Introduction to Process Technology

Second Edition

Technical Editor

Martha McKinley
McKinley Consulting
Longview, TX

Managing Director, Career Development and Employability: Leah Jewel
Director, Alliance/Partnership Management: Andrew Taylor
Director, Learning Solutions: Kelly Trakalo
Associate Content Producer: Stephany Harrington
Development Editor: Rachel Bedard
Executive Marketing Manager: Brian Hoehl
Manufacturing Buyer: Mary Ann Gloriande, LSC Communications
Cover Designer: Laurie Entringer, SPi Global
Cover Image Credit: Bim/E+/Getty Images
Editorial Production and Composition Services: SPi Global
Full-Service Project Manager: Mohamed Hameed
Printer/Binder: LSC Communications
Cover Printer: LSC Communications

Credits and acknowledgments for materials borrowed from other sources and reproduced, with permission, in this textbook appear on the appropriate page within the text.

Copyright © 2018, 2010 by Pearson Education, Inc. All rights reserved. Printed in the United States of America. This publication is protected by Copyright and permission should be obtained from the publisher prior to any prohibited reproduction, storage in a retrieval system, or transmission in any form or by any means, electronic, mechanical, photocopying, recording, or likewise. To obtain permission(s) to use material from this work, please visit http://www.pearsoned.com/permissions/

Library of Congress Cataloging-in-Publication Data

Names: Pearson (Firm), issuing body.
Title: Introduction to process technology.
Other titles: Introduction to process technology (Pearson (Firm))
Description: 2e. | Hoboken, New Jersey : Pearson, [2017] | Includes index.
Identifiers: LCCN 2017028740| ISBN 9780134808246 | ISBN 013480824X
Subjects: LCSH: Chemical processes.
Classification: LCC TP155.7 .I598 2017 | DDC 660/.28--dc23 LC record available at https://lccn.loc.gov/2017028740

ISBN 10:   0-13-480824-X
ISBN 13: 978-0-13-480824-6

Preface

The Process Industries Challenge

In the early 1990s, the process industries recognized that they would face a major staffing shortage due to the large number of employees retiring as baby boomers. Industry partnered with community colleges, technical colleges, and universities to remedy this situation. Together, they developed this series, which provides consistent curriculum content and exit competencies for process technology graduates to ensure a trained and competent staff that is ready to take over the demands of the field. The collaborators in education and industry also recognized that training for process technicians would benefit industry by reducing the costs associated with training and traditional hiring methods. Thus was born the NAPTA series for Process Technology.

To achieve consistency of exit competencies among graduates from different schools and regions, the Gulf Coast Process Technology Alliance and the Center for the Advancement of Process Technology identified a core technical curriculum for the Associate Degree in Process Technology. This core consists of eight technical courses and is taught in alliance member institutions throughout the United States. Instructors who teach the process technology core curriculum, and who are recognized in industry for their years of experience and depth of subject matter expertise, requested that a textbook be developed to match the standardized curriculum. A broad range of reviewers from process industries and educational institutions participated in the production of these materials so that the presentation of content would address the widest audience possible. This textbook is intended to provide a common national standard reference for the *Introduction to Process Technology* course in the Process Technology degree program.

This textbook is intended for use in high schools, community colleges, technical colleges, universities, and any corporate setting in which process technology is taught. Current and future process technicians will use the information within this textbook as their foundation for work in the process industries. This knowledge will make them better prepared to meet the ever-changing roles and responsibilities within their specific process industry.

What's New!

The second edition has been thoroughly updated and revised including the following

- **New** Learning Outcome alignment with NAPTA core objectives, with links from objective to text page provided!
- **All New** Full Color Art Program! With over 400 fully revised artwork and photos including new visuals of process flow for different industries, quality control, and logic diagrams.
- **New!** Key term definitions on text page with content as well as at beginning of each chapter.
- **Updated Review and New Answers Appendix!** Checking Your Knowledge questions have been updated to meet new chapter objectives and now include answers and rationales.
- **Thoroughly Updated Content!** Including growth and development of industry into the 2010s; appropriate new legislation; and new data sets.
- **New!** Emphasis on Safety throughout and inclusion of industry events that led to new safety legislation.
- **Expanded Content** on current methods of oil and gas extraction (fracking); process-auxiliary systems, and process service utilities.
- **All New** Instructor Resource Package including lesson plans, test banks, review questions, and PowerPoints.

Organization of the Textbook

This book has been divided into four parts with 27 chapters. Part 1 (Chapters 1–7) provides an overview of process technology and the various process industries. Part 2 (Chapters 8–10) describes what might be called "soft skills" that ensure success in the work environment. Part 3 (Chapters 11–13) describes concepts that are the foundation for the process technician's role in industry. Part 4 (Chapters 14–27) gives information on common types of equipment that process technicians will encounter as part of their job.

Each chapter is organized in the following way:

- **Objectives** for each chapter are aligned with the revised NAPTA curriculum and may cover one or more sessions in a course.
- **Key Terms** are a listing of important terms and their respective definitions, which students should know and understand before proceeding to the next chapter.
- The **Introduction** may be a simple introductory paragraph or may introduce concepts necessary to the development of the content of the chapter itself.
- Any of the **Key Topics** may have several subtopics. Topics and subtopics address the objectives stated at the beginning of each chapter.
- The **Summary** is a restatement of the important points in the chapter.
- **Checking Your Knowledge** questions are designed to help students do self-testing on essential information in the chapter.

- **Student Activities** provide opportunities for individual students or small groups to apply some of the knowledge they have gained from the chapter. They should generally be performed with instructor involvement.

Acknowledgements

The second edition of this series would not have been possible without the support of the entire NAPTA Board and, in particular, without the leadership and dedication of Eric Newby, Executive Director. A particular and special thank you also goes to Martha McKinley for her key role in the revision of this title. Her dedication to quality and timeliness of work is greatly appreciated.

Contributors

Gayle Cannon, Middlesex County College, Millstone Township, NJ (Art program)

Frank Huckabee, Remington College-Mobile, Mobile, AL (Instructor supplements)

Jeffrey Laube, Kenai Peninsula College, Kenai, AK (Art program, Text revisions, Instructor supplements)

Martha McKinley, McKinley Consulting, Longview, TX (Art program, Text revisions)

Reviewers

Ammar Alkhawaldeh, PhD, HCC Global Energy Institute, Houston, TX

Curtis Briggs, Development Coordinator for INEOS, League City, TX

Tommie Ann Broome, Mississippi Gulf Coast Community College, Moss Point, MS

Regina Cooper, Marathon Petroleum, Texas City, TX

Cleve Fontenot, Technical Training Manager, Corporate Learning and Development, Geismar, LA

Frank Huckabee, Remington College-Mobile, Mobile, AL

Troy Key, Phenol Reliability Supervisor, Shell Chemical, Deer Park, TX

B. Ray Player, Training Instructor, Eastman Chemical Company, Gladewater, TX

Walbert Schulpen, University of Alaska, Anchorage, AK

The following organizations and their dedicated personnel supported the development of the first edition of this textbook. Their contributions set the foundation for this revision and continue to be greatly appreciated.

Alaska Process Industry Careers Consortium
Anchorage Water and Wastewater Utility
Basell USA
BASF
BP
Chevron Texaco
ConocoPhillips
Dow Chemical
Eastman Chemical
Equistar Chemical
ExxonMobil
Formosa Plastics Corp
Heads Up Systems
Huish Detergents Inc.
Ingenious, Inc.
Kraft Foods
Marathon
Mississippi Power
Novartis
Pasadena Refining System
Shell Chemical LP
Shell Oil Products
Sherwin Alumina
TAP Safety Services
Union Carbide
Valero
Westlake Chemical Corporation

Contents

Part 1 Introduction to Process Technology

1 Process Technology: An Overview — 1

Course Overview — 2
  Individual Expectations — 2
  Program Purpose — 2
  Program Value — 2
  Industry Involvement — 2
What Process Industries Are — 2
What Process Industries Do — 3
How Process Industries Operate — 6
What a Process Technician Does — 7
Duties, Responsibilities, and Expectations of Process Technicians — 8
  Skills and Traits — 8
  Job Duties — 9
  Equipment — 10
Workplace Conditions and Expectations — 10
  Safety — 11
  Quality — 11
  Teams — 11
  Shift Work — 12
  Union Versus Nonunion Work Environment — 13
Issues and Trends in Industry — 13
  Impact of Process Technology Industries — 13
  Response of Industry to Current Issues and Trends — 14
  Future Trends in the Role of Process Technicians — 15
Responsibilities of Regulatory Agencies — 15
  Summary — 16
  Checking Your Knowledge — 17
  Student Activities — 18

2 Mineral Extraction Industries: Oil and Gas and Mining — 19

Introduction to the Oil and Gas Industry — 20
Growth and Development of the Oil and Gas Industry — 21
Sectors of the Oil and Gas Industry — 26
Duties, Responsibilities, and Expectations of Process Technicians in the Oil and Gas Industry — 30
Introduction to the Mining Industry — 30
Growth and Development of the Mining Industry — 34
Sectors of the Mining Industry — 36
Duties, Responsibilities, and Expectations of Process Technicians in the Mining Industry — 37
  Summary — 37
  Checking Your Knowledge — 38
  Student Activities — 39

3 Chemical and Pharmaceutical Industries — 40

Introduction to the Chemical Industries — 41
Growth and Development of the Chemical Industry — 41
Overview of the Chemical Industry — 44
Duties, Responsibilities, and Expectations of Process Technicians in the Chemical Industry — 45
Introduction to the Pharmaceutical Industry — 45
Growth and Development of the Pharmaceutical Industry — 46
Overview of the Pharmaceutical Industry and Drug Manufacturing Process — 48
Duties, Responsibilities, and Expectations of Process Technicians in the Pharmaceutical Industry — 49
  Summary — 50
  Checking Your Knowledge — 51
  Student Activities — 52

4 Power Generation Industry — 53

Introduction — 54
Overview of Electricity — 54
Major Sectors of the Power Generation Industry — 54
Growth and Development of the Power Generation Industry — 55
  Early Civilizations — 55
  1700s — 56
  1800s — 56
  1900s — 57
  2000–Present — 58
Overview of the Power Generation Industry — 58
  Coal, Oil, and Gas — 58
  Hydroelectric — 59
  Nuclear — 60
  Fusion Reaction — 61
  Renewable Resources — 61
Expectations of Process Technicians and Future Trends — 63
  Summary — 63
  Checking Your Knowledge — 64
  Student Activities — 65

5 Food and Beverage Industry — 66

Growth and Development of the Food and Beverage Manufacturing Industry — 67
  1900s — 67
  2000–Present — 67

Overview of the Food and Beverage Manufacturing Industry	68
Food Processing Hazards	68
Food Processing Methods	69
Food Processing Systems	70
Duties, Responsibilities, and Expectations of Process Technicians	71
Equipment	72
Workplace Conditions and Expectations	72
Environmental Regulations and Considerations	72
Summary	74
Checking Your Knowledge	74
Student Activities	75

6 Water and Wastewater Treatment Industry 76

Introduction	77
Growth and Development of the Water and Wastewater Treatment Industry	77
History of the Wastewater Treatment Industry	78
Overview of the Water Treatment Process	78
Municipal Water Treatment Process	78
Industrial Wastewater Treatment Process	79
Duties and Responsibilities of Process Technicians	80
Environmental Regulations and Considerations	81
Summary	81
Checking Your Knowledge	82
Student Activities	82

7 Pulp and Paper Industry 83

Introduction	83
Growth and Development of the Pulp and Paper Industry	84
1400s	84
1600s	85
1700s	85
1800s	85
1900s	85
2000–Present	85
Overview of the Pulp and Paper Industry	85
Pulp Mills	86
Paper Mills	88
Segments of the Pulp and Paper Industry	88
Equipment Used in the Pulp and Paper Industry	89
Summary	89
Checking Your Knowledge	89
Student Activities	90

Part 2 Skills for Process Technicians

8 Working as Teams 91

Introduction	92

Work Groups and Teams	92
Types of Teams in the Process Industries	93
Characteristics of High-Performance Teams	94
Composition	94
Technique	94
Process	95
Synergy and Team Dynamics	95
Team Tasks Versus Process	95
Stages of Team Development	96
Stage 1: Forming	97
Stage 2: Storming	97
Stage 3: Norming	97
Stage 4: Performing	97
Stage 5: Adjourning	98
Factors that Prevent Team Success	98
Resolving Conflict	99
Giving Feedback	101
Workforce Diversity	102
From the Melting Pot to the Salad Bowl	102
Terms Associated with Discussion of Diversity	102
Respecting Diversity	103
Summary	103
Checking Your Knowledge	104
Student Activities	105

9 Safety, Health, Environment, and Security 106

Introduction	108
Safety, Health, and Environmental Hazards Found in the Process Industries	109
Regulatory Agencies and Their Responsibilities	110
Environmental Protection Agency (EPA)	112
Occupational Safety and Health Administration (OSHA)	113
Department of Transportation (DOT)	113
Nuclear Regulatory Commission (NRC)	114
Regulations Affecting the Process Industries	114
29 CFR Chapter XIV Equal Employment Opportunity Act	114
OSHA 1910.119—Process Safety Management (PSM)	114
OSHA 1910.132—Personal Protective Equipment (PPE)	115
OSHA 1910.1200—Hazard Communication (HAZCOM)	115
OSHA 1910.120—Hazardous Waste Operations and Emergency Response (HAZWOPER)	116
OSHA 1910.1000 Air Contaminants	116
DOT CFR 49.173.1—Hazardous Materials—General Requirements for Shipments and Packaging	116
EPA 40 CFR Parts 239–282—Resource Conservation and Recovery Act (RCRA)	116
EPA Clean Air and Clean Water Acts	116
The Role of the Process Technician	117

Attitudes and Behaviors That Help to Prevent Accidents	117
Components of the Fire Triangle and the Fire Tetrahedron	118
Fire Tetrahedron	119
Classes of Fire	119
Consequences of Noncompliance with Regulations	120
Legal	120
Moral and Ethical	120
Safety, Health, and Environmental	120
Economic	121
Engineering Controls, Administrative Controls, and PPE	121
Correct Use of Personal Protective Equipment (PPE)	122
Head Protection	122
Face Protection	123
Eye Protection	123
Ear Protection	123
Respiratory Protection	123
Body Protection	124
Hand Protection	124
Foot Protection	124
The OSHA Voluntary Protection Program (VPP)	125
The ISO 14000 Standard	125
Physical Security and Cybersecurity	126
Summary	127
Checking Your Knowledge	128
Student Activities	129

10 Quality 130

Introduction	131
What is Quality?	132
Industry Response to Quality Issues and Trends	132
The Quality Movement And Its Pioneers	132
Dr. W. Edwards Deming	133
Joseph M. Juran	134
Philip B. Crosby	135
The Japanese Influence	135
Quality Initiatives	135
Total Quality Management (TQM)	136
ISO	136
Statistical Process Control and Other Analysis Tools	137
Six Sigma	141
Self-Directed or Self-Managed Work Teams	141
Malcolm Baldrige National Quality Award	141
Maintenance Programs	142
Total Productive Maintenance (TPM)	143
Predictive/Preventive Maintenance (PPM)	143
Process Technicians and Quality Improvement	144
Summary	145
Checking Your Knowledge	145
Student Activities	146

Part 3 Basic Knowledge for Process Technicians

11 Basic Physics 147

Introduction	150
Applications of Physics in the Process Industries	150
States of Matter	150
Key Concepts in Physics for Process Industries	151
Mass	151
Density	152
Elasticity	153
Viscosity	153
Specific Gravity	153
Buoyancy	154
Flow	154
Velocity	155
Friction	156
Temperature	156
Heat Transfer	158
Electricity	158
Force and Leverage	158
Pressure	159
Gas Laws	161
Boyle's Law (Pressure-Volume Law)	161
Charles' Law (Temperature–Volume Law)	161
Dalton's Law (Law of Partial Pressures)	162
General (or Combined) Gas Law	162
Bernoulli's Law (Bernoulli's Principle)	163
Summary	164
Checking Your Knowledge	164
Student Activities	166

12 Basic Chemistry 167

Introduction	169
Applications of Chemistry in the Process Industries	169
Elements and Compounds	169
The Periodic Table	170
Characteristics of Atoms	171
Characteristics of Compounds	172
Chemical Versus Physical Properties	173
Chemical Reactions	173
Mixtures and Solutions	175
Acids and Bases	175
Summary	176
Checking Your Knowledge	177
Student Activities	178

13 Process Drawings 179

Introduction	180
Purpose of Process Drawings	180

Common Components and Process Drawings
    Information   181
  Symbols   181
  Legend   181
  Title Block   182
  Application Block   183
Types of Process Drawings and Their Uses   183
  Block Flow Diagrams (BFDs)   184
  Process Flow Diagrams (PFDs)   184
  Piping and Instrumentation Diagrams (P&IDs)   184
  Engineering Flow Diagrams (EFDs)   186
  Plot Plan Diagrams (PPDs)   186
  Utility Flow Diagrams (UFDs)   186
  Electrical Diagrams   187
  Isometric Drawings   190
  Other Drawings   191
  Summary   192
  Checking Your Knowledge   192
  Student Activities   193

Part 4 Equipment Used in Process Technology

14 Piping and Valves   194
Introduction   195
Purpose and Function of Piping and Valves   195
Construction Materials in Piping and Valves   196
Connecting Methods for Piping and Valves   196
  Threaded (Screwed)   197
  Flanges   197
  Welds   198
  Bonds   198
Fitting Types   198
Valve Types   199
  Ball Valve   200
  Plug Valve   201
  Butterfly Valve   201
  Check Valve   201
  Diaphragm Valve   202
  Gate Valve   203
  Globe Valve   203
  Relief And Safety Valves   204
  Valve Actuators   206
Operational Hazards   206
Monitoring and Maintenance Activities   206
  Piping and Valve Symbols   207
  Summary   208
  Checking Your Knowledge   208
  Student Activities   209

15 Vessels   210
Introduction   212
Purpose of Vessels   213
Types of Tanks   214
Common Components of Vessels   215
Containment Walls, Dikes, and Firewalls   220
Reactors: Purpose and Types   220
Operational Hazards   222
Monitoring and Maintenance Activities   223
  Symbols for Vessels and Reactors   224
  Summary   224
  Checking Your Knowledge   225
  Student Activities   225

16 Pumps   226
Introduction   227
Types of Pumps   227
  Positive Displacement Pumps   228
  Dynamic Pumps   229
  Axial Pumps   231
Operational Hazards   232
  Overheating   232
  Overpressurization   232
  Cavitation   233
  Leakage   233
  Vibration   234
Monitoring and Maintenance Activities   234
  Pump Symbols   234
  Summary   234
  Checking Your Knowledge   235
  Student Activities   235

17 Compressors   237
Introduction   238
Differences Between Compressors and Pumps   238
Common Types of Compressors   239
  Positive-Displacement Compressors   239
  Dynamic Compressors   240
  Multistage Compressors   242
  Blowers   242
Operational Hazards   243
  Overpressurization   244
  Overheating   244
  Surging   244
  Leakage   244
Monitoring and Maintenance Activities   245
  Compressor Symbols   245
  Summary   246

Checking Your Knowledge	246
Student Activities	247

18 Turbines 248

Introduction	249
Types of Turbines	249
Steam Turbines	250
Gas Turbines	250
Hydraulic Turbines	251
Wind Turbines	251
Steam Turbine Components	252
Principles of Operation	253
Operational Hazards	254
Monitoring and Maintenance Activities	255
Steam Turbine Symbol	255
Summary	256
Checking Your Knowledge	256
Student Activities	256

19 Electricity and Motors 258

Introduction	259
What is Electricity?	259
Volts	260
Resistance (Ohms)	260
Amps	260
Ohm's Law	261
Watts	261
Circuits	261
Grounding	262
Electrical Transmission	262
Understanding Alternating (AC) and Direct (DC) Current	263
Which Type of Current is Used Most in the Process Industries?	264
Uses of Electric Motors in the Process Industries	265
Primary Components of a Typical AC Induction Motor	265
Principles of Operation of AC Electric Motors	266
Operational Hazards	267
Monitoring and Maintenance Activities	270
Symbols Associated with Electricity	270
Summary	271
Checking Your Knowledge	271
Student Activities	272

20 Heat Exchangers 273

Introduction	274
Purpose of Heat Exchangers	274
Types of Heat Exchangers	274
Components of a Typical Shell and Tube Heat Exchanger	277
Principles of Operation of Heat Exchangers	277
Heat Exchanger Applications	280
Relationships Among Different Types of Heat Exchangers	280
Operational Hazards	281
Monitoring and Maintenance Activities	281
Heat Exchanger Symbols	282
Summary	284
Checking Your Knowledge	284
Student Activities	285

21 Cooling Towers 286

Introduction	287
Purpose of Cooling Towers	287
Types of Cooling Towers	287
Component Parts of an Open Cooling Tower	289
Principles of Operation of Open Circuit Cooling Towers	290
Factors that Affect Cooling Tower Performance	291
Cooling Tower Applications	292
Operational Hazards	292
Monitoring and Maintenance Activities	293
Cooling Tower Symbols	293
Summary	294
Checking Your Knowledge	294
Student Activities	294

22 Furnaces 295

Introduction	296
Purpose of Furnaces	296
Components of a Furnace	297
Radiant Section	298
Convection Section	299
Principles of Operation of Furnaces	299
Furnace Designs and Draft Types	299
Furnace Designs	299
Furnace Draft Types	299
Operational Hazards	302
Monitoring and Maintenance Activities	302
Furnace Symbol	303
Summary	303
Checking Your Knowledge	303
Student Activities	304

23 Boilers 305

Introduction	306
Purpose of Boilers	306
Parts of a Boiler	306
How a Boiler Works	308

Fuels Used in Boilers	309
Boiler Types	309
Fire Tube Boiler	309
Water Tube Boiler	310
Operational Hazards	310
Monitoring and Maintenance Activities	311
Boiler Symbols	311
Summary	312
Checking Your Knowledge	312
Student Activities	312

24 Distillation — 313

Introduction	314
Purpose of Distillation	314
Distillation System Components	315
Column Sections	316
Trays	316
How the Distillation Process Works	318
Distillation Methods	318
Packed Columns	319
Operational Hazards	319
Monitoring and Maintenance Activities	320
Distillation Column Symbols	320
Summary	321
Checking Your Knowledge	321
Student Activities	322

25 Process Service Utilities — 323

Introduction	324
Types of Process Service Utilities	324
Water	324
Steam and Condensate Systems	328
Compressed Air	329
Breathing Air	330
Nitrogen System	330
Fuel Gas	331
Electricity	331
Types of Equipment Associated with Utility Systems	331
Operational Hazards	333
Water	334
Steam and Condensate	334
Compressed Air	334
Breathing Air	334
Nitrogen	334
Gas	335
Electricity	335
Monitoring and Maintenance Activities	335
Symbols Associated with Process Utilities	335
Summary	335
Checking Your Knowledge	336
Student Activities	336

26 Process Auxiliaries — 337

Introduction	338
Types of Process Auxiliaries	338
Flare Systems and Associated Equipment	339
Refrigeration Systems and Associated Components	340
Components of a Mechanical Refrigeration System	341
Lubrication Systems and Associated Components	341
Hot Oil Systems and Associated Components	342
Other Common Auxiliary Systems	343
Amine System	343
Fluidized Bed System	343
Nitrogen Header System	344
Operational Hazards	344
Monitoring and Maintenance Activities	345
Symbols for Process Auxiliaries	345
Summary	345
Checking Your Knowledge	346
Student Activities	347

27 Instrumentation — 348

Introduction	350
Purpose of Process Control Instrumentation	350
Key Process Variables Controlled by Instrumentation	350
Pressure	350
Temperature	351
Level	352
Flow	352
Analytical Variables	354
Typical Process Control Applications	354
Location	355
Function	355
Power Source	355
Other Components	356
Process Control Instrumentation Signals	356
Instrumentation and Control Loops	357
Distributed Control Systems and Their Application	358
Operational Hazards	359
Monitoring and Maintenance Activities	359
Symbols for Process Control Instruments	360
Summary	361
Checking Your Knowledge	362
Student Activities	362

Appendix I Answers for Checking Your Knowledge Questions — 363

Glossary — 369

Index — 385

PART 1 Introduction to Process Technology

Chapter 1
Process Technology: An Overview

 Objectives

Upon completion of this chapter you will be able to:

1.1 Explain the framework of the process technology course. (NAPTA History 1) p. 2

1.2 Define what process technology is and what process industries do. (NAPTA History 1) p. 2

1.3 Identify the duties, responsibilities, and expectations of the process technician. (NAPTA Overview 1, 2; Career 1) p. 7

1.4 Describe working conditions in the process industries. (NAPTA Career 1, 2) p. 10

1.5 Explain process industries' impact and industry responses to current issues and trends. (NAPTA History 2, 3) p. 13

1.6 Describe the responsibility of regulatory agencies in process industries. (NAPTA History 4) p. 15

Key Terms

Facility—also called a plant; a place that is built or installed to serve a specific purpose, **p. 6.**
Feedstock—a raw material (such as propane or ethane) or an intermediate component (such as plastic) that is used to create a product, **p. 3.**
Process—a system of people, methods, equipment, and structures that creates products from other materials, **p. 3.**
Process industries—a broad term for industries that convert raw materials, using a series of actions or operations, into products for consumers, **p. 3.**
Process technician—a worker in a process facility who monitors and controls mechanical, physical, and/or chemical changes, throughout many processes, to produce either a final product or an intermediate product made from raw materials, **p. 3.**
Process technology—processes that take quantities of raw materials and transform them into other products, **p. 3.**

Product—also called output; the desired end components from a particular process, **p. 6.**

Production—output, such as material made in a plant, oil from a well, or chemicals from a processing plant, **p. 3.**

Raw materials—also called feedstock or input. The material sent to a processing unit to be converted into a different material or materials, **p. 6.**

Unit—an integrated group of process equipment used to produce a specific product or products, **p. 6.**

1.1 Course Overview

Individual Expectations

As a process technician in the workforce, you should plan to meet certain expectations. Demonstrate leadership within your role and job assignment. Ask questions, because processes and the technology involved are constantly changing and improving. Come to work on time and be aware that personnel in many process industries cannot leave the unit until relieved by the oncoming shift or relief personnel. Expect to follow a career development plan designed specifically for each position in the process unit. Be motivated to learn. Many industries have multiple process units, which means it could take years to learn all the job positions.

Program Purpose

The Process Technology program was set in place to enhance the knowledge base of an entry level employee. The term "related instruction" is used in many process industries as the umbrella term for the knowledge needed as the base for any process. It refers to core information about principles of operation for equipment and instrumentation. Industry subject matter experts dedicated hundreds of hours to creating the Process Technology program degree. Its purpose is to prepare students for process technician positions in industry.

Program Value

Education and a degree in Process Technology are becoming more valuable to the process industries. They minimize training costs of entry level employees, create a safer work environment, and train employees in process unit specifics. It is very helpful when new employees have already attained related knowledge as preparation for work in the unit.

Industry Involvement

A series of plant incidents and accidents created an awakening in the process industries about the need for more knowledgeable entry level employees. In the past, employees were hired off the street, some having no experience, and some having worked in other industrial positions. The majority of employees had no related knowledge and skills for work around machinery, instrumentation, and other types of equipment. A core group of industry representatives partnered with local colleges in their geographical area to develop eight core technical courses of related instruction for the process industries. From that effort, the Process Technology degree was developed and continues to evolve as changes in technology dictate.

1.2 What Process Industries Are

This chapter provides you with a description of process industries and their impact on people, the economy, and the environment. It also describes what a process technician does, and the expectations and working conditions of process technicians.

Throughout this text book, the term *process industries* is used. The term **process industries** describe industries that convert raw materials, using a series of actions or operations, into products for consumers.

Generally speaking, the process industries involve **process technology**—processes that take quantities of raw materials and transform them into other products. The result might be an end product for a consumer or an intermediate product that is used to make an end product. Each company in the process industries uses a **process**, a system of people, methods, equipment, and structures, to create products from other materials. A **process technician** is a person who monitors and controls mechanical, physical, and/or chemical changes, throughout many processes, to produce either a final or an intermediate product.

Process industries share some basic processes and equipment. For example, the distillation process (discussed later in this textbook) can be used to make anything from alcoholic beverages to petroleum products, depending on the materials used. However, most of these industries have their own unique processes. For example, the way water is used in papermaking differs greatly from the way water is treated and made suitable for drinking.

The process industries are some of the largest industries in the world, employing hundreds of thousands of people in almost every country. These industries, both directly and indirectly, create and distribute thousands of products that affect the daily lives of almost everyone on the planet.

> **Process industries** a broad term for industries that convert raw materials, using a series of actions or operations, into products for consumers.
>
> **Process technology** processes that take quantities of raw materials and transform them into other products.
>
> **Process** a system of people, methods, equipment, and structures that creates products from other materials.
>
> **Process technician** a worker in a process facility who monitors and controls mechanical, physical, and/or chemical changes, throughout many processes, to produce either a final product or an intermediate product made from raw materials.

What Process Industries Do

Various industries are classified as process industries. These include the following:

- **Oil and Gas**—The exploration and **production** (output) segment locates oil and gas, then extracts them from the ground using drilling equipment and production facilities (Figure 1.1).

  The transportation segment moves petroleum from where it is found to refineries and petrochemical facilities, then takes finished products to markets.

  The refining segment of the oil and gas process industry takes quantities of hydrocarbons and transforms them into finished products, such as gasoline and jet fuel, or into **feedstock** (a raw material or intermediate component used to make something else, like plastics).

> **Production** output, such as material made in a plant, oil from a well, or chemicals from a processing plant.
>
> **Feedstock** a raw material (such as propane or ethane) or an intermediate component (such as plastic) that is used to create a product.

Figure 1.1 **A.** Drilling field for oil. **B.** Coal mining equipment such as this pickaxe and hardhat have largely been replaced by automated equipment.
CREDIT: **A.** Eugene Chernetsov/Fotolia; **B.** Adam J/ Shutterstock.

A. B.

- **Mining**—Mining is a complex process which involves the extraction and processing of rocks and minerals from the ground (see Figure 1.1). Mining products are integral to a wide range of industries, serving as base materials for utilities and power generation, construction, transportation, agriculture, electronics, food production, pharmaceuticals, personal hygiene, consumer products, precious metals, and so on.
- **Chemicals**—Chemicals play a vital role in a wide range of manufacturing processes, resulting in products such as plastic, fertilizers, dyes, detergent, explosives, film, paints, food preservatives and flavors, synthetic lubricants, and more (Figure 1.2).

Figure 1.2 A. Chemical industry. **B.** The mortar and pestle are the classic image that represents the pharmaceutical industry.
CREDIT: **A.** Kadmy/Fotolia; **B.** Sherry Yates Young/Shutterstock.

A.

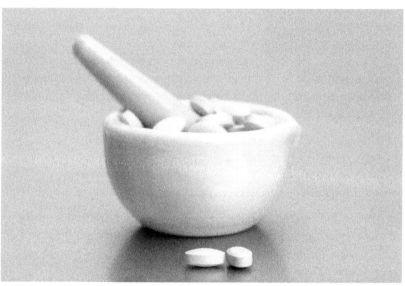

B.

- **Pharmaceuticals**—Modern drug manufacturing establishments produce a variety of products, including finished drugs, biological products, bulk chemicals, botanicals used in making finished drugs, and diagnostic substances such as pregnancy and blood glucose kits.

    Modern drugs save lives and improve the well-being of countless patients while improving health and quality of life and reducing healthcare costs.

- **Power Generation**—Power generation involves the production and distribution (Figure 1.3) of electrical energy in large quantities to industries, businesses, residences, and schools. The role electricity plays in everyday life is enormous, supplying lighting, heating and cooling,

Figure 1.3 High-voltage electrical power lines.
CREDIT: MEzairi/Shutterstock.

and power to everything from coffeepots to refineries. There are three main segments of the power generation industry: generation, transmission, and distribution. Power can be generated in a variety of ways. These include burning fuels, splitting atoms, and using water, wind, or thermal energy.

- **Food and Beverages**—The food and beverage manufacturing industry (Figure 1.4) links farmers to consumers through the production of finished food products. The products created by this industry can vary dramatically and range from fresh meats and vegetables to processed foods that can simply be heated in the microwave.

Figure 1.4 Typical products of the food and beverage industry.
CREDIT: Mariia Pazhyna/Fotolia.

- **Water and Wastewater Treatment**—Clean water (Figure 1.5) is essential for life and many industrial processes. It is through water treatment facilities that process technicians are able to process and treat water so it is safe to drink and safe to return to the environment.

Figure 1.5 Clean drinking water as represented here is a huge process industry in the United States.
CREDIT: djahan/Fotolia.

- **Pulp and Paper**—Paper (Figure 1.6) plays a huge role in everyday life. If, along with paper, you include items made from natural wood chemicals, then the pulp and paper industry creates and distributes thousands of products involved in daily life. The products include items such as packaging, documents, bandages, insulation, textbooks, playing cards, and money.

Figure 1.6 Paper processing plant.
CREDIT: Westend61/Getty Images.

Although this list of industries is not all inclusive, it covers a broad range of industries that are responsible for producing or contributing to important everyday items, including the following:

- DVD and CD media
- Electronic parts
- Telephones and cell phones
- Antihistamines, aspirin, and antibiotics
- Carpets and wallpaper
- Plastics
- Parts for cars, boats, and planes
- Gasolines and fuels
- Fertilizers
- Foods and drinks
- Boxes and packaging
- Jewelry
- Electricity
- Clean water
- Tires
- Checks, notepads, newspapers, and books
- Credit cards
- Detergents and soaps
- Paints and glues
- Sporting goods
- Cosmetics and colognes
- Cameras and film
- Toothpaste
- Cement, steel, and other building materials

It is easy to see the importance of the process industries and their impact on everyday life.

How Process Industries Operate

Although there are specific processes and products associated with each type of process industry, they have some common operations:

1. **Raw materials** (sometimes called input or feedstock) are made available to a process **facility** (or plant), a place that is built to perform a specific function.
2. The raw materials are sorted by the proper process. Most facilities have different units that perform a specific process. A **unit** is an integrated group of process equipment used to produce a specific product.
3. Units can be referred to by the processes they perform or be named after their end products.
4. The raw materials are processed using people, equipment, and methods. Safety, quality, and efficiency are key elements of the process.
5. A **product** (output), or desired component, is the result of a particular process. The product can be either an end product for consumers or an intermediate product used as part of another process to make an end product.
6. The product is distributed to consumers.

Raw materials also called feedstock or input. The material sent to a processing unit to be converted into a different material or materials.

Facility also called a plant; a place that is built or installed to serve a specific purpose.

Unit an integrated group of process equipment used to produce a specific product or products.

Product also called output; the desired end components from a particular process.

1.3 What a Process Technician Does

A process technician (Figure 1.7) is a key member of a team responsible for planning, analyzing, and controlling the production of products from the acquisition of raw materials through the production and distribution of products to customers in a variety of process industries.

A process technician monitors and controls mechanical, physical, and/or chemical changes throughout many processes to produce a final product or an intermediate product, made from raw materials.

Process technician duties can also include the acquisition and testing of raw materials, monitoring the production and distribution stages and, ultimately, shipment of products to customers in a variety of industries.

It is essential for a process technician to have the ability to work effectively in a team-based environment. Productivity within industries is accomplished by teams of people from many different backgrounds and with many different levels of education. Strong computer, oral, and written communication skills are essential for a process technician. Technicians must complete work assignments during shift, describe activities for relief personnel, maintain data logs, prepare reports, and perform other needed tasks.

The duties of a process technician include maintaining a safe work environment; controlling, monitoring, and troubleshooting equipment; analyzing, evaluating, and communicating data obtained using technology; and training others while also continuing one's own learning process.

A process technician applies quality principles to all activities performed. Particular emphasis is placed on process control of production operations and continuous improvement of those operations. While applying these principles, the process technician completes tasks safely, wears safety equipment, uses industrial safety devices, and/or promotes safety among co-workers.

The life of a process technician must be flexible since the process technician may do shift work in all types of weather. This career provides a variety of experiences for an individual looking for a challenging occupation.

Process technician duties include but are not limited to the following:

- Maintaining a safe work environment
- Controlling, monitoring, and troubleshooting equipment and instrumentation in processes
- Analyzing, evaluating, and communicating data obtained through the use of technology

A process technician ensures customer satisfaction by applying quality principles to all activities performed, with an emphasis on production, operations, and continuous improvement of those operations.

Employees in the process industries are generally rewarded for job excellence through salary increases, promotions, and bonuses. Job benefits usually include health and dental insurance, profit sharing, and retirement plans.

Figure 1.7 Process technician at work.

COURTESY of Eastman Chemical Company.

Duties, Responsibilities, and Expectations of Process Technicians

Today's process technicians must deal with more automation and computerization than workers in the past. Fewer people are now required to work as process technicians due to advances in technology. Today's workers must have more education and be more highly skilled. A two-year Associate degree is recommended.

Process technicians are required to have strong reading, writing, and verbal skills, as well as knowledge of computer skills. They also need an understanding of math, physics, and chemistry.

Process technician jobs are more complex and cross-functional (meaning workers are required to perform a wider variety of tasks) than in the past. Technicians are responsible for entire complex processes, not just simple tasks.

Companies have increased their emphasis on keeping costs low by improving safety, performing preventative maintenance, optimizing processes, and making efforts to increase efficiency.

Along with technology changes, process technicians must keep up with constantly changing governmental regulations pertaining to safety and the environment.

This section covers general responsibilities. There is a wide range of processes involved with the process industries, and companies have different methods of organizing job duties.

Skills and Traits

TRAINABLE SKILLS The skills and traits companies look for when hiring process technicians include the following:

- A high priority on safety and a firm understanding of its importance in a process unit
- Education (preferably a two-year Associate degree), including a foundation in science and math
- Technical knowledge and skills
- Interpersonal skills (relating to and working with others)
- Communication skills (reading, writing, speaking)
- Ability to work on and contribute to a team
- Computer skills
- Physical capabilities such as lifting, pulling, and climbing
- Ability to deal with change
- Technology skills
- Ability to stay current in work skills
- Ability to understand and troubleshoot a process problem

PERSONALITY TRAITS

- Strong work ethic
- Ability to deal with change
- Ability to be a flexible, self-directed team player
- Ability to appreciate a diverse workplace
- Ability to follow safety, health, environment, and security procedures and policies
- Sense of responsibility
- Positive attitude
- Respect for others
- Ability to accept criticism and feedback

Job Duties

Process technicians are expected to keep in mind the principles of cost savings, efficiency, and safety. There is a strong emphasis on these principles within the industry.

Process technicians are required to have strong communication skills (i.e., reading, writing, and verbal), a solid base of computer skills, and an understanding of biology, chemistry, and safety.

Process technicians may be responsible for a wide variety of tasks, including the following:

- Monitor and control processes and equipment
  - Sample processes
  - Inspect equipment
  - Start, stop, and regulate equipment
  - View instrumentation readouts
  - Analyze data
  - Evaluate processes for optimization
  - Make process adjustments
  - Respond to changes, emergencies, and abnormal operations
  - Document activities, issues, and changes
- Assist with equipment maintenance
  - Change or clean filters or strainers
  - Lubricate equipment
  - Monitor and analyze equipment performance
  - Prepare equipment for repair
  - Return equipment to service
- Troubleshoot and problem solve
  - Apply troubleshooting processes such as root cause analysis
  - Participate in corrective action teams
  - Use statistical tools
- Communicate and work with others
  - Write reports
  - Write and review procedures
  - Document incidents
  - Help train others
  - Learn new skills and information
  - Work as part of a team
- Perform administrative duties
  - Do housekeeping (make sure the work area is clean and organized)
  - Perform safety checks
  - Perform environmental checks
- Work with a focus on safety, health, environment, and security
  - Take safe actions and keep safety, health, environmental, and security regulations in mind at all times
  - Look for unsafe acts and correct them
  - Watch for signs of potentially hazardous situations and report or mitigate them

Equipment

Process technicians work with many different types of equipment, including the following:

- Piping and valves
- Vessels (tanks, drums, reactors)
- Pumps
- Compressors
- Furnaces/heaters
- Boilers
- Heat exchangers
- Cooling towers
- Distillation columns
- Turbines
- Motors, engines, and generators
- Instrumentation
- Auxiliary systems
- Utilities

Later chapters in this textbook explain what this equipment does, how to monitor and maintain it, potential hazards, and other information.

NOTE: Safety, health, environment (SHE), and security are constant concerns in industry, but several different acronyms are used to reference them depending on industry and region. Acronyms include EH&S (Environment, Health, & Safety), HESS (Health, Environment, Safety, Security), SHEM (Safety, Health, Environmental Management), and others.

1.4 Workplace Conditions and Expectations

The following are some workplace conditions that process technicians might expect:

- Outdoor job positions that require working in all types of weather, including extreme conditions
- Working on holidays
- Working in hot and noisy conditions
- Maintaining a drug- and alcohol-free environment
- Working shift work, both days and nights, in a facility that operates 24 hours a day, 7 days per week
- Using tools and lifting some heavy objects
- Potential for working away from home for long periods of time (eg, offshore rigs, Alaska pipeline)

Companies expect that process technicians will do all the following:

- Work safely, in compliance with government and industry regulations
- Comply with environmental regulations
- Work with a focus on business goals as directed by the plant, department, and process unit
- Look for ways to reduce waste and improve safety and efficiency
- Keep up with industry trends and constantly improve skills

Figure 1.8 Personal safety equipment.
CREDIT: THPStock/Shutterstock.

Safety

A major expectation of employers is that employees will have a proactive attitude regarding safety (Figure 1.8). Two crucial safety points that process technicians should remember are that accidents are preventable and that employees must be proactive (think ahead) before a problem occurs.

Government regulations are in place to protect worker health and safety, the community, and the environment. Many federal government agencies and regulations have an impact on the process industries: Occupational Safety and Health Administration (OSHA), Environmental Protection Agency (EPA), Department of Transportation, Department of Energy, Nuclear Regulatory Commission, U.S. Coast Guard, Mine Safety and Health Administration, and so on. State agencies also regulate certain elements within their jurisdiction.

OSHA works under the direction of the Department of Labor (DOL). This agency is charged with ensuring safe workplace conditions by education of employees, supervision, and company administration. OSHA maintains oversight of regulations and standards developed in accordance with the OSH Act.

The Process Safety Management (PSM) standard was enacted when accidents and incidents continued to occur, releasing toxic and other types of hazardous chemicals, and sometimes causing injury and death to employees. The PSM standard was established to protect employees in the private sector from the hazards of flammable, reactive, toxic, and explosive materials. The PSM standard contains fourteen mandated elements whose responsibility of enforcement falls on employers. PSM information can be found within the OSHA Standard CFR 1910.119. These points will be discussed in later chapters.

Companies use physical and cyber-security measures to protect assets and workers from internal and external threats. Process technicians must be trained and be able to recognize safety and security threats, and understand the impact on themselves and the facility where they work.

Safety rules and agencies are discussed later in this chapter and in Chapter 9 Safety, Health, Environment, and Security.

Quality

Quality is an important part of the process industries. Quality has two major definitions: (1) a product or service free of deficiencies; and (2) the characteristics of a product or service that bear on its ability to satisfy stated or implied needs.

Without quality measures, products and services could be deficient or unsatisfactory. Unsatisfactory products lead to unhappy customers, increased waste, inefficiencies, increased costs, reduced profits, and an inability to maintain a competitive edge.

Teams

Teams and teamwork are at the core of success in process technology. A team usually consists of a small number of people. People are picked for a team because they have skills that complement other team members' skills. Everyone on the team is committed to a common purpose. All team members hold the other members mutually accountable for the success of the team.

When working as part of a team, it is important to recognize and appreciate others for their contributions to the workplace while not discounting them because of their differences. It is vital to understand diversity and practice its principles in the workplace.

Chapter 8 discusses in depth the attributes and importance of teams.

Shift Work

Because most process facilities operate 24 hours a day, 7 days a week, 365 days a year, process technicians are typically shift workers. Each plant or process facility is unique in the way shifts and workdays are arranged.

Most shift work involves 8-hour or 12-hour rotations, with numerous variations in the way nonworking days are arranged (Figure 1.9). Some examples of working day arrangements follow, though not all companies address shift work patterns in the same way.

August

Sun	Mon	Tue	Wed	Thu	Fri	Sat
1	2	AM 3	AM 4	AM 5	AM 6	7
8	PM 9	PM 10	PM 11	PM 12	13	14
AM 15	AM 16	AM 17	AM 18	19	20	PM 21
PM 22	SI 23	PM 24	25	26	AM 27	AM 28
AM 29	AM 30	31 31	1	PM 2	PM 3	PM 4
PM 5	6	7	AM 8	AM 9	AM 10	AM 11

Figure 1.9 A sample shift work calendar.

- Four days on, four days off—the process technician works four consecutive days and then is off for four consecutive days. Typically, 12-hour shifts switch back and forth from day to night after days off. Eight-hour shifts may rotate to the next set of hours (e.g., midnight. to 8 a.m., 8 a.m. to 4 p.m., 4 p.m. to midnight, usually with at least one extra shift off between changes).
- EOWEO (every other weekend off)—the process technician works one weekend and is off the next weekend.

Process technicians must to be able to adjust to the schedules that a plant or process facility uses. Because of this, it is vital that process technicians understand the impact of shift work.

A person operates on a natural time clock that is different from work hours. Two mental low points can occur every 24 hours, typically between 2–6 a.m. and 2–6 p.m. The "sun up effect" may also occur, where a person wakes up when the sun rises, no matter how sleep deprived the person is.

Shift work has been compared to having permanent jet lag. Many people who work long, irregular hours tend to experience the following:

- Fatigue
- Reduced attention span
- Slowed reaction time
- Conflicting body clock and work schedule
- Reduced attentiveness

- Inability to think and remember clearly
- Increased potential for accidents

If not handled well, shift work can affect the following:

- Physical health—resulting in high rates of alcohol, drug, and tobacco use, overeating, lack of exercise, and long-term sleep disturbances
- Emotional health—resulting in increased irritability and a tendency toward depression and a lack of social life or healthy leisure activities
- Family matters—resulting in higher divorce rates, little time with children and spouse, few shared family activities, and missing out on social outings

Process technicians can reduce the impact of shift work by taking care of themselves. To maintain physical and mental health, a process technician should do the following:

- Establish as regular a schedule as possible
- Create a day-sleeping environment
- Take naps when possible
- Avoid stimulants, alcohol, and caffeine
- Eat only light snacks in the 2–6 a.m./p.m. periods
- Compensate for lower awareness

Union Versus Nonunion Work Environment

Many workers in the process industries belong to labor organizations called unions. Process technicians should understand how working conditions can vary based on union versus nonunion work environments.

Not every employee in the process industries is a union member. For example, only about 5% of the workers in the pharmaceutical and medicine manufacturing industry are union members or are covered by a union contract. This is compared with about 15% of workers throughout private industry. Companies with employees who belong to unions often have different labor practices and operating conditions than companies with nonunion employees.

A goal of a union is to help employees establish better working environments and benefits for themselves. Unions use collective bargaining to negotiate various work conditions and policies for their members. Unions usually select representatives to negotiate with employers. These negotiations seek to create agreements (work rules) that govern company practices ranging from wages and benefits, to unit staffing levels, to job assignments based on seniority.

Nonunion employees do not negotiate labor practices with their employers as union members do. They follow the labor policies the employer has established. On the other hand, in companies without unions there are no union dues and there may be more flexibility in moving from one position to the next regardless of seniority.

The labor policies that employers provide differ from company to company. Companies provide standard benefits and compensation for their employees. Adopting such policies helps to ensure employee retention and job satisfaction.

1.5 Issues and Trends in Industry
Impact of Process Technology Industries

IMPACT ON COMMUNITY Process industries impact our communities by helping the economy. Industries bring jobs into the area, create greater commerce, and increase the standard of living in part due to the higher wage bracket paid by these types of positions. New industry arriving in a community often creates a growth of housing and supporting businesses to help sustain its operation. Management representatives work with community representatives to promote and foster good relationships and transparency in all actions that could affect air and water quality in the geographical area surrounding the plant site. Great effort is taken to ensure the environmental footprint is minimized.

IMPACT ON OTHER INDUSTRIES Industries that either supply feedstock or take product from another industry can be affected as the global economy shifts and cycles. For example, if the world market is glutted with oil, the price of oil drops and oil production may need to be cut back. In a domino effect, all supporting industry related to oil production is then in less need and is forced to cut back as well. Industrial processes that need oil to create products also feel the impact. When oil prices rise, the reverse effects can occur.

IMPACT ON THE ENVIRONMENT Process industries need a ready and plentiful supply of water. The optimum location of these facilities is to have them placed close to the supply source. Due to the nature of many processes, water becomes contaminated and must be cleaned before returning to its source. Companies work closely to monitor, treat, and analyze water effluent before allowing it to leave the plant reservoirs. During periods of start-up and shutdown, many process plants need to purge and clean lines and vessels. The resulting gases are released to the atmosphere by vent or by flare, depending on volatility. Every effort is made to ensure noise pollution is kept to a minimum in order not to disturb "near neighbors" who may live and work in the surrounding area.

IMPACT ON THE ECONOMY Process industries are a boon to the economy of both global and local geographic locations of specific plants. They and their supporting businesses supply jobs, goods, and services. However, changing trends in the world market sift down to companies associated with particular products. An example is the price of crude oil per barrel. Gasoline and diesel fuels are derived from crude oil and when the barrel price escalates, the price of gasoline at the pump goes up as well. When companies have to pay more for the feedstock, it costs more to produce the fuel, and that cost is passed on to the consumer. If the global market does not adjust or return to balance, eventually entire plant sites may be closed or begin a cutback of employees. Cutbacks within companies affect all support businesses that depend on the plant site for work, including contractors, food supply, transportation, and maintenance services.

Response of Industry to Current Issues and Trends

GLOBAL COMPETITION Global competition has become a major factor in industry. Products such as steel that once were only made within the country are now shipped around the globe. Likewise, products that once were made and used locally are exported to wherever they find a market. The result for process industries is a heightened level of competition and an increased need for attention to cost containment. Work must be done quickly, safely, and efficiently using modern technology and processes. Time, quality, and cost are all affected by the impact of global competition.

SAFETY AND ENVIRONMENTAL REGULATIONS Global awareness has become a feature of our lives, including the importance of workplace safety and the long-term impact of damage to the environment. For example, after a tsunami destroyed the Fukushima nuclear power plant in 2011, satellites and television stations traced the radiation that traveled from Japan through the air and the debris that traveled by ocean current to the Pacific shores of the United States and Canada. Food and fishing industries within the damage zone were shut down until the effects from the accident were assessed and safety of products was ensured.

Global summits now create international agreements about reducing or maintaining low levels of pollution. These agreements impact industry heavily. Countries create laws that require industries to revamp old plants or build new ones to ever-stricter safety and environmental standards. Industries that do not meet these regulations are fined.

TECHNOLOGICAL ADVANCEMENTS AND TRENDS
Factors Responsible for Future Changes in Role of PT Process technicians will be part of an evolving environment in which mechanical controls will be guided by computerized instrumentation. Production processes may be overseen by fewer people as machines are

developed to perform tasks once done by workers. The result will be a greater role in monitoring processes, maintaining equipment, and troubleshooting errors that may occur.

The process industries are some of the largest industries in the world, employing over 500,000 refining and chemical processing workers alone. However, these industries have predicted a severe shortage of skilled workers as the "baby boom" generation takes retirement.

Unfortunately, records indicate that as many as 70% of current high school students do not have the basic knowledge and skills required to fill these industry positions. In the past, most processing jobs involved operating machinery. Today, much of the machinery is automated and computerized. Jobs are more complex and cross-functional than in the past, and process technicians must be able to keep up with constantly changing technology and regulations. For these reasons, education beyond the high school level will be required for all process technicians; for most positions, a two-year Associate degree is preferred for process technicians.

Future Trends in the Role of Process Technicians

The process technician of the future will need a two-year Associate degree to become familiar with basic knowledge used in various industries. Some process industries use more math or chemistry than others, but the basic principles are important to understand. The ability to use and learn new processes on computers will be an essential component of preparing for the process technician role.

The role of process technicians in industry is changing. Process technicians should expect to see the following trends in the future:

- Increased use of computers, automation, and advanced technology
- More involvement in process control and business decisions
- Increased workplace diversity
- Increased reliance on communication skills
- Continuous upgrading of skills and knowledge
- Restructuring of organizations to reduce costs and compete more effectively
- Less traditional methods of supervision
- New training methods and technologies, including virtual facilities, advanced simulators, and partnerships between education and industry
- New and updated environmental, government, and industry regulations or restrictions (e.g., on types of mining and access to land)
- More stringent water, air, and environmental quality standards
- International expansion to areas with low labor and production costs
- More efficient, safer, and cleaner operations (e.g., cleaner burning coal technologies that burn coal more efficiently and also reduce emissions)
- Market fluctuations in pricing of raw materials and feedstocks

NOTE: Pharmaceutical research and development ranks among the fastest growing manufacturing industries, so demand for this industry's products is expected to remain strong and work is likely to be stable even during periods of high unemployment. The industry is relatively insensitive to changes in economic conditions.

1.6 Responsibilities of Regulatory Agencies

Federal, state, and local standards are continually reviewed in order to provide a safe working environment for workers and a safe living environment for populations surrounding industry. The following are some of the major agencies or offices that regulate health, environmental, safety, and security issues. Responsibilities of these and other agencies are discussed in more detail in Chapter 9.

EPA (Environmental Protection Agency)—a federal agency charged with authority to make and enforce the national environmental policy.

OSHA (Occupational Safety & Health Administration)—a U.S. government agency created to establish and enforce workplace safety and health standards, conduct workplace inspections and propose penalties for non-compliance, and investigate serious workplace incidents.

DOT (Department of Transportation)—a U.S. government agency with a mission of developing and coordinating policies to provide an efficient and economical national transportation system, taking into account need, the environment, and national defense.

NRC (Nuclear Regulatory Commission)—a U.S. government agency that protects public health and safety through regulation of nuclear power and the civilian use of nuclear materials.

DHS (Department of Homeland Security)—the department of the U.S. federal government charged with protecting U.S. territory from terrorist attacks (including both physical and cyber-attacks) and providing a coordinated response to large-scale emergencies.

ATF (Alcohol, Tobacco, and Firearms)—a law enforcement agency in the United States Department of Justice that protects the community from violent criminals, criminal organizations, the illegal use and trafficking of firearms, the illegal use and storage of explosives, acts of arson and bombings, acts of terrorism, and the illegal diversion of alcohol and tobacco products.

In addition to the Federal government, local and state regulatory agencies may set regulations for industry. These laws also impact industry processes and costs.

Summary

Process technology is the set of processes that take quantities of raw materials and transform them into products. Process industries created these products for consumers using a series of actions and operations or an intermediate product that is used to make an end product. Each company in the process industries uses a system of people, methods, equipment, and structures to create products.

Process industries share some basic processes and equipment. However, most of these industries have their own unique processes.

The process industries are some of the largest industries in the world, employing hundreds of thousands of people in almost every country. Both directly and indirectly, they create and distribute thousands of products that affect the daily lives of almost everyone on the planet.

Process technicians are the workers employed in these industries who monitor and maintain equipment to support industry processes. Due to rapid technological advances, their jobs are more cross-functional than in the past. Technicians are now responsible for entire complex processes, not just single tasks. Working conditions in process industries vary widely but generally include a basic understanding of technology, ability to problem solve and work well on teams, acceptance of diversity, an expectation of continuous learning, and other factors such as shift work or the possibility of working in outdoor conditions.

Process industries affect their communities, other industries, the environment, and the economy. Global competition encourages exploration into improved processes. There is an increased emphasis on keeping costs low by improving safety, performing preventive maintenance, optimizing processes, and making efforts to increase efficiency.

Industries continually respond to changes and trends in the market and to the availability of raw materials and feedstock. Along with technology changes, process technicians must keep up with constantly changing governmental regulations pertaining to safety and the environment.

Checking Your Knowledge

1. Define the following terms:
   a. facility
   b. process
   c. process industries
   d. process technician
   e. process technology
   f. product
   g. production
   h. raw materials

2. A(n) _____ is a system of people, methods, equipment, and structures that creates products from other materials.

3. Name at least five process industries.

4. Which of the following is the best definition of a unit in the process industries?
   a. 144 items grouped together
   b. a distribution center
   c. an industrial plant that makes a particular component for use in creating products for distribution and sale
   d. an integrated group of process equipment used to produce a specific product or products for distribution and sale

5. Name at least three trainable skills that companies look for when hiring process technicians.

6. Name at least five personality traits that companies look for when hiring process technicians.

7. *(True or False)* Many process industry facilities operate 24 hours a day, 7 days a week.

8. Which are considered to be mental low periods during the day?
   a. 12 a.m. to 2 a.m. and 12 p.m. to 2 p.m.
   b. 12 p.m. to 4 p.m. and 12 a.m. to 4 a.m.
   c. 2 a.m. to 6 a.m. and 2 p.m. to 6 p.m.
   d. 10 p.m. to 2 a.m. and 10 a.m. to 2 p.m.

9. What can be done to combat the fatigue associated with shift work? (Select all that apply)
   a. Drink plenty of coffee or tea.
   b. Take naps when possible.
   c. Avoid stimulants and alcohol.
   d. Eat only light snacks in the 2–6 a.m. or 2–6 p.m. period.
   e. Work an irregular schedule.

10. Name tasks you would do to monitor a process? (Select all that apply)
    a. Sample process streams.
    b. Use statistical tools.
    c. Analyze data.
    d. Make process adjustments.
    e. Do housekeeping (make sure the work area is clean and organized).

11. What tasks are associated with equipment maintenance? (Select all that apply)
    a. applying processes such as root cause analysis
    b. lubrication
    c. housekeeping
    d. monitoring and analyzing equipment performance
    e. changing filters

12. What tasks are associated with troubleshooting and problem solving? (Select all that apply)
    a. applying processes such as root cause analysis
    b. using statistical tools
    c. doing housekeeping (making sure work area is clean and organized)
    d. participating in corrective action teams
    e. performing safety checks

13. Name ways that process industries respond to current and future trends. (Select all that apply.)
    a. Purchase feedstock at regular intervals throughout the year.
    b. Search for ways to increase efficiency.
    c. Sell and buy only within national borders.
    d. Provide a fixed 10-year plan for the industry.
    e. Plan for computerized instrumentation.

NOTE: Answers to Checking Your Knowledge questions are in Appendix I.

Student Activities

1. Which process industry sounds the most interesting to you? Why? Write a one-page paper in response.

2. What do you think are the benefits of working as a process technician? What are the drawbacks? Make a chart with one column for benefits and one for drawbacks. Write down your responses, then compare the benefits and drawbacks.

3. Think about the impact shift work would have on your life. Make a list of how it would affect your daily life.

4. Think about the skills required of process technicians. Make a list of the skills you think you already have and the skills you will need to learn.

Chapter 2
Mineral Extraction Industries: Oil and Gas and Mining

 ## Objectives

Upon completion of this chapter, you will be able to:

2.1 Explain the background, growth, and development of the oil and gas industry. (NAPTA History 1-4) p. 21

2.2 Describe the different sectors of the oil and gas industry. (NAPTA History 1, 3) p. 26

2.3 Identify duties, responsibilities, and expectations of the process technician in the oil and gas industry. (NAPTA Career 1) p. 30

Key Terms

Conversion—a process that changes the size and/or structure of the petroleum components by breaking them down, combining them, or rearranging them, **p. 29.**

Deposit—a natural accumulation of ore, **p. 31.**

Distillation—the process (also called *fractionation*) of refining a compound such as petroleum into separate components. This process uses the different boiling points of liquids to separate components (called *fractions*), **p. 22.**

Exploration—the process of locating oil and gas reservoirs by conducting surveys and studies, and drilling wells, **p. 27.**

Formulating and blending—a process that combines or mixes components and additives to produce finished products with specific performance requirements; also called *blending*, **p. 29.**

Geology—the study of Earth and its history as recorded in rocks, **p. 30.**

Hydraulic fracturing—a process using pressurized liquid to create cracks in rock formations to release oil and natural gas (commonly called *fracking*), **p. 26.**

Hydrocarbon—organic compounds that contain only carbon and hydrogen; most often found occurring in petroleum, natural gas, and coal, **p. 21.**

19

LNG—liquefied natural gas; a clear, colorless, nontoxic liquid formed when natural gas is purified and cooled to −162°C (−260°F), **p. 25.**

Metal—chemical elements that have luster (ability to reflect light) and can conduct heat and electricity (e.g., copper, bauxite, iron, lead, gold, silver, zinc, nickel, and uranium), **p. 31.**

Mine—a pit or excavation from which minerals are extracted, **p. 31.**

Minerals and gems—naturally occurring inorganic substances that have a definite chemical composition and a characteristic crystalline structure, **p. 31.**

Mining—the extraction of valuable minerals or other geological materials from the earth, **p. 31.**

Nonmetals—substances that conduct heat and electricity poorly; are brittle, waxy, or gaseous; cannot be hammered into sheets or drawn into wire (e.g., gems and precious stones, coal, gravel, sand, lime, stone, soda ash, phosphate rock, and clay), **p. 31.**

Ore—a metal-bearing mineral that is valuable enough to warrant mining (e.g., iron or gold), **p. 31.**

Petrochemical—a chemical derived from fossil fuels or petroleum products, **p. 20.**

Petroleum—a substance found in the earth, such as oil or gas, composed of chemical compounds consisting primarily of hydrogen and carbon, **p. 21.**

Plate tectonics—a theory describing movement of semi-rigid plates in the earth's crust, which cause geological changes such as volcanoes, mountains, and earthquakes, **p. 30.**

Quarry—an open excavation from which stones are extracted, **p. 31.**

Refining—the process of purifying a crude substance into other products, such as petroleum being separated into gasoline, kerosene, gas, and oil, **p. 29.**

Rocks—minerals or gems, or a combination of minerals or gems mixed with other chemical compounds, **p. 30.**

Rock cycle—the process by which rocks are formed, exposed to forces that erode or break them down, and then are reformed by forces such as heat and pressure, **p. 30.**

Rock formations—geology that is arranged or formed in a certain way; also, scenic outcroppings formed by weathering and erosion, **p. 30.**

Transportation—the oil and gas industry segment responsible for moving petroleum from wells to processing facilities and finished products to consumers. Transportation methods include pipelines, watercraft, railways, and trucks, **p. 28.**

Treatment—a process that prepares the components for additional processing and creates some final products. Treatments can remove or separate various components or contaminants at this point, **p. 29.**

Introduction to the Oil and Gas Industry

The oil and gas industry has been a keystone to economic development since the first tapping and drilling of wells in the 1800s. Oil that was once transported from wells to refineries in wooden barrels now often flows in permanent pipelines over hundreds of miles to reach ports or refineries. Technology has evolved from planting pipes into oil-rich fields to well stimulation techniques such as hydraulic fracturing. Oil and natural gas industries continue to play a major role in providing energy for industrial processes. A closely related industry is the chemical industry. Sometimes you will hear the term **petrochemical**, which refers to a chemical derived from fossil fuels or petroleum products. See Chapter 3, *Chemical and Pharmaceutical Industries*, for details.

Petrochemical a chemical derived from fossil fuels or petroleum products.

2.1 Growth and Development of the Oil and Gas Industry

Petroleum is the name given to liquid oils found in the earth. Petroleum is a substance made almost exclusively of **hydrocarbons** (organic compounds that contain only carbon and hydrogen).

Petroleum can be a solid, liquid, or gas. The main benefits of petroleum are that it contains a considerable amount of energy, and that it can be used to create many products (e.g., fuels, plastics, adhesives, paints, explosives, and pharmaceuticals).

Although no one is exactly sure how petroleum was formed, the most accepted theory is that it came from the remains of carbon-based plants and animals that fell to the sea floor millions of years ago and were buried by sediments like silt and sand. Because oxygen could not penetrate the layers of sediment covering these remains, the material did not decay but built up in layers. The heat and pressure of all these layers caused the organic material to change into petroleum and the surrounding sediments to change into sedimentary rock. The end product of these changes was oil-saturated rock, not unlike a kitchen sponge.

As geological forces changed the face of Earth over time, petroleum was forced into different formations, both onshore and offshore. These formations lay dormant for many years until they were discovered by ancient civilizations.

Petroleum a substance found in the earth, such as oil or gas, composed of chemical compounds consisting primarily of hydrogen and carbon.

Hydrocarbon organic compounds that contain only carbon and hydrogen; most often found occurring in petroleum, natural gas, and coal.

Did You Know?
The word *petroleum* comes from the Latin words *petra*, which means rock, and *oleum*, which means oil. Literally, the term *petroleum* means "rock oil," or oil that comes from rocks.

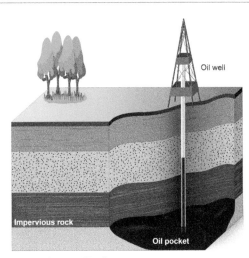

CREDIT: designua/Fotolia.

THE EARLY DAYS OF REFINING The chemical processing industry can be traced back thousands of years to civilizations including the ancient Egyptians, the Chinese, and Native Americans. In ancient times, the ability to harness and use raw petroleum products was very limited.

In 1821, a gunsmith named William Aaron Hart drilled the first natural gas well in Fredonia, New York. The gas from this 27-foot well was piped to nearby buildings and was used for fuel and lighting. In the years following Hart's experiment, many scientists and researchers continued investigating applications for coal, shale, crude oil, and natural gas.

In 1850, Samuel Kier built the first petroleum refinery in Pittsburgh, Pennsylvania. This refinery consisted of a one-barrel still that was used to distill petroleum. Later Kier added a larger five-barrel still.

Distillation the process (also called *fractionation*) of refining a compound such as petroleum into separate components. This process uses the different boiling points of liquids to separate components (called *fractions*).

In 1853, kerosene was extracted from petroleum for the first time. This was an important discovery because whale oil (the first source of lamp oil) was becoming scarce.

In 1855, Benjamin Sillman, Jr., a Yale University professor, began researching crude oil. He hypothesized that the components of the crude mixture could be separated using heat and the **distillation** process. He was correct. Sillman discovered that each component in the mixture had a unique boiling point and could, therefore, be separated out into a number of products (including naphtha, solvents, kerosene, heavy oils, and tars).

From ancient times until the mid-1800s, petroleum was collected when it seeped to the surface or collected in very shallow wells. Although this method produced a quantity of oil, it was not satisfactory for large-scale applications. In 1859, four years after Sillman's discovery, Colonel Edwin Drake (Figure 2.1) changed the face of the industry when he began exploring for oil using an old steam engine to power a drill.

Figure 2.1 Colonel Edwin Drake
CREDIT: Contributor Collection: Bettmann.

Did You Know?

- The ancient Egyptians used petroleum to help preserve mummies.
- Pitch, a tarry petroleum substance, was used extensively in ancient construction to seal off the pyramids and to waterproof Egyptian reed boats.
- The ancient Chinese sometimes found oil when they were digging for brine (salt water). They captured this oil and burned it as fuel to heat their homes.
- In ancient Persia, the streets of Babylon were said to be paved with asphalt, a solid form of petroleum.
- The ancient Persians, Sumatrans, and American Indians believed that oil had medicinal properties.
- In the 1800s, before people knew what could be done with petroleum, it was bottled throughout the United States and sold as a miracle tonic that people would rub on their skin or drink.

Drake used a steam engine-drill combination at a drill site near Titusville, Pennsylvania. The 70-foot deep well on that spot met with almost immediate success. Drake's well is considered the first commercially successful oil well.

In response to Drake's success, other oil drillers began setting down wells. Before long, the pristine Pennsylvania landscape was changed into an industrial community teeming with roughnecks, wooden oil derricks, and wagons waiting to carry raw crude to riverboats. From there, it was transported to a handful of refineries located along the East Coast.

The introduction of commercial oil wells created a need for more refineries. By 1860, the United States was home to 15 refineries, all of which primarily produced kerosene.

These early refineries consisted mainly of large iron drums with a long tube that acted as a condenser. A coal fire heated the drum, and three parts of the crude (called fractions) were boiled off. The first fraction to boil off was naphtha. The second was kerosene, and the third consisted of the heavy oils and tar, which were considered useless and were dumped into nearby rivers and streams.

Oil was transported from wells to refineries in wooden barrels (usually old whiskey barrels). Wagons or barges hauled these barrels. Barrels used to transport oil ranged greatly in size, but eventually were standardized to 42 gallons (5.6 cubic feet) around 1870. This 42-gallon standard is still the unit of measure used today.

Early attempts at transporting oil were often unsuccessful. It was quite common for oil to seep out of the barrels, with a smell of raw petroleum that would be nauseating to anyone working nearby. Furthermore, the volatility of oil made it dangerous to transport because of the likelihood of fire or explosion.

In the early days of refining, horse-drawn wagons were employed to move the crude materials from the drilling site to the refinery for processing. Sometimes the wagons would take the barrels directly to the refineries. However, in most cases, water and barges were used to complete the transport. Wagon drivers were called *teamsters* because they managed the team of horses needed to pull the wagons. They carried countless loads from the drilling site to nearby refineries, railroads, or barges.

Early oil barges were pulled by horse or depended on the river current to move. Eventually, however, barges and riverboats were transformed into tanker boats specially designed to haul oil.

Hauling oil in barrels by horse-drawn wagons and riverboats was a slow and tedious process so, in 1865, a pipeline was built from Drake's Titusville well to a nearby railroad. Early pipelines were made of wood or iron. Most designs tried to use gravity flow to push oil through the pipes and were quite inefficient in the hilly regions where oil was being drilled.

In 1865, Samuel Van Syckel built a wrought iron pipeline with steam pumps to push the oil. It also had a telegraph line running next to it, so people on each end of the pipeline could communicate about oil shipments, leaks, and more. This pipeline is considered one of the first successful oil pipelines in the United States. This new transportation system, however, was not without problems. Teamsters, who were angry at losing transport business to the pipeline, attacked the pipeline with pickaxes and caused significant damage.

In 1869, the Transcontinental Railroad was completed. For the first time, railroads stretched from the East Coast of the United States to the West Coast. Special tanker cars were built to hold the petroleum. By 1869, a wooden horizontal-style tank car with a dome was in use. The dome allowed expansion of dangerous gases. This basic style remains the industry standard today, although metal has replaced wood.

The Transcontinental Railroad helped improve the transportation of refinery products to all parts of the United States. That same year, new products made from refinery "wastes" (e.g., petroleum jelly, lubricants, candle wax, and chewing gum) were introduced into the U.S. market. However, there was still no market for the major waste product, gasoline.

In the early days, gasoline was tested as a local anesthetic. It was also used in place of kerosene but with extremely limited success because it had a tendency to set homes ablaze or, in more severe cases, to cause them to explode.

Finally, in 1885, Karl Benz invented a use for gasoline—the gasoline automobile engine. This was followed in 1892 by the diesel engine.

Between 1885 and 1886, Benz and another German named Gottlieb Daimler separately invented gasoline-powered automobile engines. With this technology, a number of companies in France, Germany, and the United States began to produce automobiles. However, they tended to be costly because only one car could be manufactured at a time.

Did You Know?

Henry Ford owed his success to Karl Benz and Gottlieb Daimler's invention of the gasoline engine.

The Ford Motor Company was founded in 1903. The first mass-produced automobile, the Model T, was not manufactured until 1908.

CREDIT: Julie Clopper/Shutterstock.

In the United States, Henry Ford created a way to mass-produce automobiles that were more reasonably priced. In 1908, the first Model T rolled off the assembly line and changed the transportation industry forever. Automobile transportation became affordable for many Americans. Tank trucks were developed to haul petroleum products where rail, watercraft, and pipelines could not go.

With the introduction of the automobile, the "waste product" gasoline became a valuable commodity. As a new market developed for this product, better processes for making gasoline were explored.

In 1897, H.L. Williams built a wharf extending 300 feet into the Pacific Ocean, erected a drilling rig, and drilled the first offshore oil well in California. When the well began to produce successfully, other wharves were built, some of them extending more than 1,200 feet out into the ocean.

WORLD WAR I AND WORLD WAR II In 1914, war broke out in Europe. World War I (then called "The Great War") was the first mechanized, chemical war. Tanks (Figure 2.2), airplanes, and chemical weapons were used that changed the nature of warfare. This brought renewed focus to the importance of the oil, gas, and chemical industries to military power.

Figure 2.2 World War I tank
CREDIT: Pecold/Fotolia.

In 1939, war broke out again in Europe and around the world. World War II spurred a huge growth in the oil, gas, and chemical industries, stimulating new advances in processing.

During the war, oil tankers were the lifeblood of both the Allied forces (i.e., the United States, Britain) and the Axis forces (Germany, Italy, Japan). German U-boats sank many Allied oil tankers as they traveled to East Coast refineries. Looking for safer alternatives, the federal government built two pipelines, the 24-inch diameter Big Inch Pipeline and the 20-inch diameter Little Big Inch Pipeline, to deliver crude oil by land from Texas to the East Coast.

Transporting oil and gas by rail and truck also was important during both world wars. The outcome of many battles was determined by the supply and performance of petroleum products.

In 1940, Standard Oil developed a new process to increase the octane in gasoline, which helped Allied transportation outperform that of the Axis powers. The Allies' logistical superiority in moving fuels around the world also is credited with winning some victories, as German offensives stalled when they ran out of fuel.

At the end of the war, millions of veterans returned home. Their return unleashed several "booms" (dramatic increases in activity): a consumer buying boom, a housing boom, an education boom, and a baby boom. During this time, the oil and gas industry enjoyed solid growth and profitability.

Increases in automobile ownership and new interstate highway systems increased the demand for gasoline products. Consumer goods manufacturing also increased to meet the demand for household goods such as plastics, synthetic fibers, and other petrochemical products.

THE 1950s TO 2000 In the 1950s, American and European companies built manufacturing locations all over the world and dominated the petroleum industry. Other countries profited little when the U.S. and Europe tapped and exported their oil resources.

In the 1960s and 1970s, foreign oil-producing countries realized they could go into business for themselves. A wave of nationalization occurred, and state-owned petroleum companies emerged. Increased production of oil caused prices to fall.

The first liquefied natural gas (LNG) plant was put into production in Algeria. **LNG** is a clear, colorless, nontoxic liquid formed when natural gas is purified and cooled to −162°C (−260°F). The cooling process shrinks the volume of the gas 600 times, making it easier and safer to store and ship. In its liquid state, LNG does not ignite. After transport, a regasification process returns the liquid to its gaseous state for use in commercial facilities and residences.

LNG liquefied natural gas; a clear, colorless, nontoxic liquid formed when natural gas is purified and cooled to −162°C (−260°F).

In 1973, the Organization of Petroleum Exporting Countries (OPEC) decided to boost prices with an oil embargo (trade restriction). The result was skyrocketing prices in the United States and Europe, lines at the gas stations (Figure 2.3), national trends toward conservation (e.g., no Christmas lights on houses), and purchases of smaller, more fuel-efficient cars.

Figure 2.3 Sign at gas station during the 1970s oil embargo.

CREDIT: Everett Historical/Shutterstock.

The oil embargo led to increased interest in alternative energy sources, such as hydroelectricity, nuclear power plants, and solar energy. In addition, exploration and production expanded into areas with harsh environments, such as Alaska, the North Sea, and the deep waters of the Gulf of Mexico.

High prices encouraged other nations (e.g., in South America, in Africa, China, and Russia) to begin producing their own oil. Eventually, the supply of oil increased and prices dropped. Concern for the environment and worker safety boomed in the 1960s and 1970s. Governments enacted legislation to provide for cleaner air and water, along with safer working environments.

In the 1980s, the price of oil dropped, greatly affecting the oil and gas industry. Exploration and production dropped off, and refining operations were optimized for better performance. During the 1980s and 1990s, increased global competition, governmental regulations, a reduction in workforce, and other factors contributed to tight markets. Many companies downsized or merged with others to remain in business.

Hydraulic fracturing a process using pressurized liquid to create cracks in rock formations to release oil and natural gas (commonly called *fracking*)

In the 1990s, a technique came into use called **hydraulic fracturing** (or fracking). Fracking is a process that uses pressurized liquid to create cracks in rock formations to release oil and natural gas. It is implemented in wells dug deep under the surface and then horizontally; it releases oil and gas that would be inaccessible by simple drilling. Fracking was first applied to shales that contain natural gas, then later to shale containing crude oil and natural gas liquids. Many wells are fracked to produce a mix of oil and gas.

2000 TO THE PRESENT   As concern for the environment escalates, there is increased pressure to minimize the footprint or impact the process industries have on the air, land, and waterways. Environmental disasters such as the Texas City oil refinery explosion of 2005 and the 2010 destruction of the Deepwater Horizon oil rig keep concerns about the safety of oil and gas in the public eye.

Governments have enacted more stringent legislation to provide for cleaner air and water, along with safer working environments. These laws have helped change the culture in many industries. Companies now seek ways to create products in a manner less dependent on local environmental resources. Process plant designs integrate hot and cold processes as a means of heating and cooling. Companies have begun recycling programs to reduce waste that was previously sent to landfills. Wastewater that was once partially purified in lagoons, or retention ponds, is now treated in onsite wastewater facilities before returning to the natural waterway. Those lagoons and retention ponds have largely been cleaned up and the top layer of soil removed to be burned in an incinerator. Waste management programs continue to be implemented, and those programs evolve to meet the requirements of local government entities.

Technological advances in plant design are still being driven by the response to air emission of hazardous pollutants in the 1990s. Plants are being either built or retrofitted to minimize the release of nitrogen oxides (NOx) and sulfur oxides (SOx). New technological designs for flare systems ensure that smoke and hydrocarbons are not released into the air. This change in regulations will continually drive company culture to ensure all processes are operated to meet those standards.

Fracking has become a major method of producing natural gas and oil. The process is controversial because of concerns related to environmental contamination and tectonics. However, it has resulted in tremendous increases in production of oil and natural gas.

2.2 Sectors of the Oil and Gas Industry

The oil and gas industry is made up of three major sectors (Figure 2.4): exploration and oil field production, transportation (e.g., pipelines, trucks), and refining. The primary focus of this chapter is the refining sector, although transportation and exploration and production also will be discussed.

Figure 2.4 Three sectors of the oil and gas industry.

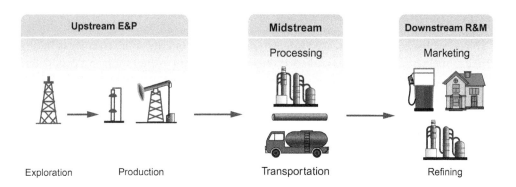

The exploration and production sector is sometimes called E&P or upstream. This sector locates oil and gas, then extracts it from the ground using drilling equipment and production facilities.

The transportation sector (also called the midstream sector) moves petroleum from where it is found to the refineries and petrochemical facilities. It also moves finished products to markets.

Petroleum consists primarily of hydrogen and carbon, or hydrocarbons. The refining sector of the oil and gas process industry (the downstream segment) takes petroleum and transforms it into finished products, such as gasoline and jet fuel, or into feedstocks (such as plastics). The downstream sector is sometimes called R&M for refining and marketing.

EXPLORATION AND PRODUCTION **Exploration** is the *upstream* process of locating oil and gas reservoirs by conducting surveys and studies, and by drilling wells. During the exploration process, scientists called geologists and geophysicists examine the earth and locate potential petroleum reservoirs on land or under water using seismic studies and other tools. After a location is identified, surveys, legal issues, and drilling rights are resolved to allow a company to drill. After a potential reservoir is detected, a well is drilled to verify the presence of hydrocarbons. Only a small percentage of wells in a new prospect are commercially viable.

After the company is ready to drill, specialized drilling equipment (called a rig) is set up at the location. Land rigs can be transported by truck, by barge, or sometimes by helicopter, while offshore rigs are built into barges, movable platforms, or ships. Figure 2.5 shows an example of an offshore rig.

Exploration the process of locating oil and gas reservoirs by conducting surveys and studies, and drilling wells.

Figure 2.5 Offshore oil rig.
CREDIT: noomcpkstic/Shutterstock.

A rig consists of a derrick (also called a tower), a power system, a mechanical hoisting system, and rotating equipment.

Through the rig, workers drill a hole in the ground using a rotating bit. The rotating equipment on the rig provides the bit with the motion it needs to cut the hole (called a *well bore*). Sections of drilling pipe (Figure 2.6) are attached to the bit, which is raised and lowered by the derrick as the well bore gets deeper.

Figure 2.6 Oil drilling pipes.
CREDIT: Pavell /Shutterstock.

As the well bore is drilled, fluid (either water or drilling fluid called mud) is circulated into the hole to cool the drill bit and remove cuttings. Casing, or pipe, is placed in the hole to keep it from collapsing. A blow-out preventer is placed on the surface above the well bore; this device prevents an uncontrolled rush of oil or gas to the surface.

Drilling continues in stages. As the well is drilled deeper, more casing is added. Testing processes (samples and electronic sensors) are used to check for the presence of petroleum. When the well's depth reaches the petroleum reservoir, a device blows holes in the casing to let the petroleum flow. Tubing is inserted into the well and a valve structure is placed at the top of the well. This valve structure allows the flow of petroleum to be controlled.

After the well is drilled, the production phase starts. Facilities and equipment are brought in to extract the petroleum to the surface and transport it to another location for processing. Traditional wells or fracking wells may be drilled.

Many state and federal agencies are involved with the regulation of oil and gas exploration and production. These agencies include Occupational Safety and Health Administration (OSHA), Environmental Protection Agency (EPA), Department of Transportation, U.S. Coast Guard, Minerals Management Service, and others.

Transportation the oil and gas industry segment responsible for moving petroleum from wells to processing facilities and finished products to consumers. Transportation methods include pipelines, watercraft, railways, and trucks.

TRANSPORTATION Transportation (*midstream* sector) is the oil and gas industry segment responsible for storing, marketing, and moving petroleum from wells to processing facilities and finished products to consumers. Transportation methods include pipelines, watercraft, railways, and trucks. Midstream operations are sometimes included in the downstream category.

Pipelines are the prime method for transporting petroleum products. They move high volumes of material product at a low cost. Pipelines account for almost two-thirds of the petroleum quantities moved in the United States. In general, pipelines tend to run east-west because a majority of the oil is in the states of Texas, Louisiana, Oklahoma, and Alaska, while demand and some of the refining capabilities are on the East Coast (Figure 2.7A).

Gas pipelines and oil pipelines are traditionally separate systems. A gas pipeline uses compressors to generate pressure as a driving force for the gas, while an oil pipeline uses pumps to generate pressure to drive liquids. When compared with other transportation methods, pipelines are one of the safest methods for transporting oil and gas.

The second most used transport method for oil and gas is watercrafts (e.g., oil tankers, as seen in Figure 2.7B). The costs for this type of transportation are moderate.

Figure 2.7 Oil transportation methods. **A.** A portion of the Alaska pipeline. **B.** Oil tanker.

A. Courtesy of Martha McKinley.
B. EvrenKalinbacak/Fotolia.

A.   B.

Tankers transport products around the world from large oil-producing regions to consumer countries. Within North America, barges and other vessels transport petroleum products through intercoastal waterways and to coastal ports. After the boom of the 1970s and the following economic downturn, the fleet of oil tankers began to diminish in size. New vessel construction remains slow as the fleet continues to age.

Pipelines replaced railways as the major method of transport during the twentieth century. The cost of rail delivery is moderate, and it is limited to the existing system of rails. The current use of rail transportation for petroleum products is fairly static, remaining at less than 5 percent of the total volume of transportation. Since rails tend to run north-south rather than east-west, they are useful in transporting petroleum products to northern and southern locales.

Trucks are the most flexible transportation method for the delivery of petroleum. They can be used to transport material to smaller distribution centers and from fields that do not have a pipeline system in place. However, the amount of material that can be transported by truck at one time is limited, so the cost of using trucks for petroleum product transportation is high.

Many state and federal agencies are involved with the regulation of oil and gas transportation. These agencies include OSHA, EPA, Department of Transportation, U.S. Coast Guard, and Minerals Management Service, among others.

REFINING Refining (*downstream* sector) is the process of purifying a crude substance into other products, such as petroleum being separated into gasoline, kerosene, gas, and heavy oil. The downstream sector includes petroleum refinement, petroleum product distribution, retail outlets, and natural gas distribution companies.

A refinery is a facility that separates various petroleum components into fuel products or feedstocks for other processes, to produce such items as plastics or pharmaceuticals (Figure 2.8). Refineries involve a series of processes to separate and change petroleum components.

Refining the process of purifying a crude substance into other products, such as petroleum being separated into gasoline, kerosene, gas, and oil.

Figure 2.8 Oil refinery.
CREDIT: FOTOimage Montreal/Shutterstock.

The refining process relies on differences in the physical properties of components in a petroleum mixture (e.g., boiling points) to separate those components. The following is an overview of the refining process:

1. Distillation (also called fractionation). The refining process begins with the distillation of petroleum into separate components. This process uses the different boiling points of liquids to separate components (called fractions). Atmospheric and vacuum towers are used during this process.
2. **Conversion**. A process that changes the size and/or structure of the petroleum components. One method called "cracking" breaks down large molecules into smaller ones. If heat is used, the process is called *thermal cracking*. If a catalyst (a substance that is used to affect the rate of a chemical reaction) is used, the process is called *catalytic cracking*. Other methods include combining components and rearranging components.
3. **Treatment**. A process that prepares the components for additional processing. This process also creates some final products. Various components or contaminants are removed or separated at this point. Chemical or physical separation can be used, along with a variety of other methods and treatments.
4. **Formulating and blending**. A process that combines or mixes components and additives to produce finished products with specific performance requirements.
5. Additional processes, which include recovering components, treating wastes and water, cooling, handling, moving, and storing.

Conversion a process that changes the size and/or structure of the petroleum components by breaking them down, combining them, or rearranging them.

Treatment a process that prepares the components for additional processing and creates some final products. Treatments can remove or separate various components or contaminants at this point.

Formulating and blending a process that combines or mixes components and additives to produce finished products with specific performance requirements; also called *blending*.

Again, many state and federal agencies are involved with the regulation of oil and gas refining. These agencies include OSHA, EPA, Department of Transportation, U.S. Coast Guard, Minerals Management Service, and others.

2.3 Duties, Responsibilities, and Expectations of Process Technicians in the Oil and Gas Industry

In addition to items described in detail in Chapter 1, process technicians in the oil and gas industry must be prepared to fulfill a variety of tasks. These range from operation of oil transfer pumps and filtration equipment to visual inspection and preparation for maintenance of piping and chutes. They work with volatile substances in potentially hazardous conditions, so safety must be a priority. Process technicians employed in these fields must be prepared for outside jobs and work in every type of weather and season. Coordination and physical strength are assets.

Process technicians must know about the federal government agencies and regulations that affect these industries: OSHA, EPA, Department of Transportation, Mine Safety and Health Administration, Minerals Management Service, Bureau of Land Management, and others.

Objectives

Upon completion of this chapter, you will be able to:

2.4 Explain the background, growth, and development of the mining industry. (NAPTA History 1-4) p. 34

2.5 Describe the different sectors of the mining industry. (NAPTA History 1, 3) p. 36

2.6 Identify duties, responsibilities, and expectations of the process technician in the mining industry. (NAPTA Career 1) p. 37

Introduction to the Mining Industry

HOW MINERALS ARE FORMED Many minerals are formed when magma (molten rock beneath the earth's surface) interacts with other liquids and gases. As magma cools, deposits of the mineral (called veins or lodes) collect according to the mineral's chemical properties. Coal, an important energy-producing mineral, is an organic material consisting primarily of hydrogen and carbon, formed millions of years ago by biological matter and geological forces.

Process technicians in the mining industry should study and understand **geology** (study of Earth and its history as recorded in rocks). Key concepts in geology include forces such as the following:

- **Rock cycle**—The process by which the different types of rocks are formed, exposed to weather and other forces that erode or break them down, and then are reformed by geological forces such as heat and pressure.
- **Plate tectonics**—A theory that explains how the planet moves, as semi-rigid plates in the earth's crust drift or flow, causing geological changes such as volcanoes, mountains, and earthquakes.
- **Rock formations**—Geology that is arranged or formed in a certain way. Certain formations, such as sandstone, are of interest to industry because they contain oil and gas.
- **Rocks**—Minerals and gems, or a combination of minerals or gems mixed with other chemical compounds; their history is intertwined with the history of ancient civilizations, specifically with regard to buildings, food and cooking, weapons, religion, tools, and medicine.

Geology the study of Earth and its history as recorded in rocks.

Rock cycle the process by which rocks are formed, exposed to forces that erode or break them down, and then are reformed by forces such as heat and pressure.

Plate tectonics a theory describing movement of semi-rigid plates in the earth's crust, which cause geological changes such as volcanoes, mountains, and earthquakes.

Rock formations geology that is arranged or formed in a certain way; also, scenic outcroppings formed by weathering and erosion.

Rocks minerals or gems, or a combination of minerals or gems mixed with other chemical compounds

Mining is the extraction of valuable minerals or other geological materials from the earth. **Minerals** are naturally occurring, inorganic substances that have a definite chemical composition and a characteristic crystalline structure. According to the International Mineralogical Association (IMA), more than 5,000 different types of minerals have been identified. Minerals are the building blocks of rocks. Rocks can be composed of a single mineral or a combination of minerals. Rocks and minerals may be just below the earth's surface or deeply buried. Minerals generally are categorized as metallic (metals) or nonmetallic (nonmetals):

- **Metals** are chemical elements that have luster (ability to reflect light) and can conduct heat and electricity (e.g., copper, bauxite, iron, lead, gold, silver, zinc, nickel, and uranium).
- **Nonmetals** are poor conductors of heat and electricity; are brittle, waxy, or gaseous; and cannot be hammered into sheets or drawn into wire (e.g., gems and precious stones, coal, gravel, sand, lime, stone, soda ash, phosphate rock, and clay).

Coal, like oil and gas, is considered an energy-producing mineral.

Minerals can be extracted from either land or water. A **mine** is a pit or excavation from which minerals are extracted. A **quarry** is an open excavation from which stones are extracted. Mining operations also take place offshore. For example, electrolysis and other techniques can be used to remove minerals from seawater.

Ore is a solid material bearing metal or other types of minerals that are valuable enough to warrant mining (e.g., iron or gold). A **deposit** is a natural accumulation of ore. Ore is extracted through mining.

Mining involves three fundamental elements: earth, air, and water. For example, during the mining process:

- Large amounts of earth are moved.
- Air (ventilation) is provided to enclosed mines.
- Water is pumped in or out, as the process requires.

Mining uses powerful equipment and forces to handle these elements. Explosives and massive machines are used to move earth. Large compressors and fans are used to provide ventilation. Huge pumps are used to move water.

Mining products (Figure 2.9) are integral to a wide range of industries. They serve as base materials for utilities and power generation, construction, transportation, agriculture, electronics, food production, pharmaceuticals, personal hygiene, and consumer products.

> **Did You Know?**
>
> More than two-thirds of the world's magnesium supply is mined from seawater.
>
>
>
> CREDIT: stringerphoto/Fotolia.

> **Did You Know?**
>
> The Fort Knox gold mine in Alaska retrieves only about 0.028 ounce of gold per ton of rocks moved.
>
>
>
> CREDIT: johnphotostock/Fotolia.

Mining the extraction of valuable minerals or other geological materials from the earth.

Minerals and gems naturally occurring inorganic substances that have a definite chemical composition and a characteristic crystalline structure.

Metal chemical elements that have luster (ability to reflect light) and can conduct heat and electricity (e.g., copper, bauxite, iron, lead, gold, silver, zinc, nickel, and uranium).

Nonmetals substances that conduct heat and electricity poorly; are brittle, waxy, or gaseous; cannot be hammered into sheets or drawn into wire (e.g., gems and precious stones, coal, gravel, sand, lime, stone, soda ash, phosphate rock, and clay).

Mine a pit or excavation from which minerals are extracted.

Quarry an open excavation from which stones are extracted.

Ore a metal-bearing mineral that is valuable enough to warrant mining (e.g., iron or gold).

Deposit a natural accumulation of ore.

Figure 2.9 One product of mining industries.

CREDIT: RTimages/Fotolia.

The following are some examples of the many impacts products from mining processes have on everyday life:

- Coal for power generation
- Zinc for pharmaceuticals
- Salt for food
- Cement for construction
- Iron for steel (e.g., for cars, ships, buildings)
- Copper for electronics
- Gold for communications and jewelry
- Graphite for pencils
- Fluoride for toothpaste
- Silver or diamonds for jewelry
- Soda ash for fertilizers
- Feldspar for ceramics.

MINING ECONOMICS AND TYPES OF MINING The economics of mining are challenging. First, companies must search for rock and mineral sites, and then obtain mining rights. Then, a considerable amount of extracting work must be done, often to obtain only small amounts of minerals. Finally, physical and/or chemical processing must be performed to prepare and refine the rocks and minerals. To add to these challenges, this process, which is often expensive and time consuming, occurs in an economic environment where prices can fluctuate rapidly.

Mining operations vary greatly depending on the type of rock or mineral being mined, the location of the deposit (above ground, near the surface, or deep below the surface), the environment, and many other factors (Figure 2.10). For this reason, this text describes mining operations in very general terms.

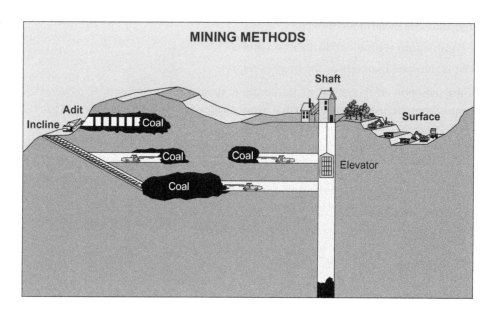

Figure 2.10 Different types of mining methods.

There are two basic types of mining: surface and underground.

Surface Mining Surface mining, also called open-pit or strip mining, is used when the rocks or minerals are located near the earth's surface (Figure 2.11). Surfacing mining is typically more cost-effective than underground mining, because it requires fewer workers to produce the same amount of ore.

Figure 2.11 Example of a surface mine.
CREDIT: vladimirnenezic/Fotolia.

These general tasks are associated with surface mining:

1. Explosives are used to move quantities of earth away from the deposit.
2. Workers use large earth-moving equipment to move away layers of dirt and rock covering the deposit.
3. When the deposit is exposed, workers use smaller earth-moving equipment to lift and load the ore from the deposit into trucks. If necessary, explosives might be used to break down the deposit further.
4. The ore retrieved is often crushed on site and then transported to a nearby facility for processing.
5. When all of the ore that can be practically extracted has been removed, the mine and its surrounding area are restored to original condition.

Quarrying operations are similar to mining operations in that stones (e.g., marble, granite, sandstone) are split into blocks from a massive rock surface.

Underground Mining The different types of underground mining are:

- Conventional. This oldest method involves cutting beneath the ore to control the direction of its fall, and using explosives to blast the ore free.
- Longwall. Like conventional mining, longwall mining uses a machine with a rotating drum to shear the ore and load it onto a conveyor; hydraulic jacks are used to reinforce the roof.
- Continuous. A machine called a continuous miner rips ore out of the ground and loads it directly onto a transportation device.

The following tasks are generally associated with underground mining:

1. Miners dig openings (tunnels) in the earth. Depending on the deposit's location in relation to the surface, tunnels may be vertical, horizontal, or sloped. Tunnels serve one of two purposes: (a) allow miners to reach the deposit, extract the mineral, and transport it to the surface; or (b) act as ventilation, allowing fresh air into the mine.
2. Supports and pillars of unmined ore are used to support the mine roof.
3. Additional interconnecting passages are dug.
4. Ore is removed from the deposit and taken to the surface using shuttle carts, rail cars, conveyor belts, or other methods.

Did You Know?

Arsenic is an element that is naturally found in geological formations, especially granite. As long as this arsenic remains bound to the granite, it causes no harm.

When the granite is disturbed, however, either through mining or natural erosion, the arsenic can be released into the soil, thereby contaminating groundwater. That is why mining companies must monitor water quality during and after mining operations.

CREDIT: horkins/Shutterstock.

Did You Know?

On August 5, 2010, 33 miners were trapped 2,300 feet underground in a mine in Northern Chile after a main shaft collapsed. After a search of 17 days, a probe into the mine was returned with a note stating that all 33 miners were alive and unhurt. It took more than 5 weeks and international support for a bore hole to reach the area where miners were trapped, and another month for the hole to be widened and secured. On October 13, a rescue pod was lowered 33 times, returning all miners safely to the surface.

5. Ore is crushed (usually on site) and then hauled away for processing.
6. When all of the ore that can practically be extracted has been removed, the mine and its surrounding area are restored to original condition.

After underground mining operations have ceased, the mining company must check groundwater to make sure it is not contaminated and must take actions to ensure that abandoned mines do not collapse.

2.4 Growth and Development of the Mining Industry

Many early civilizations were familiar with metals and other minerals (Figure 2.12). Because of this, the early history of civilization often is described in different ages, based on the types of minerals or metals they used (e.g., stone, bronze, iron, steel).

- Stone Age (around 2 million years ago)—The earliest technological period in human culture when metal was unknown, and tools were made of stone, wood, bone, or antlers.
- Copper Age (around 5000 BC)—A phase in the development of human culture in which the use of early metal tools (made of copper) appeared alongside the use of stone tools.
- Bronze Age (around 2500 BC)—The period in history characterized by the development of bronze and its use in the creation of weapons and tools.
- Iron Age (around 1500 BC)—The period in history characterized by the development of iron and its use in the creation of weapons and tools.
- Steel Age (around AD 200)—The period in history characterized by the manufacture and use of steel in tools, weapons, and construction materials.

Figure 2.12 Helmet and sword.
CREDIT: steliangagiu/Shutterstock.

A variety of other metals and minerals were used during these ages as well. For example, the Egyptian pharaohs used gold, silver, and quarried stones to create jewelry and ornamentation for buildings, furnishings, and monuments. The Chinese used jade and ceramics, and the Greeks had marble statues and cosmetics made from minerals.

Initially, these early civilizations found metals and minerals on the earth's surface, but they soon discovered ways to obtain metals from the ground when surface supplies became scarce. During these ancient times, the techniques for mining and processing minerals changed very little.

In the sixteenth century, coal was discovered to be a good source of energy (i.e., heat). Because of this, the demand for coal increased. Within a few hundred years, it became the fuel source for the Industrial Revolution.

INDUSTRIAL REVOLUTION The Industrial Revolution, which started in England in the 1800s and spread to the United States shortly thereafter, changed the nature of manufacturing and sparked an increase in consumer products. Factories, powered by coal, began mass-producing products that previously were made by hand. Workers flocked to cities where factories provided jobs.

Coal also helped the growing transportation industry, which was required to deliver products to new markets. Railroads and ships benefited from the power of coal.

The discovery of gold and other precious minerals also spurred population shifts, such as gold rushes in California and Alaska.

WORLD WARS I AND II In the early and mid-1900s, two world wars increased the need for mining and minerals. Coal, along with oil and gas, provided fuel to the war efforts, while metals such as steel (iron with low carbon content) and brass (essentially copper and zinc) were used to make vehicles and weapons.

Following World War II, a boom in the economy meant more minerals were needed for consumer products. As the construction business skyrocketed, for example, cement, stones, and metals were needed. Also, manufacturing increased, leading to a rise in the production of products such as electronics and appliances. The face of transportation changed, as more automobiles were produced, air travel became common, and ships transported goods around the world.

1960s TO 2000 In the 1960s, coal was used to satisfy about half the world's energy needs. By the 1970s, oil and gas had become the world's primary energy source, fulfilling two-thirds of total energy demand.

Concern for the environment and worker safety boomed in the 1960s and 1970s. Governments enacted legislation to provide for cleaner air, water, and land, along with safer working environments (Figure 2.13).

Figure 2.13 Process industry flare.
CREDIT: George Spade /Fotolia.

During the 1990s, production of both minerals and coal increased. Metal prices were volatile, leading to fluctuating production. Nonmetallic mineral prices were more stable, leading to fewer production changes.

Changes in technology made mining techniques safer and more automated (e.g., some mining operations use lasers and robotics in their processes). Productivity has increased. Old sources of minerals that had been considered unprofitable to mine are now being revisited because of more efficient recovery techniques.

2000 TO PRESENT In this second decade of the twenty-first century, the mining process is being challenged to utilize advanced technologies to ensure the sustainability of the industry. Fatal accidents in the past brought to the world's attention the need for changes in occupational health to maintain and promote the safety of employees. Mining has been seen as a detriment to the environment, and changes in government regulations challenge the mining industry to minimize the footprint left by the process of ore removal.

Mining is now a global industry, very competitive in its search to broker partnerships with other countries. Technologies have been developed to allow mines that were once thought to be exhausted to be explored again for greater resources. Advanced technologies and an emphasis on safety will be the focus of the mining industry of the future.

2.5 Sectors of the Mining Industry

In general, the mining process involves the following sectors:

1. Searching for rock and mineral deposits
2. Extracting the rocks and minerals
3. Processing the rocks and minerals
4. Transporting the rocks and minerals to consumers

PROCESSING OPERATIONS   After the ore is mined, it must be processed. Processing plants are usually located next to or near a mine or quarry. The job of the processing plant is to remove the target metals from the ore, then process it to meet certain specifications. Figure 2.14 provides a process diagram of the mining industry.

Figure 2.14 A. Overview of processes in the mining industries.

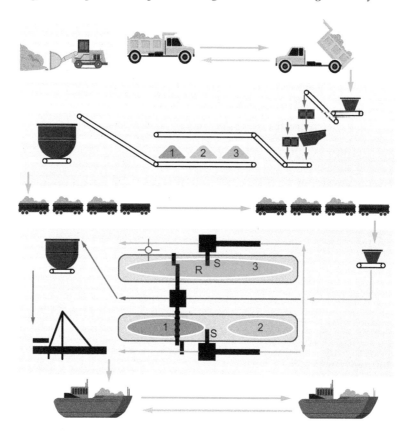

The following are some common methods associated with processing:

- Separating and classifying. (Figure 2.15A) Sorting minerals for processing. Machines called cyclones typically use centrifugal force to sort minerals.
- Crushing. Using physical impact to reduce the size of ore. Crushing is typically a dry operation performed in three stages.
- Grinding. Using impact and abrasion to further reduce the size of crushed rock and ore. Grinding can be dry or wet, and can take place in three stages.
- Sizing. Dividing crushed and ground minerals into different size categories (Figure 2.15B).
- Chemical processing. Using different solutions to process the minerals. For example, during leaching, ore is mixed with chemical solutions, solvents, or other liquids in order to separate the different material. Thickeners and clarifiers are added to condition the minerals.

- Filtering. Using a porous medium to separate solids from liquids. In this method, a filter retains solids while allowing liquids to pass through.
- Washing. Using water or other fluids to rinse the minerals and carry away impurities.

Figure 2.15 **A.** Ore conveyor machine. **B.** Rock sorting mechanism.

A. Courtesy of Martha McKinley. **B.** Werner Stoffberg/Shutterstock.

2.6 Duties, Responsibilities, and Expectations of Process Technicians in the Mining Industry

Process technicians in the mining industry will be expected to work in a variety of working conditions, including working around noisy equipment and dust and performing duties that require physical exertion and stamina.

Besides the usual equipment involved in process units, process technicians in mining also might be asked to work with the following equipment:

- Pipes and chutes
- Excavation equipment
- Supports and jacks

Process technicians also might work with explosives, jackhammers, power tools, heavy lifting equipment, earth-moving equipment, and various transportation equipment (e.g., carts, conveyors, and railcars), cutting tools, and other related equipment.

Along with the responsibilities discussed in Chapter 1, safe practice must be a top priority for process technicians in the mining industry.

Summary

Oil and Gas Industry

Many segments make up the oil and gas industry, including oil field exploration and production (the upstream segment), transportation (midstream segment), and refining (downstream segment). The upstream segment locates oil and gas and extracts them from the ground using drilling equipment and production facilities. The midstream segment moves petroleum from where it is found to the refineries and petrochemical facilities, and then takes finished products to markets. The downstream segment of the oil and gas process industry transforms petroleum into finished products such as gasoline or into feedstocks such as plastics that are used to make other products.

The history of the oil and gas process industry stretches from ancient civilizations to the present. Early uses for oil varied from heating to medicine to waterproofing. In the nineteenth century, early refineries and oil wells were started in the United States, mainly producing kerosene for heating and lighting. The invention of the automobile, two world wars, and a population boom after World War II led to tremendous expansion in the use of petroleum products. This resulted in increased air and water pollution. Because of this, environmental, health, and safety regulations grew in importance during the 1960s and 1970s. New technologies such as wind and solar power were also developed.

Changes in availability and production have resulted in mergers, downsizing, and newer processes such as fracking. Companies now operate more efficiently and effectively for lower cost. Changes include increased automation and computerization. Fewer workers are needed to run processes, but these people need more skill and education than workers in the past. A two-year associate degree is preferred for process technicians.

Process technicians must respond continually to environmental, health, and safety legislation related to recycling and reducing pollutants into air, land, and water. This requires keeping informed, communicating among teams, and learning new skills in automation and technology.

Mining Industry

The general mining process involves searching for rock and mineral deposits, extracting the rocks and minerals, processing the rocks and minerals, and providing the rocks and minerals to consumers. Mining products serve as base materials for utilities and power generation, construction, transportation, agriculture, electronics, food production, pharmaceuticals, personal hygiene, consumer products, and other industries.

Minerals are inorganic substances that occur naturally, with different chemical compositions and properties. Minerals generally fall into two categories: metallic (e.g., iron, copper) and nonmetallic (e.g., coal, stone).

Mining involves extraction and processing of rocks and minerals from the ground, either from a mine (pit) or a quarry (open excavation). Mining operations also take place offshore.

Processing plants are usually located near a mine or quarry. Rock and other impurities are removed from the ore, which then is processed to meet certain specifications.

From ancient times to 1800s, not much changed in the way rocks, minerals, and gems were excavated and processed. Then the Industrial Revolution changed the nature of manufacturing and sparked an increase in consumer products made in factories powered by coal.

The two world wars increased the need for mining and minerals. Coal, along with oil and gas, provided fuel to the war efforts, while metals such as steel (iron with low carbon content) and brass (essentially copper and zinc) were used to make vehicles and weapons.

Pollution from fossil fuel industry led to concern for the environment in the 1960s and 1970s. Governments enacted legislation to provide for cleaner air, water, and land, along with safer working environments. Changes in technology have made mining techniques safer and more automated. As a result, productivity has increased.

Process technician jobs are more complex and cross-functional than in the past. Technicians are responsible for entire complex processes, not just simple tasks. Increased emphasis is placed on keeping costs low by improving safety, performing preventative maintenance, optimizing processes, and increasing efficiency.

Checking Your Knowledge

1. Define the following key terms:
   a. deposit
   b. exploration
   c. geology
   d. hydrocarbon
   e. metal
   f. mineral
   g. mining
   h. nonmetal
   i. ore
   j. petroleum
   k. quarry
   l. refining
   m. transportation

Oil and Gas Industry

2. _____ is a substance found in the earth, such as oil or gas, composed of chemical compounds consisting primarily of hydrogen and carbon.

3. Put the following events in order, from earliest to latest:
   a. automobile invented
   b. first refinery built in the United States
   c. improvements in refining during World War II
   d. first successful commercial oil well drilled in Pennsylvania

4. Name three segments of the oil and gas industry.

5. What did the earliest refineries produce?
   a. kerosene
   b. propane
   c. gasoline
   d. diesel

6. What is the most common method of transporting petroleum?
   a. watercraft
   b. pipelines
   c. railcars
   d. trucks

7. What is the process of purifying a crude substance into other products called?
   a. exploration
   b. transmutation
   c. fractional distillation
   d. refining

Mining Industry

8. _____ are inorganic substances that occur naturally, with different chemical compositions and a definite crystalline structure.

9. What are the two primary types of minerals?
   a. coal and metal
   b. hard and soft
   c. sedimentary and igneous
   d. metallic and nonmetallic

10. What are the two main types of mining?
    a. surface and strip
    b. strip and open pit
    c. surface and underground
    d. crushing and stripping

11. What event sparked the use of coal?
    a. the Crusades
    b. the Industrial Revolution
    c. World War II
    d. the 1970s energy crisis

12. In the 1970s, coal provided about what portion of the world's energy needs?
    a. one-half
    b. seventy-five percent
    c. ten percent
    d. one-third

13. What unique equipment might process technicians in the fossil fuel industries be expected to use?
    a. pumps
    b. hoses
    c. pipes and chutes
    d. personal protective equipment

NOTE: Answers to Checking Your Knowledge questions are in Appendix I.

Student Activities

1. Which event or period in time had the biggest impact on the oil and gas industry? Why? Write at least five paragraphs in response.

2. Tell which of the following oil and gas segments sound the most interesting to you and explain why. Write several paragraphs in response.
   a. exploration and production
   b. transportation
   c. refining

3. Explain which type of mineral has the greatest impact on modern living, metals or nonmetals? Write at least five paragraphs in response.

4. Pick one of the following tasks and research some specific details about it for the oil and gas industry or the mining industry. Use the Internet, library, or other resources for your research. Write at least five paragraphs in response.
   a. monitoring and controlling process
   b. maintaining equipment
   c. troubleshooting and problem solving
   d. performing administrative activities
   e. performing assigned duties with a focus on safety, health, environment, and production
   f. returning equipment to service

Chapter 3
Chemical and Pharmaceutical Industries

Objectives

Upon completion of this chapter, you will be able to:

3.1 Explain the growth and development of the chemical industry. (NAPTA History 1-4) p. 41

3.2 Provide an overview of the chemical industry sectors. (NAPTA History 1) p. 44

3.3 Identify the duties, responsibilities, and expectations of the process technician in the chemical industry. (NAPTA Career 1) p. 45

Key Terms

Alchemy—a medieval practice that combined occult mysticism and chemistry, **p. 41.**
Antibiotics—substances derived from mold or bacteria that inhibit the growth of other microorganisms (e.g., bacteria or fungi), **p. 47.**
Apothecary—a person who studies the art and science of preparing medicines; in modern times we call these individuals pharmacists, **p. 46.**
Biologicals—products (e.g., vaccines) derived from living organisms that detect, stimulate, or enhance immunity to infection, **p. 47.**
Chemical—a substance with a distinct composition that is used in, or produced by, a chemical process, **p. 41.**
Commodity chemicals—basic chemicals that are typically produced in large quantities and in large facilities. Most of these chemicals are inexpensive and are used as intermediates, **p. 41.**
Compounding—mixing two or more substances or ingredients to achieve a desired physical form, **p. 46.**
Drugs—substances used as medicines or narcotics, **p. 46.**
Intermediates—substances that are not made to be used directly, but are used to produce other useful compounds, **p. 41.**
Petrochemical—a chemical derived from fossil fuels or petroleum products, **p. 45.**
Pharmaceuticals—manmade or naturally derived chemical substances with medicinal properties that can be used to treat diseases, disorders, and illnesses, **p. 45.**

Specialty chemicals—chemicals that are produced in smaller quantities, are more expensive, and are used less frequently than commodity chemicals, **p. 41.**

Synthetic—a substance resulting from combining components, instead of being naturally produced, **p. 42.**

Introduction to the Chemical Industry

A **chemical** is a substance with a distinct composition that is used in, or produced by, a chemical process. Chemicals play a vital role in a wide range of manufacturing processes. Consider plastic, one of the most common chemical products. Plastics are found in almost every aspect of modern society. They are used in computers, cars, hospitals, houses, appliances, electronics, food packaging, games, toys, manufacturing, aircraft, schools, businesses, sporting goods, and more.

Table 3.1 lists some of the items produced by the chemical industry.

Chemical a substance with a distinct composition that is used in, or produced by, a chemical process.

Table 3.1 Examples of Products Produced Through Chemical Manufacturing

Industry	Examples of Chemical Products
Food and beverage	sweeteners, preservatives, flavorings, packaging
Medical	pharmaceuticals, plastics (e.g., syringes, tubes)
Construction	paints, coatings, adhesives, colored glass
Agriculture	fertilizers, pesticides
Clothing	dyes, fibers
Transportation	artificial rubber, improved fuels, oils, lightweight materials
Cleaning	detergents, bleaches
Military	explosives, armor plating
Personal hygiene	soaps, perfumes, beauty aids
Amusement and hobbies	photographic materials, plastic toys and games, sporting goods

Chemical manufacturing requires an understanding of chemistry, manufacturing processes, and the special equipment and structures used to process the chemicals.

Some chemical companies produce products that are sold directly to consumers, although most create feedstocks that are used to create other products. Some chemicals are considered commodity chemicals, and others are considered specialty chemicals.

Commodity chemicals are basic chemicals, typically produced in large quantities and in large facilities. Most of these chemicals are inexpensive and are used as **intermediates** (substances that are not made to be used directly, but are used to produce other useful compounds).

Specialty chemicals are chemicals that are produced in smaller quantities, are more expensive, and are used less frequently than commodity chemicals.

Commodity chemicals basic chemicals that are typically produced in large quantities and in large facilities. Most of these chemicals are inexpensive and are used as intermediates.

Intermediates substances that are not made to be used directly, but are used to produce other useful compounds.

Specialty chemicals chemicals that are produced in smaller quantities, are more expensive, and are used less frequently than commodity chemicals.

3.1 Growth and Development of the Chemical Industry

IN THE BEGINNING The use of chemicals can be traced back to early civilizations that existed thousands of years ago. For example, in 7000 BC Persian artisans produced glass using refined alkali and limestone. Roughly 1,000 years later, Phoenicians produced soap. Early Greek, Chinese, and Egyptian cultures also contributed greatly to advances in chemistry and chemicals. These cultures sought ways to cure diseases, prolong life, and change common metals into precious metals such as gold and silver. Through their studies and experiments, a pseudoscience called alchemy was born. **Alchemy** was a medieval practice that combined occult mysticism and chemistry (e.g., some alchemists tried to blend nonprecious metals to create gold, while others tried to create elixirs of immortality).

Alchemy a medieval practice that combined occult mysticism and chemistry.

42   Chapter 3

Arabs added to the study and practice of alchemy. During this learning process, they also created small amounts of chemical, medicinal elixirs and a compound similar to plaster of Paris that could be used to heal broken bones.

CREDIT: arkalyk/Fotolia.

> **Did You Know?**
>
> The word *chemistry* is supposedly derived from the word *khemela*, a variation of the Greek word *khumos* or juice. Khemela is said to relate to the art of extracting juices, which was how early chemistry was viewed.

1000s–1600s   Europeans picked up the practice of alchemy from the Arabs during the Crusades, around the eleventh century. About the same time, the Chinese invented black powder, an early explosive.

Between the eleventh and sixteenth centuries, many alchemy practices, which previously had been closely guarded secrets passed verbally from mentor to student, were written down by scholars and scientists. In 1597, the German alchemist Andreas Libau published what is considered the first chemistry textbook, *Alchemia*.

1600s–1800s   In the 1630s, Pilgrims in Massachusetts were using chemicals to tan leather and were making saltpeter, a crucial component in gunpowder.

In the eighteenth century, alchemy began to transition from mysticism to a serious science. A Flemish physician, Jan Baptista Van Helmont, described how air and vapors were similar in appearance but had different properties. He called air and vapors *chaos*, a Greek word for the material used to create the universe. In Flemish, *chaos* is pronounced similarly to *gas*, which is what we still collectively call air, vapors, and similar substances.

Large-scale chemical manufacturing began around the time of the Industrial Revolution, from the late 1700s through the late 1800s, when England and the United States transformed from farming and trading economies into large-scale manufacturing economies.

In the 1800s, consumer demand rose for items such as glass, paper, explosives, matches, fertilizers, soaps, and textiles. Manufacturing of these items required large quantities of chemicals, including soda ash, sulfuric acid, and potash. In the 1820s, James Muspratt began mass-producing soda ash in Britain. Soda ash is used for making soap, glass, and other products. Some years later, synthetic dyes for textiles were produced from coal tars. A **synthetic** is a substance created by combining components, instead of being naturally produced.

Synthetic   a substance resulting from combining components, instead of being naturally produced.

As early as 1839, English citizens were petitioning to have production of certain chemicals such as soda ash banned, because manufacturing emissions killed the trees, vegetation, and animal life around the factories and caused serious ailments.

Around 1890, German companies starting producing sulfuric acid in large quantities. Other companies began to mass-produce caustic soda and chlorine.

WORLD WAR I   At the start of the 1900s, synthetic fertilizers revolutionized agriculture by improving crop yields. By the late nineteenth and early twentieth centuries, the United States and Britain had become dominant players in chemical manufacturing, but Germany was more advanced in basic chemistry research.

In 1914, World War I (called "The Great War") broke out in Europe. WWI was the first major conflict that involved mechanized and chemical warfare. The countries involved in this war all used chemical weapons, primarily in the form of mustard gas, to kill the enemy. The war brought renewed focus on the importance of chemical and refining industries to military power. Extensive chemical research was conducted on chemical weapons during WWI.

Rayon (dubbed "artificial silk") was the first manufactured fiber, though not truly synthetic because it is derived from cellulose. Nylon was the first synthetic fiber made solely from petrochemicals. Nylon was first produced by Du Pont in 1928 for women's stockings.

CREDIT: ScandinavianStock/Shutterstock.

> **Did You Know?**
>
> Women's nylon stockings were in short supply during World War II because of nylon shortages (the nylon was needed for parachutes and other equipment). Because of this, nylon stockings became a black market item, and the price skyrocketed.

WORLD WAR II In the late 1930s and early 1940s, another world war (WWII) broke out and, once again, chemicals played an important role in the conflict. The war spurred a huge growth in the refining and chemical processing industries and stimulated many new process advances (Figure 3.1).

Figure 3.1 World War II created a need for new products such as synthetic rubber.

CREDIT: konstantinks/Shutterstock.

The chemical and refining industries made significant contributions to U.S. victories during World War II. By the time the United States entered the war in 1941, Japan had captured 90 percent of the rubber-producing nations in Asia. This created a need for rubber substitutes for a variety of products (e.g., tires, shoes, raingear, and tents). In 1940, the first synthetic rubber tire was produced in the United States from butadiene and styrene. Also during the war, the nylon industry was entirely taken over for the war effort, for parachutes, tents, rope, and other uses.

1960–2000 Concern for the environment and worker safety boomed in the 1960s and 1970s (Figure 3.2). Governments enacted legislation to provide for cleaner air and water, along with safer working environments.

Figure 3.2 Concern for clean air and water led to regulations that have improved air and water quality in the United States.

CREDIT: osmar01/Fotolia.

A range of federal government agencies and their regulations now affect the chemical industry. These agencies include Occupational Safety and Health Administration (OSHA), Environmental Protection Agency (EPA), Department of Transportation, U.S. Coast Guard, and more. State agencies also regulate certain elements within their jurisdiction.

During the 1980s, the U.S. chemical industry was greatly affected by increased global competition and governmental regulations. The price of oil dropped, affecting the petrochemical business. Chemical operations were optimized for better performance, and many companies downsized and/or merged with others to remain in business.

The need to minimize risks from chemical manufacturing and processing has driven many of the advancements in chemical manufacturing processes, equipment, and construction of today's chemical plants.

2000–PRESENT In the twenty-first century, global competition has escalated as countries find the need to enhance their economies. China surpassed the United States in 2011 in production of commodity chemicals. In order to compete in the world market, American companies have built plant sites in countries around the world. These sites are built with oversight to minimize cost in relation to raw materials, water supply, and access to a skilled labor market. To take advantage of easy transportation, many companies build refineries close to the sea, importing or exporting raw materials or products.

The twenty-first century will see a focus on research and development in order to expand in the world market. The continued push to lessen the impact on the environment will drive companies to run tighter controls on all processes. A sample of a chemical process diagram is shown in Figure 3.3.

Figure 3.3 Chemical processes require close supervision to limit the potential negative impact to communities and the environment. This diagram illustrates an amine system, used to convert sour gas into sweet gas.

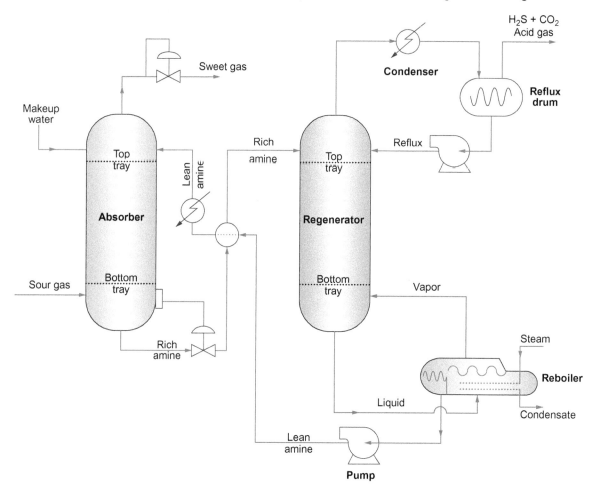

3.2 Overview of the Chemical Industry

The chemical industry is generally viewed as having seven different sectors:

1. Basic chemicals
2. Synthetic materials (e.g., resins, artificial rubber, fibers)
3. Agricultural chemicals
4. Paints, coating, and adhesives
5. Cleaning materials
6. Pharmaceuticals and medicines
7. Various chemical products.

All of the industry sectors except for pharmaceuticals are addressed generally in this section. Because of its size, the pharmaceutical industry is covered separately.

An industry closely related to the chemical industry is the oil and gas industry. Within the industries, the term *petrochemical* is often used. The term **petrochemical** refers to a chemical derived from fossil fuels or petroleum products. Petroleum was discussed in more detail in Chapter 2.

Petrochemical a chemical derived from fossil fuels or petroleum products.

3.3 Duties, Responsibilities, and Expectations of Process Technicians in the Chemical Industry

Process industries are one of the largest industries in the world. More than 500,000 workers are employed in just the refining and chemical processing industries. The population of Generation Xers, born between 1965 and 1979, is roughly 48 million, about 25 million less than the 73 million baby boomers born between 1946 and 1965. The retirement of the baby boomer generation, whether at the age of 62 or 67, will leave a large gap in the availability of skilled workers to replace them. This gap provides an opportunity for newly trained technicians.

The fundamental processes used in chemical manufacturing include many common skills and technical knowledge: storing and transporting materials, heating and cooling fluids, filtering, distillation, and other process systems. No matter what part of the chemical industry a person works in, these skills will be required. Each production process uses similar systems comprising machines, equipment, and structures including tanks, columns, pipes, valves, and pumps.

As with other process industries, process technicians in the chemical industry are required to have strong communication skills in reading, writing, and speaking and good computer skills. Process technicians in the chemical industry require a foundation in math, physics, and chemistry.

Objectives

Upon completion of this chapter, you will be able to:

3.4 Explain the growth and development of the pharmaceutical industry. (NAPTA History 1-4) p. 46

3.5 Provide an overview of the drug manufacturing process. (NAPTA History 1) p. 48

3.6 Identify the duties, responsibilities, and expectations of the process technician in the pharmaceutical industry. (NAPTA Career 1) p. 49

Introduction to the Pharmaceutical Industry

Pharmaceuticals are manmade or naturally derived chemical substances with medicinal properties that are used to treat diseases, disorders, and illnesses. Modern drug companies produce finished drugs as well as a variety of other products, including biological products, bulk chemicals, and botanicals that are used in making finished drugs. These companies also produce substances that are used for diagnostic purposes, such as chemicals used for blood glucose monitoring and pregnancy test kits.

Pharmaceuticals manmade or naturally derived chemical substances with medicinal properties that can be used to treat diseases, disorders, and illnesses.

Clearly, modern drugs save lives and can improve the well-being of countless patients. They can improve health and quality of life, extend life span, and reduce healthcare costs by keeping people out of healthcare facilities such as hospitals, emergency rooms, and nursing homes.

The process for developing new drugs is both time-consuming and expensive. From discovery to market, the process can take as long as 10 years and involve countless failures before a new safe and effective medicine is approved.

3.4 Growth and Development of the Pharmaceutical Industry

Civilizations have been using pharmaceuticals since before the dawn of history. Early humans, for example, used leaves, mud, and other naturally occurring substances to treat ailments. Their knowledge about pharmaceuticals was very limited, however, and their methods were crude. Discoveries were based purely on trial and error.

Apothecary a person who studies the art and science of preparing medicines; in modern times we call these individuals pharmacists.

2000 BC An **apothecary** is a person who studies the art and science of preparing medicines (in modern times, we call these people pharmacists). The earliest apothecary records date back to ancient Babylonia in 2600 BC (Figure 3.4). These medical texts, which were recorded on clay tablets, listed the symptoms of the illness, the prescribed treatment, and the method for creating the remedy.

Figure 3.4 The mortar and pestle were used to grind seeds, roots, and other plant parts that have medicinal properties.

CREDIT: Sherry Yates Young/Shutterstock.

Around 2000 BC, Chinese emperor Shen Nung also began investigating the medicinal properties of hundreds of products, specifically medicinal plants and herbs. During his investigation, Nung identified more than 365 **drugs** (substances used as medicines or narcotics).

Drugs substances used as medicines or narcotics.

1000 BC Although Egyptian medicine dates to 2900 BC, many of the pharmaceutical records were not created until much later. One of the best-known and most important pharmaceutical records is the "Papyrus Ebers," a collection of more than 800 prescriptions and 700 drugs.

AD 200 One of the most revered names in the professions of pharmacy and medicine is a Roman scientist named Galen, who practiced and taught medicine and pharmacy. His principles for **compounding** (mixing two or more substances or ingredients to achieve a desired physical form) and preparing medicines ruled the Western world for 1500 years. Many of Galen's original procedures for compounding are still in use today. In fact, in Europe, drug products are often referred to as Galenicals.

Compounding mixing two or more substances or ingredients to achieve a desired physical form.

> **Did You Know?**
>
> Galen created the original formula for cold cream, a cream used cosmetically to soften and clean the skin.
> The cold cream formula we use today is very similar to Galen's original formula.

CREDIT: Terry Morris/Fotolia.

1600s–1700s Pharmaceutical manufacturing as an industry began in the 1600s, but did not get underway until the mid-1700s. Early manufacturing processes began in Germany and Switzerland, mostly as an outgrowth of the chemical industry, and then spread to England, France, and the United States.

1800s Quality control for drugs and standards for education and manufacturing were not put in place until the 1800s.

In the 1820s, the first *United States Pharmacopoeia* was created. This book was the first book of drug standards to gain national acceptance. The name *pharmacopoeia* means "the art of the drug compounder." Inside this book are directions for identifying and preparing compound medicines.

The end of the 1800s marked the introduction of **biologicals**, products (e.g., vaccines) derived from living organisms that detect, stimulate, or enhance immunity to infection. When Emil von Behring and Emile Roux announced the effectiveness of diphtheria antitoxin in 1894, scientists in the United States and Europe rushed to put this new discovery into production. As a result, the lives of thousands of children were saved.

In addition to biologicals, chemically manufactured compounds also came into existence. The chemical manufacture of medicinal compounds began in the late 1800s with the creation of antipyrine, a substance used to reduce fever.

Biologicals products (e.g., vaccines) derived from living organisms that detect, stimulate, or enhance immunity to infection.

1900s–2000 Interest in pharmaceutical research began in the late 1930s and early 1940s when it became clear that organic chemistry could be used to synthesize complex chemical structures that could be used to make drugs. The field has continued to grow ever since.

In the mid-1900s, the antibiotic era began. **Antibiotics** are substances derived from mold or bacteria that inhibit the growth of other microorganisms (e.g., bacteria or fungi). While antibiotics were not new (Louis Pasteur first observed them in the late 1800s), the mass production of them was.

Sparked by pressures from World War II, pharmaceutical manufacturers were tasked with rapidly adapting their processes so they could mass-produce the antibiotic penicillin. This mass production saved many lives, allowed for a wider distribution of the drug, and cut the cost of production to 1/1000th of the original cost.

Antibiotics substances derived from mold or bacteria that inhibit the growth of other microorganisms (e.g., bacteria or fungi).

2000–PRESENT A major event affecting pharmaceuticals occurred in 2003, when scientists announced the completion of gene sequencing, the makeup of DNA. This event has aided advancement of identifying markers for disease to help find better treatments and cures. In 2006, stem cell research was found to be successful in the discovery of cloning, or mapping the cell structure. Stem cells can be programmed to become any type of cell in the body.

New drugs and therapies have been developed for human immunodeficiency virus (HIV), cancer, and human papilloma virus (HPV). These drugs are all produced by pharmaceutical manufacturers. Today, our understanding of diseases and mapping of the human genome promise rapid advances that were once unthinkable.

3.5 Overview of the Pharmaceutical Industry and Drug Manufacturing Process

Modern drug manufacturing establishments produce a variety of products. The process for developing new medicines is lengthy and expensive. Before a drug can be marketed to the public, it must go through many stages of testing. The U.S. Food and Drug Administration (FDA) was charged with regulating the approval of new medicines in 1906. Since then, a very well-regulated and well-defined process has evolved. Through this process, new medicines are tested to establish safety and efficacy. The stages or phases of this testing are listed in Table 3.2.

Table 3.2 Stages of the Drug Manufacturing Process

Stage	Description
1. Preclinical studies	Studies are conducted on nonhuman subjects that provide information pertaining about how the body processes the drug and any potential side effects.
2. Clinical evaluation	Documents are submitted to the FDA requesting approval to test the new product in a select group of people.
3. Clinical trials	FDA-approved studies are conducted on human, volunteer test subjects; the studies are used to determine the safety and effectiveness of the drug.
4. Regulatory filing	Documents are submitted to the FDA requesting marketing approval.
5. FDA approval and post-approval monitoring	Approved medication is prescribed, and patients are monitored for any adverse reactions.
6. Manufacturing	Drugs are produced; manufacturing facilities receive FDA inspection; samples are submitted to the FDA for testing; approved drugs are manufactured and distributed to consumers.

The drug manufacturing process requires a significant amount of time and financial investment. As the candidate drug(s) move through each of the stages, the potential number of compounds decreases. According to some estimates, for every 5,000 compounds that enter preclinical testing, only five will continue on to clinical trials in humans, and only one will be approved for marketing in the United States.

QUALITY Pharmaceutical manufacturing is a highly regulated business operating under a strict set of guidelines published by the FDA. These guidelines are often referred to as *good manufacturing practices*, or GMPs. GMPs were mandated by Congress in 1938

Did You Know?

The estimated average cost to develop a new drug (including the cost of failures and capital costs) reached $2.6 billion between 2000 and the early 2010s.

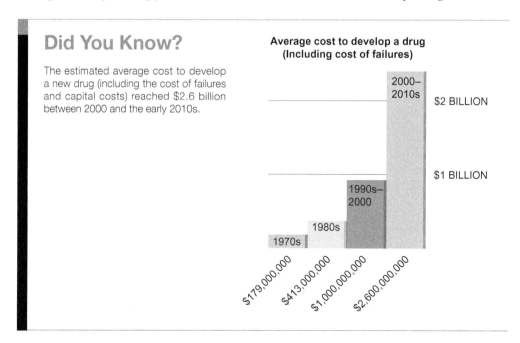

Average cost to develop a drug (Including cost of failures)

after a number of deaths occurred that were attributable to poisonous ingredients being mistakenly used in pharmaceutical manufacturing processes. As a result of these deaths, quality control and quality assurance organizations became a required and vital part of the pharmaceutical industry. Many production workers are assigned to quality control and quality assurance functions on a full-time basis.

Throughout the production process (Figure 3.5), inspectors, testers, sorters, samplers, and weighers ensure consistency and quality (e.g., tablet testers inspect tablets for hardness, chipping, and weight to assure conformity with specifications).

Figure 3.5 Generic flow chart of the pharmaceutical industry.

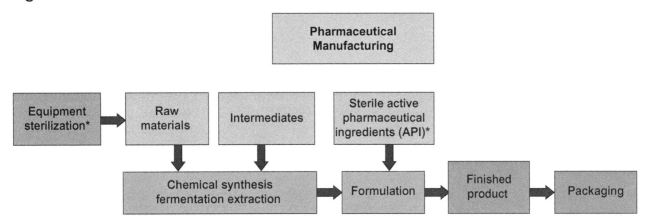

3.6 Duties, Responsibilities, and Expectations of Process Technicians in the Pharmaceutical Industry

Process technicians in the pharmaceutical industry perform a wide variety of duties, which include operating drug-producing equipment and inspecting products.

Transportation and material-moving employees package and transport the finished drugs to other facilities.

Workers in the production division might rotate through different processes (e.g., assemblers and fabricators), while others might specialize in one part of the production process (e.g., compacting machine setters, filling machine operators, and milling machine tenders).

Many production facilities are highly automated. In some processes, milling and micronizing machines (machines that pulverize substances into extremely fine particles) are used to reduce the size of bulk chemicals. These finished chemicals are then combined and processed further in mixing machines. Once mixed, these formulations are made into capsules, pressed into tablets, or made into solutions.

In order to create these products, process technicians must be able to operate a variety of equipment (e.g., pumps, valves, conveyors, and milling machines), and must have strong technical knowledge, technical and interpersonal skills, and computer skills.

JOB DUTIES Process technicians in the pharmaceutical field might be asked to perform a wide variety of job duties. In addition to duties stated in Chapter 1, duties for an employee in the pharmaceutical field can include:

- Handling chemical substances
- Following chemical procedures
- Adhering to the good manufacturing practices regulations that govern the manufacture of drugs and drug products.

EQUIPMENT The pharmaceutical manufacturing industry uses a wide variety of equipment, including the following:

- Milling and micronizing machines
- Capsule filling, sealing, and stamping machines
- Conveyors
- Centrifuges
- Bottle filling, labeling, and packaging machines.

Some of these types of equipment are explained in more detail in the equipment chapters of this textbook.

WORKPLACE CONDITIONS AND EXPECTATIONS In pharmaceutical manufacturing plants, strong emphasis is placed on keeping equipment and work areas clean because of the potential for contamination (GMP regulations). Most work areas are indoors, air conditioned, and access controlled. Ventilation systems are used to protect workers from dust, fumes, and odors. With the exception of work performed by material handlers and maintenance workers, many jobs require less physical effort than that required in some other process industries.

Like other industries, the pharmaceutical industry places a heavy emphasis on continuing education for employees. They provide classroom training in safety, environmental and quality control, and technological advances. Many companies encourage production workers to take courses related to their jobs at local schools and technical institutes or to enroll in online courses. Continuing college education is sponsored and encouraged. College courses in chemistry, biology, mathematics, and related areas are particularly valuable for highly skilled production workers who operate sophisticated equipment.

Summary

Chemical Industry

Chemicals play a vital role in a wide range of manufacturing processes. The chemical industry affects our everyday life in the production of food and beverage, medical, construction, agriculture, clothing, transportation, cleaning, military, personal hygiene, and amusements and hobbies.

Chemical manufacturing requires an understanding of chemistry, manufacturing processes, and the special equipment and structures used to process the chemicals.

The chemical industry is generally viewed as having seven different segments: basic chemicals; synthetic materials; agricultural chemicals; paints, coating, and adhesives; cleaning materials; pharmaceutical and medicines; and various chemical products.

The history of the chemical industry is long and varied. Ancient civilizations discovered many uses for chemicals, including making glass, soap, and black powder. The pseudoscience of alchemy developed, with a focus on ways to change less valuable metals into valuable ones. In the late 17th and 18th centuries, alchemy became a true science, chemistry, with many scholars and scientists contributing to its advancement. In the 1800s, consumer demand rose for items such as glass, paper, explosives, matches, fertilizers, soaps, and textiles.

World War I was the first major conflict that involved mechanized and chemical warfare. Extensive chemical research on chemical weapons was conducted during the war. In the late 1930s and early 1940s, World War II broke out and once again chemicals played an important role in the conflict. The war spurred a huge growth in refining and chemical process industries and stimulated many new advances in processes.

Concern for the environment and worker safety boomed in the 1960s and 1970s. A range of federal government agencies and their regulations began to have an impact on the chemical industry. These agencies include OSHA, EPA, Department of Transportation, U.S. Coast Guard, and so on. State agencies also regulate certain elements within their jurisdiction.

During the 1980s, the U.S. chemical industry was greatly affected by increased global competition and governmental regulations. The price of oil dropped, affecting the petrochemical business, and chemical operations were optimized for better performance. Today's process technicians deal with more automation and computerization than past workers in the industry.

Process technician jobs are more complex and cross-functional (meaning workers are required to perform a wider variety of tasks) than in the past. There is an increased emphasis on keeping costs low by improving safety, performing preventative maintenance, optimizing processes, and making efforts to increase efficiency.

Along with technology changes, process technicians must keep up with constantly changing governmental regulations addressing safety and the environment.

Pharmaceutical Industry

Pharmaceuticals have been used since early times, when knowledge of pharmaceuticals was very limited and the methods of identification were very crude. While the area of pharmaceutical research improved throughout the centuries, it did not become standardized until the 1800s.

Modern drug manufacturing establishments produce a variety of pharmaceutical products. The process for developing new drugs is both time-consuming and expensive. Mapping of the human genome has led to many new developments in the pharmaceutical industry and promise of future breakthroughs.

During the production of pharmaceutical products, quality is paramount. Process technicians in the pharmaceutical industry must possess strong technical knowledge, technical and interpersonal skills, computer skills, and physical abilities (depending on the job). They also must be able to work safely and efficiently and stay current with government rules and regulations that are always subject to change.

Most work areas are indoors, air conditioned, and access controlled. Strong emphasis is placed on keeping equipment and work areas clean because of the potential for contamination (GMP regulations). Pharmaceutical and medicinal product manufacturing ranks among the fastest growing industries. Demand for this industry's products is expected to remain strong in the future. The industry is also relatively insensitive to changes in economic conditions, so work is likely to be stable, even during periods of high unemployment.

Checking Your Knowledge

1. Define the following terms:
   a. alchemy
   b. chemical
   c. commodity chemicals
   d. intermediate
   e. pharmaceuticals
   f. specialty chemicals
   g. synthetic
   h. antibiotics
   i. apothecary
   j. biologicals
   k. compounding
   l. drugs

Chemical Industry

2. _____ refers to chemicals derived from petroleum products.
3. Name four segments of the chemical industry.
4. What chemical weapon was used extensively in World War I?
   a. anthrax
   b. sarin
   c. German gas
   d. mustard gas

5. Which of the following was the first synthetic fiber?
   a. silk
   b. rayon
   c. nylon
   d. dacron

6. Companies expect process technicians to perform with a focus on _____, health, security, and environment (choose the term below that fits best).
   a. effectiveness
   b. statistics
   c. safety
   d. technology

Pharmaceutical Industry

7. The introduction of the *United States Pharmacopoeia* in the 1820s was important because it:
   a. Marked the introduction of biologicals (vaccines).
   b. Encouraged scientists to conduct more pharmaceutical research.
   c. Was the first book of drug standards to gain national acceptance.
   d. Named all the major manufacturers of drugs in the country.

8. Which world event encouraged pharmaceutical manufacturers to adapt their processes rapidly so they could mass-produce antibiotics?
   a. World War I
   b. World War II
   c. the Cold War
   d. the Industrial Revolution

9. What event of this century has had the greatest impact on the pharmaceutical industry?
   a. Synthesis of rubber
   b. Cancer research
   c. Mapping of human genome
   d. Development of HPV vaccine

NOTE: Answers to Checking Your Knowledge questions are in Appendix I.

Student Activities

1. Use the Internet to find out more about the history of the chemical industry, including one discovery, the scientists involved, and the impact of the discovery. Write a one-page summary of your findings.

2. Make a list of common items around your home, school, or workplace that are chemical products. See how many you can identify.

3. List and describe each of the stages of the drug manufacturing process.

4. Use the Internet or other resources to research current drug manufacturing processes. Write a two to three-page paper describing what you learned.

Chapter 4
Power Generation Industry

 Objectives

Upon completion of this chapter, you will be able to:

4.1 Name the different sectors of the power generation industry. (NAPTA History 1) p. 54

4.2 Describe the growth and development of the power generation industry. (NAPTA History 1-4) p. 55

4.3 Describe the different processes and materials used for power generation. (NAPTA History 1, Green Tech 1) p. 58

4.4 Identify expectations for process technicians and future trends in the power generation industry. (NAPTA History 3, Green Tech 1) p. 63

Key Terms

Alternating current (AC)—an electrical current that reverses direction periodically. This is the primary type of electrical current used in the process industries and in residential homes, **p. 56**.

Ampere—(or *amp*) unit of electric current, **p. 56**.

Atom—the smallest particle of an element that still retains the properties and characteristics of that element, **p. 57**.

Cogeneration station—a utility plant that produces both electricity and steam that can be used for heating and cooling, **p. 58**.

Direct current (DC)—an electrical current that travels in the same direction through a conductor, **p. 56**.

Electricity—a flow of electrons from one point to another along a pathway, called a conductor, **p. 54**.

Fission—the process of splitting the nucleus, the positively charged central part of an atom, which results in the release of large amounts of energy, **p. 57**.

Generator—a device that converts mechanical energy into electrical energy, **p. 55**.

Geothermal—a power generation source that uses steam produced by the Earth to generate electricity, **p. 62**.

Hydroelectric—a power generation source that uses flowing water to generate electricity, **p. 59**.

Nuclear—a power generation source that uses the heat from splitting atoms to generate electricity, **p. 60.**

Ohm—term used to describe electrical resistance, **p. 56.**

Solar (CSP)—Concentrated Solar Power; a power generation source that uses the power of the sun to heat water and generate electricity, **p. 61.**

Solar (PV)—Photovoltaics solar power; generation of power when light excites semiconducting materials (solar panels) to generate electricity, **p. 61.**

Turbine—a machine for producing power; activated by the expansion of a fluid (e.g., steam, gas, air, or water) on a series of curved vanes on an impeller attached to a central shaft, which is used to create rotational mechanical energy, **p. 58.**

Volt—the derived unit for electrical potential, electrical potential difference (voltage), and electromotive force, **p. 56.**

Watt—(or *wattage*) a measure of electrical power, **p. 56.**

Wind power—a power generation source that uses flowing air currents to push against blades of giant wind turbines, **p. 62.**

Introduction

This chapter provides you with a history of the power generation industry and its impact on people, the economy, and the environment. This chapter also describes the tasks and working conditions of process technicians, as well as future trends and how they affect these technicians.

Overview of Electricity

Electricity is a flow of electrons from one point to another along a pathway, called a conductor. Electricity flows in a continuous current from a high potential point (i.e., the power source) to a point of lower potential (i.e., homes or businesses) through a conductor (e.g., a wire). High voltage electricity is transmitted from the power plant to the power grid, a system of wires that channel the electricity to where it is needed.

When electricity leaves a power plant, it is usually very high voltage. Substations are used to step the power down to a lower, safer voltage. The substation then distributes the electricity through feeder wires to a transformer, a device that steps down the voltage again. A service drop (or wire) runs from the transformer to homes or businesses.

Once inside a home or business, electricity is used as energy to do work (e.g., light a bulb or operate a motor). Insulators, circuit breakers, fuses, and other safety devices are used to make the power transmission process as safe as possible.

Electricity a flow of electrons from one point to another along a pathway, called a conductor.

4.1 Major Sectors of the Power Generation Industry

Power generation involves producing and distributing electrical energy in large quantities to serve the requirements of industry, businesses, residences, schools, and more. Electrically powered equipment harnesses the natural forces of electricity. We use electricity to power everything from coffee makers to refineries.

Every process industry requires power to operate. Because of this, process technicians working in industries outside of power generation should at least understand the fundamentals of this vital industry.

There are three main sectors in the power generation industry:

- Generation
- Transmission
- Distribution

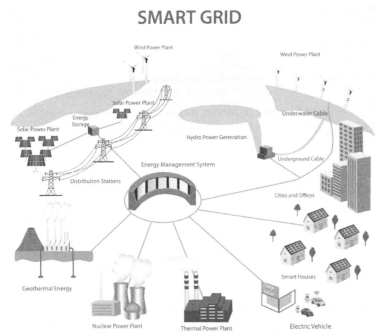

Figure 4.1 Electricity is produced by various methods and distributed from power plants to substations to homes and businesses.

CREDIT: Monica Odo/Shutterstock.

Generating plants produce electricity by converting mechanical energy or heat using a variety of methods (Figure 4.1). These methods include burning a fuel (coal, oil, or gas), using water (hydroelectric power), splitting atoms (nuclear power), or harnessing resources such as the sun, wind, or heat from the earth.

These sources of power are used to operate turbines and **generators**, devices that convert mechanical energy into electrical energy. From the generating plant, the electricity travels to a transmission facility that moves the power over high voltage lines to different regions. Distribution facilities then carry the power to end users, such as industrial or residential consumers.

Generator a device that converts mechanical energy into electrical energy.

For more information about electricity, refer to Chapter 19, Electricity and Motors.

4.2 Growth and Development of the Power Generation Industry

Electricity occurs naturally in the world, a fact that is demonstrated during a lightning storm or when you shuffle your feet across the carpet and then touch a doorknob and receive a shock. However, power must be generated in large quantities and transformed to several different levels of voltage in order to run equipment. Electrically powered equipment is a way of harnessing these natural forces of electricity. Since the early days of civilization, people have been working to harness electricity and other sources of power.

Early Civilizations

The basis of power generation began with early civilizations that sought to control and use various energy sources to do useful work. Civilizations more than 2,000 years old used the power of water to turn wheels that ground wheat into flour. The ancient Greeks conducted experiments in static electricity by rubbing amber against a cloth. A Greek inventor, Heron, created a simple amusement device that used the power of steam to rotate an object.

Did You Know?

The terms *amp*, *volt*, *ohm*, and *watt* are all named after scientists who studied power and electricity.

Term	Named After
Ampere or Amp	André-Marie Ampere
Volt	Alessandro Volta
Ohm	Georg Simon Ohm
Watt	James Watt

Eventually, around the sixteenth century, people started burning coal, an organic energy-releasing mineral consisting mainly of hydrogen and carbon, to provide heat. Windmills were also used to tap the power of wind and convert it to mechanical energy to grind wheat into flour and operate pumps.

Between the seventeenth and twentieth centuries, there were many famous discoveries pertaining to the properties of electricity. For example, in the mid-1700s, Benjamin Franklin performed experiments with kites to prove that lightning was electricity. Contrary to popular belief, Franklin did not "discover" electricity. He just showed that it existed in nature.

1700s

Watt (or *wattage*) a measure of electrical power.

In 1763, James Watt created the first practical steam engine. His invention helped spur the Industrial Revolution in England and the United States, since manufacturing could create products more quickly because of the use of machines like the steam engine. The electrical term **watt** (or *wattage*, a measure of electrical power) is named after James Watt. Electric bills to this day are calculated in kilowatt usage.

Volt the derived unit for electrical potential, electrical potential difference (voltage), and electromotive force.

In the late 1700s, Alessandro Volta discovered that when moisture comes between two different metals, electricity is generated. Using this discovery, he created a battery called the voltaic pile. This battery was constructed from thin sheets of copper and zinc separated by moist pasteboard. During his experiments, Volta demonstrated that electricity flows like water current. He also showed that electricity could be made to travel from one place to another, through conductors. The term **volt** (the derived unit for electrical potential, electrical potential difference [voltage], and electromotive force) derives from his name.

1800s

Ampere (or *amp*) unit of electric current.

In the early 1800s, André-Marie Ampere, a French scientist, experimented with electromagnetism and discovered a way to measure electrical currents. The term **ampere** (or *amp*) is used to describe units of electric current.

Ohm term used to describe electrical resistance.

Also in the early 1800s, Georg Simon Ohm, a German physicist, discovered a law relating to the intensity of an electrical current, electromotive force, and resistance. His name, **Ohm**, is used to describe electrical resistance.

Direct current (DC) an electrical current that travels in the same direction through a conductor.

In the mid-1800s, an English scientist named Michael Faraday devised a way to generate electrical current on a larger, more practical scale. Faraday expanded on earlier experiments and demonstrated that magnetism could produce electricity through motion. He showed that moving a magnet inside a coil of copper wire produced an electrical current. His experiment resulted in the first electric generator.

Alternating current (AC) an electrical current that reverses direction periodically. This is the primary type of electrical current used in the process industries and in residential homes.

In the late 1800s, English scientist Joseph Swan developed incandescent lighting. American Thomas Edison made a similar discovery in the United States several months after Swan. Swan and Edison created a company together to produce the first practical incandescent lighting.

Thomas Edison also created a direct current generator to power the lights in his lab. **Direct current (DC)** is an electrical current that always travels in the same direction. Edison's generator, powered by Watt's steam engine, signaled the start of practical, large-scale electrical power generation. However, Edison's DC power had problems with long-distance generation and voltage regulation. Nikola Tesla (who had worked for Edison) devised another system for electrical generation: alternating current. **Alternating current (AC)** is an electrical current that reverses direction periodically.

George Westinghouse, an American inventor and industrialist, developed Tesla's alternating current power system as an option to DC power. The AC system was more efficient for

Did You Know?

In 1882, Edison provided lighting to a street in New York.

That same year in Wisconsin, a water-wheel on a river was used to supply hydroelectric power to two paper mills and a house.

CREDIT: salajean/Fotolia.

sending electricity over hundreds of miles with little power loss. Alternating current became the standard for usage in industrial and residential applications.

1900s

In 1916, the German-American physicist Albert Einstein published his *Theory of Relativity*, which revolutionized physics. His theory also spurred research into developing atomic energy. *Atomic energy* is the energy released during a nuclear reaction. In a nuclear reaction, **atoms** (the smallest particles of an element that still retain the properties and characteristics of that element) are split, and heat is produced.

During the Great Depression of the 1920s and 1930s, U.S. President Franklin Delano Roosevelt instituted sweeping public works programs, which helped build the Hoover Dam and established the Tennessee Valley Authority. These electric power generation projects created jobs and brought power to new regions of the United States.

During the 1930s, three German physicists, Otto Hahn, Lise Meitner, and Fritz Strassman, performed the first successful human-initiated nuclear fission. **Fission** is the process of splitting the nucleus of an atom (the positively charged central part), which results in the release of large amounts of energy.

During the late 1930s and early 1940s, many nations experimented with nuclear energy, trying to build nuclear reactors and then atomic weapons. While the German and Japanese programs were two of the most ambitious, the United States was the first to produce atomic weapons. In 1951, the United States began generating energy with a nuclear reactor. The Soviet Union opened its first commercial nuclear reactor for power generation in 1954. Although generally considered a safe means of producing electricity, the nuclear energy industry did suffer two highly visible accidents in the late 1970s and mid-1980s. These occurred at Three Mile Island in the United States and at Chernobyl in the Soviet Union.

Increased concern for the environment and for worker safety in the 1960s and 1970s led to legislation to provide for cleaner air and water as well as safer working environments. As mentioned in the Oil and Gas section of Chapter 2, OPEC's oil embargo in the early 1970s created skyrocketing prices, lines at gas stations, national trends toward conservation, and development of smaller, more fuel-efficient cars. It also spurred interest in, and development of, alternative energy sources such as hydroelectricity, wind, nuclear power plants, and solar energy.

Atom the smallest particle of an element that still retains the properties and characteristics of that element.

Fission the process of splitting the nucleus, the positively charged central part of an atom, which results in the release of large amounts of energy.

Did You Know?

According to the U.S. Nuclear Regulatory Commission, the 1979 accident at the Three Mile Island nuclear power plant near Middletown, Pennsylvania—in which a cooling malfunction caused partial meltdown of a reactor core—was one of the most serious accidents in U.S. commercial nuclear power plant operating history.

Fortunately, there were no injuries from this event. However, it did bring about major changes in U.S. emergency response planning, operator/technician training, and other areas associated with nuclear power generation.

CREDIT: A. L. Spangler/Shutterstock.

In the 1990s, power generation companies faced regulations that affected the structure and competition within the industry. Companies were forced to reorganize their operations and reduce costs in order to compete more effectively.

2000–Present

This century has seen developments in all forms of power generation. Renewable energy sources such as hydropower, wind power, geothermal energy, and solar energy have evolved. They demonstrate the potential to provide a cleaner environment. The use of biomass fuels has grown, but its high moisture content and costs associated with its transportation make it cost-prohibitive for use in a wide-scale market. There have been major fluctuations between pressure to emphasize renewable sources of power and pressure to use the nation's fossil-fuel resources.

Destruction of a nuclear reactor at the Fukushima power plant in Japan in 2011 focused attention once again on the waste products of the nuclear industry. Because of contamination from that plant, large areas in and near Japan will remain off limits to fishing and food production for decades.

The effects of industrial and power generation pollution are now more obvious globally. We can see images of people in developing countries living in smog like the United States experienced 50 or 60 years ago, before enacting clean air legislation. We can read data about carbon dioxide in the atmosphere and temperature changes in the oceans. Worldwide competition for natural resources continues. Between the years of 2004 and 2013, the United States led the world market in investment in renewable energy sources. Global factors will have a continuing effect on power production and power generation in the future.

4.3 Overview of the Power Generation Industry

A variety of power sources are used to generate electricity. Generating plants produce electricity by converting mechanical or heat energy into electricity using a variety of methods. They include the following:

- Burning fuels (coal, oil, gas, fuel oil)
- Using water (hydroelectric)
- Splitting atoms (nuclear)
- Harnessing other resources (the sun, wind, ocean, or heat from the earth)

The transmission and distribution facilities are the same, regardless of the power source.

Coal, Oil, and Gas

Coal, oil, and gas continue to be major fuel sources for generating electricity in the United States. Many power generation facilities burn coal, oil, or gas to heat water and turn it into steam (Figure 4.2). The steam causes a turbine to rotate. A **turbine** is a rotary machine that converts a fluid (e.g., water, steam, gas) into mechanical energy. The turbine has a wheel or rotor that turns as a fluid flows past it. The wheel or rotor turns a shaft attached to a generator. The generator uses rotating magnets inside coils of copper wire to generate electricity.

One type of power generation facility is called a **cogeneration station**. This type of facility can provide heat, cooling, and electricity to consumers. When energy is converted, heat is produced as a byproduct. Cogeneration facilities use this heat, in the form of steam, to provide industrial or residential heating and cooling, instead of venting it to the atmosphere. These facilities are more efficient than conventional power generation operations.

Turbine a machine for producing power; activated by the expansion of a fluid (e.g., steam, gas, air, or water) on a series of curved vanes on an impeller attached to a central shaft, which is used to create rotational mechanical energy.

Cogeneration station a utility plant that produces both electricity and steam that can be used for heating and cooling.

Figure 4.2 Fossil fuel burning power plant.
CREDIT: Cracker Clips/Fotolia.

However, since heat energy is lost over long distances, these facilities must be located near an area that can use the heat.

Hydroelectric

Water is used to generate electricity for almost one-quarter of the world's electricity requirements. **Hydroelectric** power plants harness the energy of rushing water and use it to turn turbines that are connected to generators, creating electricity (Figure 4.3).

Typically, in a hydroelectric system, a dam is built across a river. In certain situations, though, facilities use tidal forces from the ocean as the water source.

Hydroelectric turbines are very large, weighing upward of 170 tons. In a dam-based hydroelectric facility, water is channeled into a narrow passageway (called a *penstock*) in order to increase the force of the water's movement. Water pressure builds up as it is channeled to the turbines. The power of the water is transferred to the turbine blades. The water is then released through an outtake (called a *tailrace*) downstream of the dam.

Hydroelectric power is considered a renewable power source, since the water is not consumed or used up during the energy generating process. However, there are limitations in finding waterways suitable for damming.

Hydroelectric a power generation source that uses flowing water to generate electricity.

Figure 4.3 **A.** Hydroelectric power plant. **B.** Process diagram for hydroelectric power.
CREDIT: **A.** Goce Risteski/Fotolia.

A.

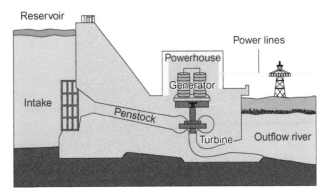

B.

Nuclear

Nuclear a power generation source that uses the heat from splitting atoms to generate electricity.

Nuclear power plants produce heat by splitting atoms in a device called a reactor (Figure 4.4). Within a reactor, atoms are split using a process called fission.

Figure 4.4 Nuclear power plant.

CREDIT: Spacekris/Fotolia.

Fission is the process of splitting the nucleus, the positively charged central part of an atom, which results in the release of large amounts of energy. Fission produces a tremendous amount of heat. This heat is used to boil water and create steam. Steam is then used to rotate a turbine that is connected to the shaft of a generator. When the generator turns, electricity is created. Figure 4.5 is a process diagram of the nuclear power production process.

Figure 4.5 Nuclear power plant process diagram.

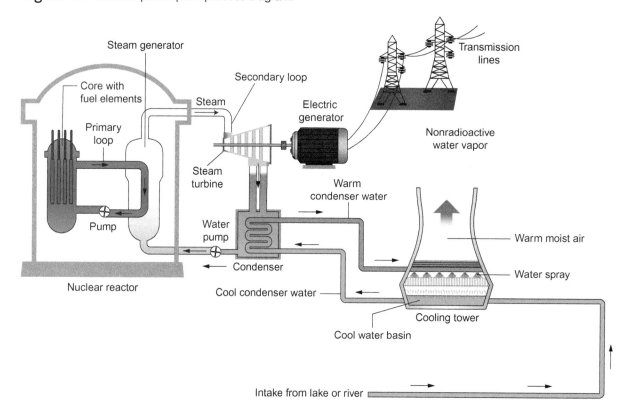

Uranium and plutonium are radioactive metallic elements used as fuel for nuclear reactors. During nuclear reactions, a small amount of fuel produces large amounts of energy. For example, on some Navy ships (e.g., nuclear-powered aircraft carriers, and submarines), a pound of uranium is roughly equivalent to 1 million gallons of gasoline.

In a nuclear reactor, the fuel is arranged in long rods and grouped into bundles. These bundles are placed in water contained in a pressurized vessel. The water also acts as a coolant during this reaction.

As the bundles undergo a slow, controlled chain reaction (i.e., a splitting of the atoms), heat is released. The heat from the chain reaction turns the water into radioactive steam. Some types of reactors use this steam to drive a turbine directly, while other types use this steam to heat a separate water loop which is used to drive a turbine (this prevents the turbine from coming into direct contact with the radioactive steam).

Shielding around the reactor and an outer building made of concrete (built strong enough to withstand the crash of a jet airplane) prevent the escape of radiation and radioactive steam.

Nuclear power plant operations are much cleaner than coal-, oil-, or gas-powered plants. However, they produce radioactive wastes that must be stored and monitored. These wastes remain hazardous for centuries.

Fusion Reaction

Fusion energy is an actively researched process in which two light atoms, such as hydrogen or helium, are fused together. The fusion creates an atom slightly lighter than the original two atoms. The difference in mass is converted to energy.

Producing a fusion reaction is a complex task requiring the atoms to be heated to millions of degrees Celsius. The fuel is turned into a hot ionized gas—plasma—which can be contained within a magnetic field without touching the inside of the reactor.

In a power plant, heavy isotopes of hydrogen fuse to form helium and an energetic neutron. The helium is contained within the plasma, but the neutron shoots into the "blanket" surrounding the plasma, heating it up. The heat produced drives a steam turbine to generate electricity.

Two roadblocks to development of fusion power are the need for extreme heat and the rapid degeneration of materials undergoing massive bombardment from high energy particles.

Renewable Resources

Other resources, such as the sun, wind, and heat of the Earth, can be used to generate electricity. Although these alternative energy sources come from renewable sources (in other words, the resource is not consumed during the power generation process), they account for only a small percentage of the world's electricity production. However, strong demands from customers and communities have gradually allowed renewables to move out of the "alternative" category and into the mainstream.

Solar energy is used to generate power through two main methods. **Concentrated solar power (CSP)** uses large reflective panels to collect and focus the sun's energy (Figure 4.6). This energy can be used to heat water to run a steam turbine connected to an electrical generator. **Photovoltaic (PV) solar power** creates electricity directly using special devices made from semiconducting materials that are called *solar panels* or *solar cells*.

Recent years have seen a huge increase in both residential and commercial solar energy applications. Costs have also declined because of changes in such "soft" costs as labor, improved supply chain, and reduced overhead. Legislative efforts to streamline permits have also benefited the industry. Solar installations at U.S. corporations and businesses, as of 2016, were enough to offset 1.1 million metric tons of carbon dioxide emissions each year.

A major stumbling block for solar and wind farm electricity generation has been the absence of a practical means of storing power when demand is low and deploying it when usage spikes. Tesla's lithium-ion batteries for electric cars and their evolution to modular unit megafactory production offer potential for wide-scale application in the power generation industry.

Solar (CSP) Concentrated Solar Power; a power generation source that uses the power of the sun to heat water and generate electricity.

Solar (PV) Photovoltaic solar power; generation of power when light excites semiconducting materials (solar panels) to generate electricity.

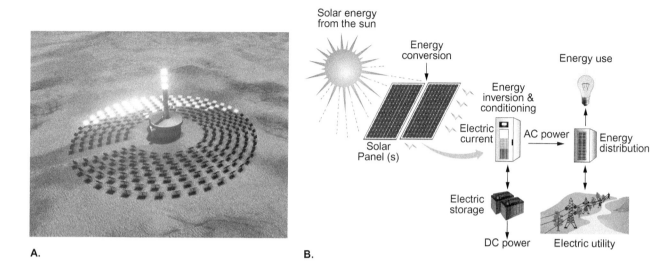

Figure 4.6 A. Solar power plant. **B.** Process diagram of the solar power generation process.
CREDIT: A. SSSCCC/Shutterstock.

A.

B.

Wind power a power generation source that uses flowing air currents to push against blades of giant wind turbines.

Wind power uses flowing air currents to push against the blades of giant windmill-like turbines (rotary machines that convert fluid pressure into mechanical energy). These wind turbines are usually gathered together on a site called a wind farm (Figure 4.7). As the turbine rotates, it turns a shaft on a generator that converts mechanical energy into electrical energy.

Figure 4.7 Power generating windmills on a wind farm.
CREDIT: Jim Parkin/Fotolia.

Wind energy capacity at the end of 2016 was 81 gigawatts (GW). Like solar, it has achieved both residential and commercial acceptance. The increase in the wind industry has been significant:

- 2000 2.53 GW produced across 4 states
- 2010 40.18 GW produced across 27 states
- 2013 60.72 GW produced across 34 states
- 2020 (projected) 113.43 GW produced across 37 states
- 2050 (projected) 404.25 GW produced across 48 states

Geothermal a power generation source that uses steam produced by the Earth to generate electricity

Geothermal power is like a steam-powered facility, except it uses the naturally occurring steam from the Earth to power turbines (Figure 4.8). Deep wells are drilled to tap underground sources of water that are superheated by hot geological forces (such as lava). The water is then cooled and re-injected into the ground. Ocean waters can be used as a thermal energy source. Ocean thermal energy conversion (OTEC) is a process using the difference in temperature between deep cold seawater and surface tropical water for the production of electricity.

Figure 4.8 Geothermal power plant.

CREDIT: Dmitry Naumov/Fotolia.

4.4 Expectations of Process Technicians and Future Trends

Process technicians working in the power generation industry control the equipment that generates and distributes electricity. Their work can be enhanced by understanding the process as a whole that provides power to consumers.

Some process technicians in this industry must obtain licensing from a government agency based on the job duties they perform (e.g., reactor operators at a nuclear power plant). Technicians must pass certain tests to obtain certification, and they need to recertify on an annual or biannual basis.

A range of federal government agencies and their regulations affect the power generation industry, including OSHA, EPA, Department of Energy, Nuclear Regulatory Commission, and the U.S. Coast Guard. State agencies also regulate certain elements within their jurisdiction.

Scientific advances continue to be made in "green" technology. The nation still relies on fossil fuels for much of its power generation. Politics will continue to play a significant role in determining the direction of the power generation industry.

Summary

For thousands of years, civilizations have harnessed resources of nature, such as water, steam, and heat.

Starting in the seventeenth century, many famous discoveries about the properties of electricity were made. Inventions such as the electric generator, the first practical steam engine, and incandescent lighting changed the world. The direct current (DC) system was followed by the alternating current (AC) system. AC is a type of electric current that reverses direction periodically and can transfer electricity over hundreds of miles with little power loss. AC is the primary type of electrical current used in industrial and residential applications.

Research into atomic energy (splitting atoms to unleash tremendous energy) led to development of nuclear reactors for electricity and also to development of atomic weapons. Although good producers of energy, nuclear reactors create waste products that take centuries to break down.

Pollution and increasing health issues in the United States in the 1960s and 1970s led to government legislation to provide for cleaner air and water and safer working environments.

An energy crisis in the 1970s spurred interest in alternative energy sources, such as hydroelectricity, nuclear power plants, solar, wind, and thermal energy.

There are three main sectors in the power generation industry: generation, transmission, and distribution. (1) Power generation involves producing and distributing electrical energy in large quantities to serve the requirements of industry, businesses, residences, schools, and so on. Every process industry requires power to operate turbines and generators that generate electricity. (2) From the generating plant, the electricity travels to a transmission facility that moves the power over high-voltage lines to serve different regions on that grid. (3) Distribution facilities then carry the power to end users, such as industrial or residential consumers.

Process technicians must deal with more automation and computerization than past workers in the industry. Because of this, they must have more education and be more highly skilled, with a two-year Associate degree preferred for process technicians.

Process technician jobs are more complex and cross-functional than in the past. Technicians are responsible for entire complex processes and should expect to continue learning throughout their career.

There is an increased emphasis on keeping costs low by improving safety, performing preventative maintenance, optimizing processes, and making efforts to increase efficiency.

Process technicians are required to have strong communication skills, good computer skills, and an understanding of math, physics, and chemistry. Management is mandated to ensure all employees are trained and updated on the most current government regulations.

Checking Your Knowledge

1. Define the following terms:
   a. alternating current (AC)
   b. atom
   c. ohm
   d. cogeneration station
   e. direct current (DC)
   f. volt
   g. fission
   h. generator
   i. geothermal
   j. hydroelectric
   k. nuclear
   l. solar
   m. turbine
   n. wind power

2. _____ is a flow of electrons from one point to another along a pathway, called a conductor.

3. Which Greek inventor created a steam-powered device more than 2,000 years ago?
   a. Hericles
   b. Aristophanes
   c. Casiclees
   d. Heron

4. Put the following events in order, from earliest to latest:
   a. first generator created
   b. atomic bomb tested
   c. water power used to turn wheat into flour
   d. voltaic pile battery created

5. Who is credited with inventing the first electric generator?
   a. Edison
   b. Faraday
   c. Swan
   d. Franklin

6. *(True or False)* Benjamin Franklin invented electricity.

7. Which of the following is not a major sector of the power distribution industry?
   a. distribution
   b. generation
   c. distillation
   d. transmission

8. Which of the following sources is used most often for generating electricity?
   a. solar
   b. fossil fuels
   c. nuclear fission
   d. water

9. Name one benefit and one drawback of nuclear power.

10. How may employment in the nuclear power generation industry differ from other types of power generation jobs?
    a. Process technicians may need to wear personal protective equipment.
    b. Process technicians may need to retire early due to radioactive exposure.
    c. Process technicians will not do equipment maintenance.
    d. Process technicians may need to obtain specialized licensing from a government agency.

11. Which of the following is a federal government agency that regulates the power generation industry?
    a. Department of State
    b. OSHS
    c. U.S. Navy
    d. EPA

NOTE: Answers to Checking Your Knowledge questions are in Appendix I.

Student Activities

1. If you could travel back in time, which event, period, or person in the history of the power generation industry would you choose to visit and why? Write at least five paragraphs in response.

2. Pick one of the power generation segments (coal, oil, or gas; hydroelectric; nuclear; solar [CSP and PV]; wind; or geothermal), and write at least a one-page research report on its benefits, drawbacks, and future.

3. Consider the responsibilities and expectations of process technicians working in the nuclear power generation industry. What aspects would be difficult? What aspects would be rewarding? Discuss this with a classmate or write a paper explaining your answers.

4. Think about the future of the power generation industry. Which skills would you need to learn in order to adapt to the changes and trends? Make a list of these skills.

Chapter 5
Food and Beverage Industry

Objectives

Upon completion of this chapter, you will be able to:

5.1 Explain the growth and development of the food and beverage manufacturing industry (NAPTA History 1) p. 67

5.2 Explain the food manufacturing process and the most common products produced by the industry (NAPTA History 1) p. 68

5.3 Identify the duties, responsibilities, and expectations of the process technician in the food and beverage manufacturing industry (NAPTA Career 1) p. 71

5.4 Describe the responsibility of various regulatory agencies and their impact on the food and beverage manufacturing industry. (NAPTA History 4) p. 72

Key Terms

Agglomerating—gathering materials into a mass (instant coffee crystals are made using this process), **p. 69.**

Cooking and frying—using heat (typically hot oil) to process foods such as chicken strips and potato chips, **p. 69.**

Disinfecting—destroying disease-causing organisms (e.g., through washing, irradiation, or ultraviolet exposure), **p. 69.**

Distillation—a method for separating a liquid mixture using the different boiling points of each component, **p. 70.**

Drying—removing moisture from items such as rice (using a conveyor belt moving through a hot air tunnel) or sugar (dried in a turning drum), **p. 69.**

Evaporating—removing moisture from items such as milk and coffee (to create powdered milk or coffee), **p. 69.**

Manufacturing—making a product from raw materials by hand or with machinery (e.g., cooking, decorating, grinding, milling, and mixing), **p. 69.**

Mixing—stirring materials to blend them (such as instant drink mixes), **p. 69.**

Pasteurization—heating foods to kill organisms, such as bacteria, or make them less likely to cause disease (milk and orange juice are treated this way), **p. 70.**

Preserving—preparing foods for long-term storage (e.g., canning, drying, freezing, salting), **p. 69.**

Reverse osmosis—method for processing water by forcing it through a membrane through which salts and impurities cannot pass (purified bottled water is produced this way), **p. 70.**

Roasting and toasting—using heat (from an oven) to process types of food such as cereal flakes and coffee, **p. 69.**

5.1 Growth and Development of the Food and Beverage Manufacturing Industry

From the beginning of civilization, people have been concerned about the quality of their foods. They have been so concerned, in fact, that dietary guidelines and restrictions became an integral part of daily culture and religious beliefs.

As science progressed, people began to understand germs and the role they play in food safety. This led to improved food handling guidelines and storage techniques.

Over the years, many of these techniques and guidelines have become government rules and regulations aimed at protecting public health.

The following sections provide a few examples of the legislation that has been put in place to ensure food safety and quality.

1900s

In 1906, Congress passed the original Food and Drug Act. This act prohibited interstate commerce (transportation and/or exchange between states in the United States) of misbranded and adulterated foods, drinks, and drugs.

The Meat Inspection Act also was passed in 1906. This act was created in reaction to Theodore Roosevelt's investigation of Chicago meatpackers. This investigation discovered that conditions in meatpacking plants were unsanitary and that poisonous preservatives and dyes were being used in foods intended for human consumption. As a result of this act, cleanliness standards were established for slaughterhouses and processing plants. All cattle, sheep, horses, swine, and goats were required to pass an inspection by the U.S. Food and Drug Administration prior to and after slaughter.

In 1949, the Food and Drug Administration (FDA) published the first industry guidebook, which addressed toxic chemicals in food.

In 1958, the Food Additives Amendment was enacted. This amendment required manufacturers of new food additives to establish the safety of these additives. The Delaney proviso prohibited the approval of any food additive shown to induce cancer in humans or animals.

The Nutrition Labeling and Education Act was enacted in 1990. This act required all packaged foods to bear nutrition labeling, and all health claims and nutrition information to be standardized and consistent with terms defined by the Secretary of Health and Human Services (Figure 5.1).

> ### Did You Know?
> Some scientists believe the Salem Witch Trials of 1692 were triggered by incidences of convulsive ergotism, a disorder caused by the ingestion of grain contaminated with ergot fungus.
> Symptoms of convulsive ergotism include muscle spasms, delusions, and hallucinations.

CREDIT: Andrey Burmakin/Fotolia.

2000–Present

This century has seen great efforts to stem the rising tide of childhood obesity by changing labeling on many food and beverage products. Perhaps the greatest change in the American diet during this period has been the decline of the consumption of soda. In May 2016, the Nutrition Facts Label was revised by the FDA using new scientific information to help consumers understand the link between food choices and chronic diseases such as obesity and heart disease.

There has been an increased emphasis on, and controversy about, crops grown using genetically modified organisms (GMOs).

Climate change will affect global food production in the coming decades. The food and beverage industry will focus on the need to reduce greenhouse gas emissions and deforestation to prevent the loss of crops.

Figure 5.1 Food nutrition label.

5.2 Overview of the Food and Beverage Manufacturing Industry

The food and beverage manufacturing industry links farmers to consumers through the processing of fruits, vegetables, grains, meats, dairy products, and other finished goods. After processing, these finished goods are delivered to grocers and wholesalers, who then supply them to households, restaurants, or institutional food services (Figure 5.2).

The job roles in the food processing industries can vary widely. For example, process technicians are involved with the following:

- Slaughtering and processing meat
- Processing milk and other dairy products
- Canning and preserving fruits and vegetables
- Producing grain products such as flour, cereal, and pet foods
- Making bread and other baked goods
- Manufacturing sugar, candy, or other confectionery products
- Brewing or producing beverages such as beer or soft drinks
- Processing fats and oils into products such as shortening, margarine, and cooking oil
- Preparing packaged seafood.

While this list is long, it is not comprehensive. Process technicians can be involved in a wide variety of other tasks.

Food Processing Hazards

A substantial percentage of foodborne illnesses are caused by pathogens (disease-causing microorganisms) that originated in contaminated environments or in animals before slaughter. Process technicians need to be aware of these types of organisms and how they are spread. They also need to know proper handling techniques and methods for preventing the spread of disease.

Figure 5.2 Process flow diagram for lager production.

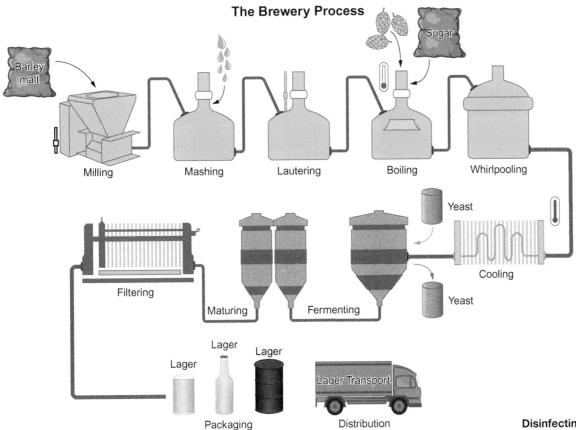

Food Processing Methods

Process technicians should be familiar with the food processing methods used in their plants. Figure 5.3 provides a general process flow diagram of one food production process. These methods can include techniques for the following:

- **Disinfecting**—destroying disease-causing organisms (e.g., through washing, irradiation, or ultraviolet exposure)
- **Preserving**—preparing for long-term storage (e.g., canning, drying, freezing, salting)
- **Manufacturing**—making a product from raw materials by hand or with machinery (e.g., cooking, decorating, grinding, milling, and mixing)
- **Drying**—removing moisture from items such as rice (using a conveyor belt moving through a hot air tunnel) or sugar (dried in a turning drum)
- **Agglomerating**—gathering materials into a mass (instant coffee crystals are made using this process)
- **Roasting and toasting**—using heat (from an oven) to process types of food such as cereal flakes and coffee
- **Cooking and frying**—using heat (typically hot oil) to process types of food such as chicken strips and potato chips
- **Mixing**—stirring materials to blend them (such as instant drink mixes)
- **Evaporating**—removing moisture from items such as milk and coffee (to create powdered milk or coffee)

Disinfecting destroying disease-causing organisms (e.g., through washing, irradiation, or ultraviolet exposure).

Preserving preparing foods for long-term storage (e.g., canning, drying, freezing, salting).

Manufacturing making a product from raw materials by hand or with machinery (e.g., cooking, decorating, grinding, milling, and mixing).

Drying removing moisture from items such as rice (using a conveyor belt moving through a hot air tunnel) or sugar (dried in a turning drum).

Agglomerating gathering materials into a mass (instant coffee crystals are made using this process).

Roasting and toasting using heat (from an oven) to process types of food such as cereal flakes and coffee.

Cooking and frying using heat (typically hot oil) to process foods such as chicken strips and potato chips.

Mixing stirring materials to blend them (such as instant drink mixes).

Evaporating removing moisture from items such as milk and coffee (to create powdered milk or coffee).

Figure 5.3 Process flow diagram for cookie production.

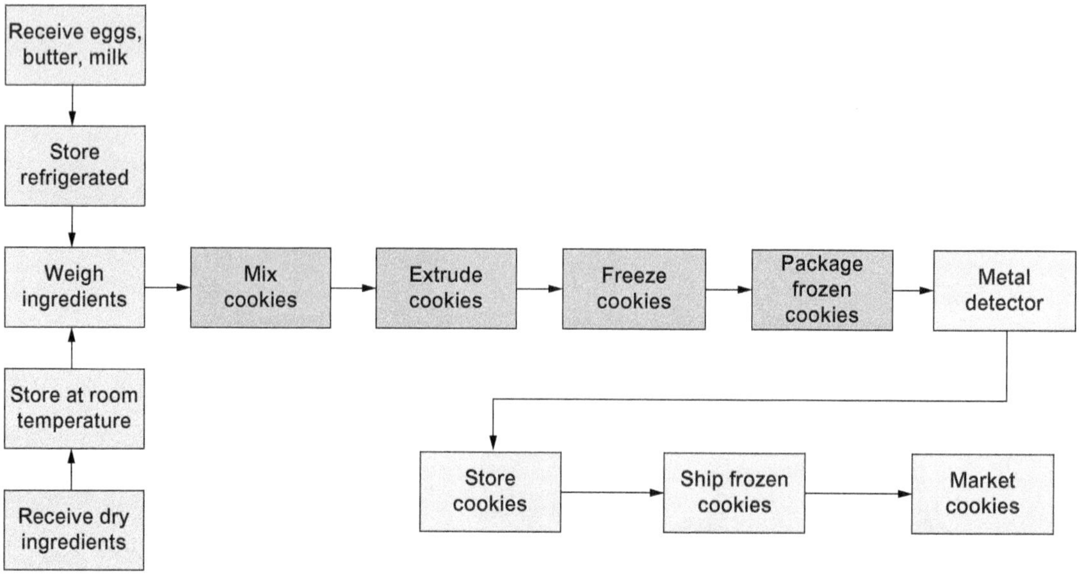

Reverse osmosis processing water by forcing it through a membrane through which salts and impurities cannot pass (purified bottled water is produced this way).

Pasteurization heating foods to kill organisms, such as bacteria, or make them less likely to cause disease (milk and orange juice are treated this way).

Distillation a method for separating a liquid mixture using the different boiling points of each component.

- **Reverse osmosis**—method of processing water by forcing it through a membrane through which salts and impurities cannot pass (bottled water is produced this way)
- **Pasteurization**—heating foods to kill organisms, such as bacteria, or make them less likely to cause disease (milk and orange juice are treated this way)
- **Distillation**—a method for separating a liquid mixture using the different boiling points of each component.

When food and beverages are processed and ready for packaging, these products are prepared for distribution using containers made from materials such as glass, plastic, paper, and metal. Many of these packaging products are created by other process industries (e.g., chemical, pulp, and paper). Food can be packaged under vacuum, low-oxygen environment, carbon dioxide environment, or hot or cold environment (Figure 5.4).

Quality standards apply to all aspects of food and beverage manufacturing, including raw materials, equipment, preparation, packaging materials, and shipping and transportation. These standards are regulated by government agencies and company policy.

Food Processing Systems

Some food processing systems have lines that are set up on a factory floor and are disassembled when the run is finished. Other systems have permanent machines that receive a constant product feed and that run 24 hours a day, 7 days a week.

Figure 5.4 Food processing line.
CREDIT: Trading Ltd. Stock/Shutterstock.

Food processing can require devices such as augers, extruders, conveyers, heaters, fryers, cookers, and freezers. Every food process is different. For example, a soft drink manufacturing plant might use the following equipment:

- Water purification and storage equipment
- Concentrate and sweetener storage containers
- Blending machines
- Food quality hosing and pipes
- Carbonators
- Canning and bottling machines
- Packing machines used to pack bottles into crates
- Bottling and can rinsing machines
- Carbonated storage and delivery systems
- Sealing machines
- Labeling machines

Some of these machines can be simple to operate, while others are much more complex.

Consider the following steps in a typical soda can bottling process:

1. The water being used to make the beverages must first be treated so the final product has a uniform taste regardless of where it was manufactured.
2. Water, sweetener, and concentrate are blended into a syrup.
3. Carbon dioxide (CO_2) is added to the syrup until it becomes carbonated.
4. The soda cans are rinsed and then placed in a filler machine for filling.
5. After the cans have been filled with soda, a machine called a *seamer* attaches the metal lid to the can.
6. The cans are then sent to a labeling machine if the container is not already marked with a label.
7. The final product is sent to the retailer.

5.3 Duties, Responsibilities, and Expectations of Process Technicians

The food manufacturing industry has many different types of workers. According to 2016 statistics of the Department of Labor (DOL), more than half of the employees in the food manufacturing industry are production workers. In addition to tasks described in Chapter 1, workers in food production jobs require manual dexterity, good hand-eye coordination and sometimes physical strength.

Food and beverage manufacturing facilities use equipment such as boilers, refrigeration equipment, air compressors, fired vessels, heat exchangers, and so on. The use of automation has increased. Workers still are needed to maintain and operate mixers, blenders, slicers, bottling machines, ovens, and other similar machinery. Some process technicians are responsible for cleaning, maintaining, lubricating, and repairing machines.

Other production workers use their hands or hand tools to do their jobs. For example, cannery workers and meat processors use knives and saws to process meat. Fruit and vegetable processors use their hands to sort, grade, wash, trim, peel, or slice. Bakery technicians use mixing machines, ovens, and other types of equipment to mix and create baked goods. Other technicians use their hands to decorate or apply artistic touches to prepared foods.

Equipment

Process technicians in the food and beverage industry also might be asked to work with a wide variety of equipment including the following:

- Blenders
- Grinders
- Mixers
- Ovens
- Slicers
- Conveyors
- Bottling machines
- Labeling machines

Food processing technicians work with hand tools and mechanized types of equipment.

Workplace Conditions and Expectations

Workplace conditions will vary from industry to industry. Some jobs require minimal physical effort and are in climate-controlled facilities. Other jobs occur in facilities without climate control and require heavy lifting or other forms of physical exertion. For instance, employees in chicken processing plants work in sub zero freezers. Specially designed clothing is mandatory to prevent frostbite. Some food processing divisions, especially chicken and meats, use flash freezers. Not only is special personal protective equipment (PPE) required, but special safety precautions and procedures must be followed to prevent entrapment in the freezer. In all settings, attention to sanitation and safety are key priorities.

5.4 Environmental Regulations and Considerations

The United States has one of the world's safest food supplies, primarily because it is closely monitored and highly regulated by federal, state, and local authorities. The main authorities involved with food handling and safety include the following:

- Food and Drug Administration (FDA)
- Centers for Disease Control and Prevention (CDC)
- U.S. Department of Agriculture (USDA)
- U.S. Environmental Protection Agency (EPA)
- National Oceanic and Atmospheric Administration (NOAA)
- Bureau of Alcohol, Tobacco and Firearms (ATF)
- State and local governments

Each of these authorities has a unique set of products that it oversees. Table 5.1 lists these products.

When monitoring and maintaining the safety of various food products, each of these authorities has its own unique set of responsibilities. Table 5.2 lists some of these responsibilities.

Table 5.1 U.S. Food Regulating Agencies and the Products They Oversee

Agency	Oversees
Food and Drug Administration (FDA)	All domestic and imported food sold in interstate commerce (except for that overseen by the USDA), eggs (excluding egg products)
	Bottled water
	Wine beverages with less than 7% alcohol
Center for Disease Control (CDC)-	All foods
United States Department of Agriculture (USDA)	Domestic and imported meat, poultry, and egg products
	Processed egg products (generally liquid, frozen, or dried pasteurized)
Environmental Protection Agency (EPA)	Drinking water
National Oceanic Atmospheric Association (NOAA)	Fish and seafood products
Bureau of Alcohol Tobacco and Firearms (ATF)	Alcoholic beverages except wine beverages containing less than 7% alcohol
State and local governments	All foods within their jurisdictions

Table 5.2 Roles of U.S. Food Regulating Agencies in Food Safety

Agency	Responsibilities
FDA	Enforces food safety laws governing domestic and imported food, except meat and poultry, by: <ul><li>Inspecting food production establishments and food warehouses, and collecting and analyzing samples for physical, chemical, and microbial contamination</li><li>Reviewing safety of food and color additives before marketing</li><li>Reviewing animal drugs for safety to animals that receive them and humans who eat food produced from the animals</li><li>Monitoring safety of animal feeds used in food-producing animals</li><li>Developing model codes and ordinances, guidelines, and interpretations and working with states to implement them in regulating milk and shellfish and retail food establishments, such as restaurants and grocery stores. An example is the model Food Code, a reference for retail outlets and nursing homes and other institutions on how to prepare food to prevent food-borne illness.</li><li>Establishing good food manufacturing practices and other production standards, such as plant sanitation, packaging requirements, and Hazard Analysis and Critical Control Point programs</li><li>Working with foreign governments to ensure safety of certain imported food products</li><li>Requesting manufacturers to recall unsafe food products and monitoring those recalls</li><li>Taking appropriate enforcement actions</li><li>Conducting research on food safety</li><li>Educating industry and consumers on safe food handling practices</li></ul>
CDC	<ul><li>With local, state, and other federal officials, investigates sources of foodborne disease outbreaks</li><li>Maintains a nationwide system of food-borne disease surveillance. Designs and puts in place rapid electronic systems for reporting food-borne infections. Works with other federal and state agencies to monitor rates of and trends in food-borne disease outbreaks. Develops state-of-the-art techniques for rapid identification of food-borne pathogens at the state and local levels.</li><li>Develops and advocates public health policies to prevent food-borne diseases</li><li>Conducts research to help prevent food-borne illness</li><li>Trains local and state food safety personnel</li></ul>
Food Safety and Inspection Service	Enforces food safety laws governing domestic and imported meat and poultry products by: <ul><li>Inspecting food animals for diseases before and after slaughter</li><li>Inspecting meat and poultry slaughter and processing plants</li><li>With USDA's Agricultural Marketing Service, monitoring and inspecting processed egg products</li><li>Collecting and analyzing samples of food products for microbial and chemical contaminants and infectious and toxic agents</li><li>Establishing production standards for use of food additives and other ingredients in preparing and packaging meat and poultry products, plant sanitation, thermal processing, and other processes</li><li>Making sure all foreign meat and poultry processing plants exporting to the United States meet U.S. standards</li><li>Seeking voluntary recalls by meat and poultry processors of unsafe products</li><li>Sponsoring research on meat and poultry safety</li><li>Educating industry and consumers on safe food-handling practices</li></ul>
EPA	Foods made from plants, seafood, meat, and poultry <ul><li>Establishes safe drinking water standards</li><li>Regulates toxic substances and wastes to prevent their entry into the environment and food chain</li><li>Assists states in monitoring quality of drinking water and finding ways to prevent contamination of drinking water</li><li>Determines safety of new pesticides, sets tolerance levels for pesticide residues in foods, and publishes directions on safe use of pesticides</li></ul>

Table 5.2 Roles of U.S. Food Regulating Agencies in Food Safety (*Continued*)

Agency	Responsibilities
NOAA	▪ Through its fee-for-service Seafood Inspection Program, inspects and certifies fishing vessels, seafood processing plants, and retail facilities for federal sanitation standards
ATF	▪ Enforces food safety laws governing production and distribution of alcoholic beverages ▪ Investigates cases of adulterated alcoholic products, sometimes with help from FDA
State and local governments	▪ Work with FDA and other federal agencies to implement food safety standards for fish, seafood, milk, and other foods produced within state borders ▪ Inspect restaurants, grocery stores, and other retail food establishments, as well as dairy farms and milk processing plants, grain mills, and food manufacturing plants within local jurisdictions ▪ Embargo (stop the sale of) unsafe food products made or distributed within state borders

Summary

Food and beverage quality is integral to our safety, our daily culture, and even our religious beliefs. Various techniques and government guidelines have been enacted since 1906 to protect public health. Food processing methods include disinfecting, preserving, manufacturing, drying, agglomerating, roasting and toasting, cooking and frying, mixing, evaporating, reverse osmosis, pasteurization, and distillation.

The food and beverage manufacturing industry links farmers to consumers through the processing of fruits, vegetables, grains, meats, dairy products, and other finished goods. Finished goods are delivered to grocers and wholesalers, who supply them to households, restaurants, or institutional food services.

Foodborne illnesses often are caused by pathogens originating from contaminated environments or animals.

The United States has one of the world's safest food supplies, primarily because it is closely monitored and highly regulated by federal, state, and local authorities. Government agencies such as the Food and Drug Administration and the U.S. Department of Agriculture oversee a wide range of food and beverage products.

Checking Your Knowledge

1. Define the following terms:
   a. agglomerating
   b. cooking and frying
   c. distillation
   d. drying
   e. evaporating
   f. manufacturing
   g. mixing
   h. pasteurization
   i. preserving
   j. roasting and toasting

2. When did the U.S. Congress pass the original Food and Drug Act?
   a. 1906
   b. 1954
   c. 1865
   d. 1990

3. A substantial percentage of food-borne illnesses are caused by pathogens found in _____ or _____.

4. Disinfecting is the process of:
   a. preparing for long-term storage
   b. making a product from raw materials
   c. removing moisture
   d. destroying disease-causing organisms

5. Name the process that involves forcing water through a membrane to remove impurities.
   a. evaporation
   b. reverse osmosis
   c. agglomeration
   d. distillation

6. Name three government agencies that regulate the food and beverage industry.

7. Which government agency oversees fish and seafood products?
   a. ATF
   b. USDA
   c. NOAA
   d. NCAA

8. Which government agency oversees domestic and imported meat and poultry products?
   a. EPA
   b. USDA
   c. DOT
   d. NRC

9. Which statement would require further teach about process technicians in the food and beverage industry?
   a. Process technicians may need to work in subzero temperatures.
   b. Technicians can set up all products the same way for production.
   c. Process technicians may work with augers, extruders, augers, and heat exchangers.
   d. Processing technicians sometimes work with hand tools.

10. True or false: Food processing systems in a plant may be temporary or permanent.

NOTE: Answers to Checking Your Knowledge questions are in Appendix I.

Student Activities

1. Using library resources or the Internet, research the history of food manufacturing. Identify five items or events that you believe have had a significant impact on food quality today. Write a two-page paper that lists and describes each of these items.

2. Select two government agencies that affect the food and beverage manufacturing industry. Using the Internet or other resources, write a one-page paper about what food- and/or beverage-related regulations these agencies oversee and how they affect the industry and consumers.

3. Divide into teams of four students. Select one of the following food or beverage items and develop the food supply chain from origin to consumer, including the money train. Place your food supply chain in a flow diagram and present to your classmates.
   a. chocolate
   b. beer
   c. chicken
   d. wine
   e. soda
   f. potatoes
   g. peanuts
   h. cotton
   i. newspaper

Chapter 6
Water and Wastewater Treatment Industry

Objectives

Upon completion of this chapter, you will be able to:

6.1 Describe the growth and development of water treatment and wastewater treatment. (NAPTA History 1, 2) p. 77

6.2 Explain the water treatment and wastewater treatment processes. (NAPTA History 3) p. 78

6.3 Identify process technician duties and responsibilities associated with water and wastewater treatment. (NAPTA Career 1) p. 80

6.4 Identify regulations associated with water and wastewater treatment, and explain the purpose of each. (NAPTA History 4) p. 81

Key Terms

Absorb—to draw inward, **p. 80.**
Adsorb—to stick together, **p. 80.**
Desalination—the removal of salts and minerals from a target substance such as saltwater, **p. 78.**
Disinfection—a process of killing pathogenic organisms, **p. 77.**
Dissolved solids—solids that are held in suspension indefinitely, **p. 80.**
Filtration—the process of removing particles from water, or some other fluid, by passing it through porous media, **p. 77.**
Pathogen—a disease-causing microorganism, **p. 77.**
Potable water—water designated as ingestible for human consumption or food preparation, **p. 77.**
Process water—water used in industrial processes that may contain inorganic and organic compounds not suitable for release to the environment, **p. 78.**
Process water sewer—water collection system of drains surrounding and under process equipment, which is directed to a wastewater treatment plant, **p. 79.**
Settleable solids—solids in wastewater that can be removed by slowing the flow in a large basin or tank, **p. 80.**
Suspended solids—solids that cannot be removed by slowing the flow, **p. 80.**
Turbidity—cloudiness caused by particles suspended in water or some other liquid, **p. 77.**

Introduction

It is well understood that the availability of clean water is essential for life. What is sometimes overlooked is the importance of clean water to many industrial processes. Through the operation of water treatment facilities, process technicians are able to process and treat water so it is safe to drink, protects public health, and is clean and available for industry.

Wastewater treatment facilities are also necessary to eliminate the spread of pollution and disease. From the use of early lagoons to modern complex treatments, the spread of pollution and waterborne diseases has been controlled.

In order to work in a water or wastewater treatment facility, process technicians must be able to perform a variety of tasks and be familiar with the rules and regulations pertaining to water and wastewater treatment. Some states require process technicians to obtain certification prior to working in a wastewater treatment facility.

6.1 Growth and Development of the Water and Wastewater Treatment Industry

The importance of good drinking water was recognized early on in history. However, it took centuries for people to understand that their senses (taste, sight, and smell) were inadequate when it came to judging water quality.

During the eighteenth century, the Europeans devised some of the earliest water treatment systems. These systems used **filtration** (removing particles from water by passing it through porous media) to improve the taste of the water and remove suspended particles. However, filtration did not address the problem of biological contaminants such as typhoid, dysentery, and cholera.

Filtration the process of removing particles from water, or some other fluid, by passing it through porous media.

The impact of biological contaminants in water was not known until the second half of the nineteenth century, when infectious diseases were first recognized and the ability to spread or transmit these diseases through water was demonstrated. Scientists became increasingly focused on biological contaminants and disease-causing microorganisms (**pathogens**) and their impact on public water supplies.

Pathogen a disease-causing microorganism.

During their research, scientists discovered that **turbidity** (cloudiness) in water was not only unappealing but also an indicator of a potential health risk. Turbidity can be caused by suspended particles that have the potential to harbor pathogens. Based on this information, new and improved water treatment systems were designed to reduce turbidity and pathogenic organisms.

Turbidity cloudiness caused by particles suspended in water or some other liquid.

By the beginning of the twentieth century, water treatment had improved. In 1906, chlorine was first introduced as a means of water **disinfection** (a process of killing pathogenic organisms). Chlorine helped eliminate waterborne diseases in drinking water. Since then, scientists and engineers have continued to develop new ways to process water more quickly and effectively, and at a lower cost.

Disinfection a process of killing pathogenic organisms.

When fresh (**potable**) water is scarce, such as in the desert or in places where water has been contaminated or compromised in some way due to an act of nature, the only way to

Potable water water designated as ingestible for human consumption or food preparation.

Did You Know?

Excessive water contamination can occur when rain washes debris from nearby public trails and pastures into water sources.

CREDIT: watcherfox/Fotolia.

> **Did You Know?**
>
> The use of chlorine as a water disinfection agent is considered one of the greatest discoveries of the last millennium.
>
>
>
> **CREDIT:** Dmitry Naumov/Fotolia.

Desalination the removal of salts and minerals from a target substance such as saltwater.

provide safe fresh water is by desalination. **Desalination** is the process of removing salts and minerals from saltwater, typically sea water. There are many process types for this operation depending on location, water source, and the supply of fresh water needed in the location. In 2015, there were over 18,000 desalination plants operating worldwide.

History of the Wastewater Treatment Industry

As water is used, wastewater is formed. Examples are dirty dishwater, laundry and biological wastes, and process water. **Process water** is water used in industrial processes that may contain inorganic and organic compounds not suitable for release to the environment. In the beginning, wastewater was dumped into streets or nearby streams. Over time, stone-lined ditches, and later elaborate collection systems, were developed to carry wastes to receiving streams. However, the end result was the same. Wastes dumped upstream were carried to communities downstream, and diseases were spread.

Process water water used in industrial processes that may contain inorganic and organic compounds not suitable for release to the environment.

With the discovery of disease-causing organisms came the diversion of wastewater to treatment plants (Figure 6.1). These facilities treat the water before returning it to the environment (e.g., streams, rivers, and oceans). Today's increased recycling efforts encourage the use of treated wastewater for irrigation purposes that are not for human consumption, such as for watering golf courses or crops.

6.2 Overview of the Water Treatment Process

Municipal Water Treatment Process

The water treatment process begins with water being pumped from wells, rivers, streams, lakes, and oceans. This water is sent to a water treatment plant through a series of pumps and valves. Water sources must always be protected against accidental contamination.

Well water is usually low in turbidity and often does not need to be filtered before being disinfected. Water from other sources, however, usually requires filtration.

Once inside the treatment plant, water is filtered and chemically treated. After disinfection, the water then can be distributed to residential, agricultural, and industrial customers. Figure 6.2 provides a process diagram of the water treatment industry.

Water quality analysis and testing is essential when producing water for distribution. Different quality standards are in place, depending on how the water is used. For example, water intended for human consumption must be:

- Free of corrosive properties
- Free of dissolved substances
- Free of toxic contam inants
- pH controlled
- At the proper temperature and pressure range

Figure 6.1 Water treatment plant.
COURTESY of Eastman Chemical Company.

Figure 6.2 Process flow diagram for a water treatment plant.

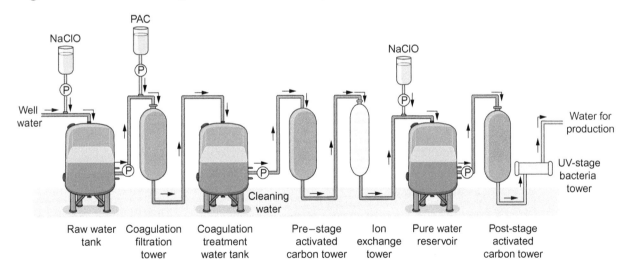

Industrial Wastewater Treatment Process

Many industrial processes use process water for heating, cooling, washing equipment, or as part of the process stream to help or enhance its completion to product. The water must be cleaned before returning it to the local waterway. Wastewater treatment plants capture this water, typically from the **process water sewer** or lift stations. In extreme run-off situations, they capture rain water for treatment prior to its return to the waterway. These plants are

Process water sewer water collection system of drains surrounding and under process equipment, which is directed to a wastewater treatment plant.

strictly regulated by the Environmental Protection Agency via the Clean Water Act and the Safe Drinking Water Act.

Collection systems carry residential and industrial wastewater, as well as storm water to a wastewater treatment facility. Large rocks, sticks, and other debris are removed early in the treatment process to avoid damaging pumps and other equipment. The first stage (known as primary or mechanical treatment) in the wastewater treatment facility is usually a settling pond or clarifier. The heavier **settleable solids** (solids in wastewater that can be removed by slowing the flow in a large basin or tank) are removed and, in many instances, dewatered. The resulting sludge is then burned in a high temperature kiln or buried in a local landfill.

In secondary (known as biological) treatment, the wastewater is adjusted for temperature and pH and then further treated to remove **suspended solids** (solids that cannot be removed by slowing the flow) and some **dissolved solids** (solids that are held in suspension indefinitely). Microorganisms, such as those found on rocks in a stream, are used to **adsorb** (stick together) and **absorb** (draw inward) the dissolved solids for removal. The wastewater continues through the treatment facility and may be disinfected with chemicals (such as sodium hypochlorite or calcium hypochlorite) before being discharged into the environment.

Some water treatment facilities must provide a third stage, known as *tertiary treatment*, to further treat the wastewater before discharging it. The most common tertiary treatment option is filtration, which removes fine suspended solids.

The EPA regulates the National Pollution Discharge Elimination System (NPDES), which issues permits for discharging the wastewater into the environment. Each permit outlines the specific water qualities that must be monitored and the conditions for discharge.

6.3 Duties and Responsibilities of Process Technicians

Process technicians in water and wastewater treatment plants are responsible for safely controlling the equipment and processes required to remove or destroy harmful materials, chemical compounds, and microorganisms from water. These technicians are also responsible for controlling pumps, valves, and other equipment used throughout the various treatment processes.

Other duties process technicians may be asked to perform include the following:

- Reading and interpreting meters and gauges to make sure that plant equipment and processes are working properly
- Using a variety of instruments to sample and measure water quality
- Operating chemical feeding devices and adjusting the amount of chemicals in the water
- Using computers to monitor equipment, store the results of sampling, make process-control decisions, schedule and record maintenance activities, produce reports, and troubleshoot malfunctions
- Sampling area waterways upstream of the plant and downstream of the effluent return to the waterway.

The specific duties of a water treatment technician depend on the type and size of the plant. In a small plant, one technician might be responsible for controlling all of the machinery, performing tests, keeping records, handling complaints, and performing repairs and maintenance. In a larger plant with more employees, technicians might be more specialized and monitor only one process.

From time to time, water treatment technicians must work during emergency situations, which can arise from within the facility (e.g., a chemical leak) or be the result of something external (e.g., a heavy rainstorm causing water volume to exceed the plant's treatment capacity). Technicians must be trained to deal with these types of situations and must be able to work under extreme pressure to correct problems as quickly as possible. Because working conditions can be dangerous, technicians must always exercise caution.

Settleable solids solids in wastewater that can be removed by slowing the flow in a large basin or tank.

Suspended solids solids that cannot be removed by slowing the flow.

Dissolved solids solids that are held in suspension indefinitely.

Adsorb to stick together.

Absorb to draw inward.

In addition to working in emergency situations, treatment technicians must be able to perform physically demanding work, indoors and outdoors, in various locations. Because of the presence of hazardous conditions, such as slippery walkways, noise, dangerous gases, and open tanks, process technicians must always pay close attention to safety and follow all facility procedures.

6.4 Environmental Regulations and Considerations

Prior to the 1970s, little was done to protect the environment from the hazardous and sometimes lethal effects of pollution. However, in 1970 a major movement was started in an attempt to educate and inform the masses about environmental issues and their impact. Several pieces of legislation were enacted as a result of that movement. Two of these items were the Clean Water Act of 1972 and the Safe Drinking Water Act of 1974.

The Clean Water Act of 1972 implemented a national system of regulation on the discharge of pollutants, while the Safe Drinking Water Act of 1974 established standards for drinking water. As a result of these two acts, industrial facilities sending their wastes to municipal treatment plants must now meet certain minimum standards to ensure that the wastes have been adequately pretreated so they will not damage the municipal treatment facility.

Municipal water treatment plants also must meet stringent standards. The list of drinking water contaminants regulated by these statutes has continued to grow over time. Because of this, plant technicians must be familiar with the guidelines established by federal, state, and local authorities, and know the impact these regulations have on their plant.

In order to ensure that water treatment technicians are familiar with the various regulations, each technician must receive training about the many aspects of water treatment, and must pass an examination to certify that they are capable of overseeing treatment plant operations. There are different levels of certification, depending on the type of treatment facility, municipal or industrial, and also upon the technician's experience and training.

Summary

The importance of good drinking water was recognized early in history. The need for advanced treatment techniques was identified in the second half of the nineteenth century when infectious diseases and their ability to spread through water was demonstrated. Scientists discovered that turbidity (cloudiness caused by suspended particles) in water was not only unappealing but also an indication of a potential health risk.

By the beginning of the twentieth century, water treatment improved. Techniques such as chlorination and filtration reduced the rates of waterborne diseases. Since then, scientists and engineers have continued to develop new ways to process water more quickly and effectively, and at a lower cost.

Clean water is essential for life, as well as for many industrial processes. Through water treatment facilities, process technicians are able to treat water so that it is safe to drink and public health is protected. Desalination of ocean water is a growing industry, providing potable water to communities in a variety of settings.

Environmental concerns in the late twentieth century led to the creation of regulations to limit pollution by limiting discharges into receiving waters. Wastewater treatment facilities are the primary method for limiting water contaminants and pollutants. Mechanical, biological, and filtration methods may be used in water and wastewater treatment plants.

Process technicians in water treatment facilities must safely perform a variety of tasks within the rules and regulations that protect water quality and ensure public health.

Checking Your Knowledge

1. Define the following terms:
   a. absorb
   b. adsorb
   c. disinfection
   d. dissolved solids
   e. filtration
   f. pathogen
   g. settleable solids
   h. suspended solids
   i. turbidity

2. The impact of biological contaminants in water was not known until the:
   a. Seventeenth century
   b. Eighteenth century
   c. Nineteenth century
   d. Twentieth century

3. Which of the following duties might a water treatment technician be required to perform?
   a. reading and interpreting meters and gauges
   b. using instruments to sample and measure water quality
   c. operating chemical feeding devices and adjusting chemical levels
   d. using computers to monitor equipment
   e. all of the above

4. Which environmental statute established standards for drinking water?
   a. the Clean Water Act of 1972
   b. the Water Safety Act of 1974
   c. the Safe Drinking Water Act of 1974
   d. the Water Protection Act of 1973

5. Primary water treatment does the following:
   a. removes suspended solids
   b. adsorbs dissolved solids
   c. removes settleable solids
   d. removes dissolved solids

6. Wastewater treatment that adjusts for temperature and pH and then uses microorganisms to remove solids is called:
   a. primary treatment
   b. turbidity treatment
   c. secondary treatment
   d. tertiary treatment

NOTE: Answers to Checking Your Knowledge questions are in Appendix I.

Student Activities

1. Use the Internet, library, or other resources to research the water or wastewater treatment process. Write a two-page paper describing what you learned.

2. Contact the local water or wastewater treatment authority and arrange a tour of a water treatment facility. Write a three-page paper describing what you saw and what you learned.

3. Working in teams of three, prepare a PowerPoint or flip chart presentation that represents the steps of one of the following:
   a. desalination of seawater
   b. wastewater treatment for a municipality
   c. wastewater treatment of a plant process sewer

Chapter 7
Pulp and Paper Industry

 Objectives

Upon completion of this chapter, you will be able to:

7.1 Describe the growth and development of the pulp and paper industry. (NAPTA History 1) p. 84

7.2 Explain pulp and paper industry processes. (NAPTA History 3) p. 85

7.3 Explain the different segments of the pulp and paper industry. (NAPTA History 1) p. 88

Key Terms

Cellulose—the principal component of the cell walls in plants, **p. 86.**
Fiber—a long, thin filament, either plant-based or manmade, resembling a thread, **p. 85.**
Fourdrinier machine—a papermaking machine, developed by Henry and Sealy Fourdrinier, which produces a continuous web of paper, **p. 85.**
Mill—a facility where a raw substance is processed and refined to another form, **p. 85.**
Pulp—a cellulose fiber material, created by mechanical and/or chemical means from various materials (e.g., wood, cotton, recycled paper), from which paper and paperboard products are manufactured, **p. 85.**

Introduction

Most paper starts as trees. A process similar to the one created almost 2,000 years ago is used to turn trees into a wide variety of paper products.

Despite the high-tech advancements of the last few decades and the emergence of the "paperless society," paper continues to play a huge role in everyday life. Products made from paper, as well as items made using natural wood chemicals, account for the thousands of products produced by the pulp and paper industry. These products are found everywhere and are used daily around the world. Consider just a few of paper's many different uses:

- Business and industry—printer paper, checks, envelopes, documents, business cards, financial statements, catalogs, and boxes
- Medical and health—bandages, medical charts, gowns and masks, sterile filters, prescription pads, and tissues

- Construction—cement bags, fiberboard, insulation, sandpaper, wallpaper, and tar paper
- Education—textbooks, notebooks, class schedules, library cards, folders, tests, and photocopies
- Household—money, books, newspapers and magazines, telephone books, napkins, toilet paper, artwork prints, food packaging, and greeting cards
- Recreation—CD covers, bumper stickers, jigsaw puzzles, party supplies, event tickets, playing cards, and board games

7.1 Growth and Development of the Pulp and Paper Industry

More than 2,000 years ago, civilizations began to write down important traditions, ceremonies, religious practices, rules, and decrees. Everything from papyrus (writing material made from water reeds), to silk, to clay was used to record this information.

> **Did You Know?**
>
> The Chinese invented toilet paper almost 1,200 years ago.
>
>
>
> **CREDIT:** pixelrobot/Fotolia.

Although many civilizations claim to have created paper or paper like substances, the first recorded efforts of papermaking involved a Chinese scholar named Ts'ai Lun. Lun ground up bark, linen, and hemp and mixed them with water. He spread this mixture onto a cloth-covered bamboo frame and then left the frame in the sun to dry. The finished product was paper. This same basic process can be used today to make homemade paper.

For hundreds of years, the papermaking process remained a secret known only to the Chinese. But starting in the 800s, the process spread to the rest of Asia, Africa, and Europe. The process was labor-intensive and consumed a significant amount of raw materials, so paper was made sparingly and was used for only the most important documents.

1400s

In these early years, a considerable amount of paper was made from rags and cloth scraps, and all documents were still being written by hand. Because of this, paper continued to be used in a limited way. However, in the mid-1400s, a German named Johannes Gutenberg created a printing machine. His invention allowed documents to be printed more quickly with less labor.

Soon, books were being printed and distributed widely. Paper became even more important and more uses for it were found. People became better educated because of the increased availability of books and other printed materials, thereby contributing to the Renaissance period in history, often referred to as the Age of Enlightenment. In addition, science and the arts flourished during this time.

> **Did You Know?**
>
> Watermarks, which have been used since the thirteenth century when the Italians first introduced them, are translucent designs embossed into a piece of paper during its production.
>
> These designs, which are visible when a sheet of paper is held up to the light, are used to identify the paper and the papermaker.
>
>
>
> **CREDIT:** Paul Paladin/Shutterstock.

1600s

In about 1690, a paper **mill** (a facility where a raw substance is processed and refined to another form) was established in the United States near Philadelphia, Pennsylvania, on a tributary of the Wissahickon Creek. Water was essential to the papermaking process and to transportation of the finished goods.

Mill a facility where a raw substance is processed and refined to another form.

1700s

In the 1700s, René de Réaumur, a French scientist, observed the paper wasp consuming wood and then spitting the mush out to make a nest. Based on his observations, Réaumur suggested that paper could be created using wood. The basic formula for making paper—wood fiber combined with water and energy—was established.

In the following decades, many people began to create inventions for turning wood into **pulp**, a cellulose fiber material that could be used to create paper.

Pulp a cellulose fiber material, created by mechanical and/or chemical means from various materials (e.g., wood, cotton, recycled paper), from which paper and paperboard products are manufactured.

1800s

In the early 1800s, Nicholas Louis Robert created the first papermaking machine. Two brothers, Henry and Sealy Fourdrinier, improved the design and invested large sums of money to develop the machine. These inventions allowed paper to be mass-produced in a continuous web. Today, most modern papermaking machines are based on the same principles as the **Fourdrinier machine**.

Various improvements in papermaking came over the next few decades. The use of paper and paper products increased as more consumer products requiring packaging were manufactured and distributed, along with books, newspapers, magazines, and more. Papermaking became a science as well as an art.

Fourdrinier machine a papermaking machine, developed by Henry and Sealy Fourdrinier, which produces a continuous web of paper.

1900s

Concern for the environment and worker safety boomed in the 1960s and 1970s. Governments enacted legislation to provide for cleaner air, water, and land, along with safer working environments.

From the 1980s on, the industry has struggled with price declines, the need for increased automation, labor issues, and growing worldwide competition.

2000–Present

The advent of electronic media brought a decline in the market for hard copies of periodicals, including books, magazines, and newspapers. The effect on the paper-producing industry has been harsh. The housing crash of 2008–2009 created a backlash of loss for the paper industry, and sawmills closed at an alarming rate. Paper technology today is ever changing to meet the needs of society.

Research is underway to develop wood pulp into many items not thought to be possible in the last decade. "Smart paper," which changes color to show when a product has lost its freshness, is being developed. Printed electric circuits are being made of paper to replace heavy circuit boards. "Scratch and sniff" applications on materials are a way of advertising and of using the senses for learning.

7.2 Overview of the Pulp and Paper Industry

Although many different materials can be used to make paper, wood is the primary material. Wood is composed of **fibers** (long, thin substance resembling threads) made up of material

Fiber a long, thin filament, either plant-based or manmade, resembling a thread.

Cellulose the principal component of the cell walls in plants.

called **cellulose** (principal component of the cell walls of plants). Fibers are held together with a natural adhesive called *lignin*. The papermaking process involves separating and rearranging those fibers.

Paper also can be made from other materials such as these:

- Recovered paper, although paper can be recycled a limited number of times before it cannot be processed anymore
- Linen
- Cotton
- Synthetics (manmade materials).

Trees are harvested from tree farms. Two categories of trees are used to make paper:

- Hardwood—trees such as oak and maple produce short fibers, which make smooth but weak paper that is suitable for writing
- Softwood—trees such as pine and spruce produce long fibers, which make strong but rough paper that is suitable for packaging (e.g., cardboard)

Papermaking operations can combine hardwood and softwood fibers to make strong, smooth papers. Almost every part of a harvested tree is used for one purpose or another. Along with the wood itself, natural chemicals from a tree, such as resin and oils, are also used to make products. These products include toothpaste, roofing shingles, car wax, crayons, clothing, sports helmets, and film stock.

Did You Know?

U.S. dollar bills are composed of 25% linen and 75% cotton. Red and blue synthetic fibers of various lengths are distributed evenly throughout the paper.

Prior to World War I, these fibers were made of silk.

CREDIT: alswart/Fotolia.

Pulp Mills

After the trees are harvested, they are trimmed to logs and transported to a pulp mill where they are washed and debarked. During the pulping process, the wood fibers must be separated into individual strands (Figure 7.1). This can be achieved either mechanically or chemically, or both:

- If the fibers are separated mechanically, a grinder processes the wood.
- If the fibers are separately chemically, a chipper processes the wood. From there, it is processed in a device called a digester, where it is treated with boiling chemicals. The chemicals are then removed using an extractor.

The type of process used is determined by the strength, appearance, and use of the paper that will be made from the pulp. Figure 7.2 provides a process flow diagram for the paper industry.

The fibers are washed, screened, and bleached. The resulting product is a mushy solution of water and fiber called pulp (see Figure 7.1). Considerable amounts of water and energy are required to make pulp. Sometimes, fillers, additives, and dyes are added to the pulp to make the finished paper glossy, absorbent, water-resistant, or colored.

Pulp and Paper Industry **87**

Figure 7.1 Wood pulp in a pulp processing plant.
CREDIT: lichaoshu/Fotolia.

Figure 7.2 Process flow diagram of paper industry.

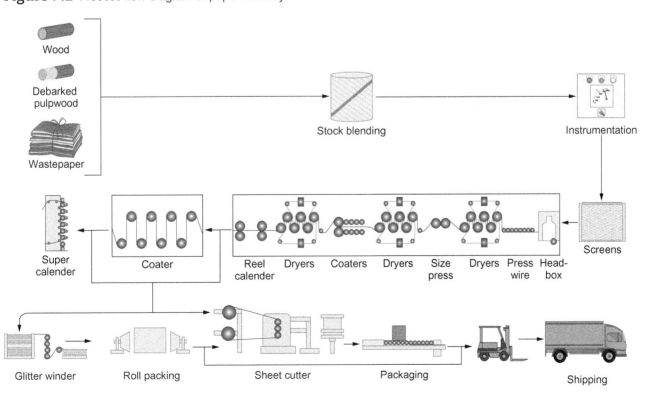

Paper Mills

The papermaking process produces many different types of paper. The quality of the paper and its characteristics will vary depending on the final application. Paper can vary in strength, weight, brightness, *opacity* (inability to be seen through), softness, smoothness, thickness, and more.

When the pulp is ready, it is spread out as a wet mixture, called a slurry, onto a giant screen. This screen passes on a conveyor from what is called the wet end of the process to the dry end. On the wet end, water is removed from the slurry using gravity and vacuums. The remaining fiber bonds into a watery sheet of paper. On the dry end, this sheet is pressed between soft, heated rollers to remove any remaining water.

Additional treatment of the paper, such as the application of coatings, is then carried out. Once complete, the resulting dry, uniform paper is stored as giant rolls.

7.3 Segments of the Pulp and Paper Industry

The segments of the pulp and paper industry include production, conversion, and distribution. Production of pulp or paper, described above, varies depending on the needs of the consumer and the base materials used (e.g., wood versus recycled newsprint). Weight, density, and color can vary widely.

Once the paper is placed on rolls, it can be converted into final products (Figure 7.3). Raw paper can be:

- Cut
- Folded
- Coated
- Glued
- Screened (printed)
- Embossed (stamped with a raised pattern)

The types of processing or conversion depend on the end product (e.g., paper napkins, stationery, boxes, or file folders).

The final products then are packaged and distributed to consumers.

Figure 7.3 Giant roll of finished paper.

CREDIT: Westend61/Getty Images.

Equipment Used in the Pulp and Paper Industry

As a process technician in the pulp and paper industry, you will work with all different types of equipment. Some types, like the following, are common to most process industries:

- Pumps (see Chapter 16)
- Compressors (see Chapter 17)
- Furnaces (see Chapter 22)

You also will work with the following:

- Reactors
- Evaporators

As with all process industries, factors such as safety awareness, teamwork, and continuous learning are expectations of process technicians in the work environment.

Summary

Most paper starts as trees. The first recorded efforts of papermaking involved a Chinese scholar called Ts'ai Lun who ground up bark, linen, and hemp and mixed them with water, spreading the mixture out on a cloth-covered bamboo frame to dry. Most paper is still made from trees. Using a process that is very similar to the one created almost 2,000 years ago, trees are turned into a wide variety of paper products.

Paper continues to play a huge role in everyday life. Along with paper, if you include items made using natural wood chemicals, then the pulp and paper industry creates and distributes thousands of products used daily worldwide.

In the mid-1400s, Johannes Gutenberg created the first printing machine, an invention that allowed documents to be printed more quickly and with less labor than ever before. This invention created a huge demand for paper, but it was not until the 1800s that Henry and Sealy Fourdrinier patented a machine that allowed paper to be mass-produced. Most modern papermaking machines are still based on the same principles as the Fourdrinier machine.

Environmental and safety concerns dominated the 1960s and 1970s, and subsequent legislation affected the profitability of the pulp and paper industry. From the 1980s on, the pulp and paper industry has struggled with price declines, the need for increased automation, labor issues, and growing worldwide competition. The advent of social media and paperless communication in the twenty-first century has had a strong impact on this industry. The pulp and paper industry has shifted focus to developing new and innovative ways of using paper products.

Because of advances in technology, fewer people are needed to produce pulp and paper than in the past. However, these people must have more education and be more highly skilled than past workers. A two-year Associate degree is preferred for process technicians.

Checking Your Knowledge

1. Define the following terms:
   a. cellulose
   b. fiber
   c. mill
   d. pulp

2. Which ancient civilization is generally credited with inventing paper?
   a. the Japanese
   b. the Chinese
   c. the Greeks
   d. the Egyptians

3. Put the following events in order, from earliest to latest:
   a. printing machine invented
   b. papermaking process invented
   c. paper mill built in America
   d. papermaking machine invented

4. What is the name associated with modern papermaking machines?
   a. Fourdrinier
   b. Justinier
   c. Caitinier
   d. Courtinier

5. When was the first paper mill built in America?
   a. 1690
   b. 1720
   c. 1836
   d. 1965

6. _____ is a component of wood fiber used to make paper.

7. Name three types of material, other than wood, that can be used in the papermaking process.

8. *(True or False)* Natural chemicals from wood are also used to make products.

9. List three steps associated with the papermaking process.

NOTE: Answers to Checking Your Knowledge questions are in Appendix I.

Student Activities

1. Conduct a scavenger hunt at home or work, keeping a list of all paper and paper-related products you can find. Limit your search to 30 minutes or less. Then, review the list and count the number of products you found.

2. Pick one of the following three topics relevant to the papermaking process, and then research the topic using the Internet, library, or other resources. Write a one-page summary of what you learned.
   a. early papermaking efforts (Egyptians, Chinese)
   b. the invention of the printing machine
   c. the invention of the papermaking machine

3. Which skills do you think are the most important for a process technician working in the pulp and paper industry? Make a list and discuss it with a classmate.

4. Working in teams of four, research the types of trees that are generally used to make wood pulp. Locate any of those trees native to your geographical area and identify them as being softwood or hardwood. Create a team report to be delivered at your next class meeting.

PART 2 Skills for Process Technicians

Chapter 8
Working as Teams

Objectives

Upon completion of this chapter, you will be able to:

8.1 Describe the differences between work groups and teams. (NAPTA Teams 1) p. 92

8.2 Describe the different types of teams encountered in the process industries. (NAPTA Teams 2) p. 93

8.3 Identify the characteristics of a high-performance or effective team. (NAPTA Teams 3, 4) p. 94

8.4 Describe the steps or stages through which a team evolves (forming, storming, norming, performing, and adjourning). (NAPTA Teams 5) p. 96

8.5 Identify factors that contribute to the unsuccessful results of a team, including:
- Inability to achieve the defined outcome
- Problems in working together and achieving full synergy. (NAPTA Teams 6) p. 98

8.6 Define workforce diversity and its impact on workplace relations:
- In a team environment
- In a work group (coworkers). (NAPTA Teams 7) p. 102

Key Terms

Criticism—a serious examination and judgment of something; criticism can be positive (constructive) or negative (destructive), **p. 101.**

DESCC conflict resolution model—a model for resolving conflict, comprising the following steps: describe, express, specify, contract, and consequences, **p. 99.**

Diversity—the presence of a wide range of variation in qualities or attributes; in the workplace, it also can refer to antidiscrimination training, **p. 102.**

Ethnocentrism—belief in the superiority of one's own ethnic group; belief that others should believe and interpret things exactly the way you do, **p. 103.**

Feedback—evaluative or corrective information provided to the originating source about a task or a process, **p. 101.**

Prejudice—attitude toward a group or its individual members based on stereotyped beliefs, **p. 103.**

Process—method for doing something, generally involving tasks, steps, or operations that are ordered and/or interdependent, **p. 95.**

Self-managed team—also called *self-directed team*; a small group of employees whose members determine, organize, plan, and manage their day-to-day activities and duties under reduced or no supervision, **p. 95.**

Stereotyping—maintaining beliefs about individuals or groups based on opinions, habits of thinking, or rumors, which lead to generalizations about all members of a group, **p. 102.**

Synergy—the total effect in which a whole is greater than the sum of its individual parts, **p. 95.**

Task—a set of actions that accomplish a job, **p. 95.**

Team—a small group of people, with complementary skills, committed to a common set of goals and tasks, **p. 93.**

Team dynamics—interpersonal relationships; ways in which workers get along with each other and function together, **p. 95.**

Work group—a group of people organized by logical structures within a company, having a designated leader, and performing routine tasks, **p. 92.**

Introduction

In the process industries, companies place people into groups of different types and sizes to work together and to achieve industry goals. Industries organize personnel into work groups, teams, or both. Teams often function within the overall industry to accomplish particular ends, and teams have different phases that will be described in this chapter. No matter what the organization of the workplace is, principles of cooperation and acceptance of diversity must apply. This chapter describes some of the important aspects of diversity in the workplace.

8.1 Work Groups and Teams

Many companies divide workers into work groups, usually based on a logical clustering of people within the overall hierarchy (structure) of the company. For example, a company might divide workers into regions, divisions, departments, shifts, units, and so on. These are collectively called **work groups**. Work groups organize staff members with similar skills to handle typically routine tasks. A supervisor usually heads up each work group, handling task assignments, monitoring worker performance, and resolving conflicts (Figure 8.1). Work groups in an industry can be large or small. An example of a work group in the pharmaceutical industry would be medication packers. Another example would be the collected group of shift personnel who operate a process unit.

Work group a group of people organized by logical structures within a company, having a designated leader, and performing routine tasks.

Figure 8.1 A supervisor usually heads up each work group.

CREDIT: yustus/Fotolia.

Work groups might be subdivided into teams. Not all work groups are teams, however.

A **team** usually consists of a small number of people who are selected because they have skills that complement those of other team members. Also, a team is committed to a common set of goals and tasks (Figure 8.2). For example, when planning a plant turnaround (also called a shutdown), management might put together a team consisting of a process technician, an instrument technician, a mechanical technician, a mechanical services contractor, vendor representatives, a work scheduler, and a safety, health, and environmental representative.

Team a small group of people, with complementary skills, committed to a common set of goals and tasks.

Figure 8.2 Team concept.
CREDIT: Rawpixel.com/Shutterstock.

Teams generally are formed to handle specific projects or tasks; they operate differently from a work group. Teams can be formed for a limited time to complete a project, or they can be ongoing, without a defined start or stop date or specific project. Team leadership is shared. The lead person and team composition might change as a project progresses through its individual phases. In effective teams, each team member treats others with respect, is committed to a common purpose, and is responsible for specific tasks and projects. In addition, each team member holds the other members mutually accountable for the success of the team.

When working as part of a team, it is important for process technicians to recognize and appreciate others for their contributions to the workplace and not to discount them because of their differences. It is vital for workers to understand diversity and to practice its principles in the workplace.

The following chart compares the defining characteristics of work groups and teams.

Work Groups	Teams
■ One leader	■ Shared leadership
■ Purpose and tasks decided by supervisor	■ Purpose and tasks selected by team leaders
■ Members answer to supervisor	■ Members answer to each other
■ Supervisor resolves conflicts between members	■ Members resolve conflicts

8.2 Types of Teams in the Process Industries

Process technicians often are required to work as part of one or more teams. Teams can vary based on factors such as organization, processes, skills, tasks, and deadlines. Some common types of teams used in the process industries include:

- Audit
- Commissioning
- Investigation and troubleshooting

- Maintenance
- Operations
- Process or quality improvement
- Safety
- Turnaround/shutdown
- Cross discipline (e.g., management, operations, and maintenance)

8.3 Characteristics of High-Performance Teams

In the early 1990s, two researchers showed that high-performance teams are effective teams. Jon Katzenbach and Douglas Smith researched teams in high-performing organizations, such as Motorola, Hewlett-Packard, the Girl Scouts, and the U.S. military (during Operation Desert Storm) to identify characteristics of the highest-performing teams.

Their research resulted in valuable findings and showed that high-performance teams were similar in three main areas:

1. Composition—how the team chose its members
2. Technique—how the team approached its tasks
3. Process—how the team operated

Composition

With regard to composition and how teams chose their members, Katzenbach and Smith learned the following about high-performance teams:

- No team started out with all the needed skills—they had to learn them along the way.
- The higher-performing teams were made up of fewer than 10 members.
- Team members possessed a mix of complementary skills in areas such as:
  - Technical or functional expertise
  - Problem solving and decision-making skills
  - Interpersonal skills (relating to others)

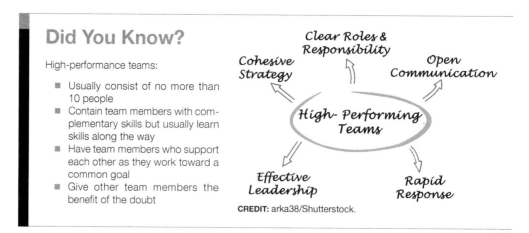

Did You Know?

High-performance teams:

- Usually consist of no more than 10 people
- Contain team members with complementary skills but usually learn skills along the way
- Have team members who support each other as they work toward a common goal
- Give other team members the benefit of the doubt

CREDIT: arka38/Shutterstock.

Technique

With regard to technique, high-performing teams approached their tasks similarly:

- Spent time getting consensus (agreement) about team purpose
- Specified measurable goals, objectives, achievements, and deadlines

- Specified goals that helped the team focus on the task at hand and allowed it to enjoy "small wins" that strengthened commitment and motivation
- Monitored team progress toward achieving its goals

High-performing teams are often **self-managed teams**. That is, they are a small group of employees whose members are self-organizing and who determine, plan, and manage their day-to-day activities and duties under reduced or no supervision. These groups are sometimes called self-directed teams.

Self-managed team also called *self-directed team*; a small group of employees whose members determine, organize, plan, and manage their day-to-day activities and duties under reduced or no supervision.

Process

Relating to process, or how high-performance teams operate, Katzenbach and Smith's research showed that team members:

- Did equivalent amounts of work, with no "free riders"
- Were open about individual members' skills and chose the best fit for the task
- Responded constructively to views expressed by others
- Recognized the interests and achievements of other team members

Synergy and Team Dynamics

When speaking of teams, it is important to understand the concepts of synergy and team dynamics.

Synergy is the result of the whole being greater than the sum of the individual parts. For example, steel is stronger than the metals that go into it (iron and carbon). In the same way, each member of a team provides talents, perspectives, and skills that make the team stronger than any of its individual members.

Team dynamics (sometimes called *interpersonal dynamics*) is a term to describe how team members get along with each other and function together. As new members are added or old members are removed, team dynamics change. These changes can be either positive or negative.

Both synergy and team dynamics are vital to the success of teams. Strong synergy and positive dynamics can result in high performance. Poor synergy and negative dynamics can result in unachieved outcomes and an overall lack of success. Process technicians working on teams should be aware of the need for synergy and do their part to improve or maintain positive team dynamics.

Synergy the total effect in which a whole is greater than the sum of its individual parts.

Team dynamics interpersonal relationships; ways in which workers get along with each other and function together.

Team Tasks Versus Process

When working on tasks as a team, it is important to understand the difference between task and process. A **task** is a set of actions that accomplish a job by a mutually agreed-upon deadline. A **process** is a method for doing something that generally involves tasks, steps, or operations that are sequential or interdependent.

When approaching tasks, teams must do two things: (1) decide how to accomplish the task; and (2) establish a process for accomplishing the task that will ensure the team can work effectively.

The following might be a typical process for accomplishing a task:

- Deciding what exactly is to be accomplished
- Identifying ideas about how to accomplish the task
- Selecting the best idea
- Achieving mutual buy-in
- Dividing up the responsibilities

Task a set of actions that accomplish a job.

Process method for doing something, generally involving tasks, steps, or operations that are ordered and/or interdependent.

- Setting deadlines
- Tracking progress toward achieving the goal
- Supporting fellow team members and their efforts
- Sharing in the rewards and recognizing individual and team accomplishments

Although this process sounds simple, a team can be unsuccessful if it does not properly address both the task and its related process.

Ineffective teams often have problems with their members and do not address these problems through their processes. Differences in work methods, goals, personalities, and personal beliefs can block progress, even when everyone shares a common goal. In contrast, highly effective teams rely on their processes to work through people problems while still accomplishing tasks and meeting objectives.

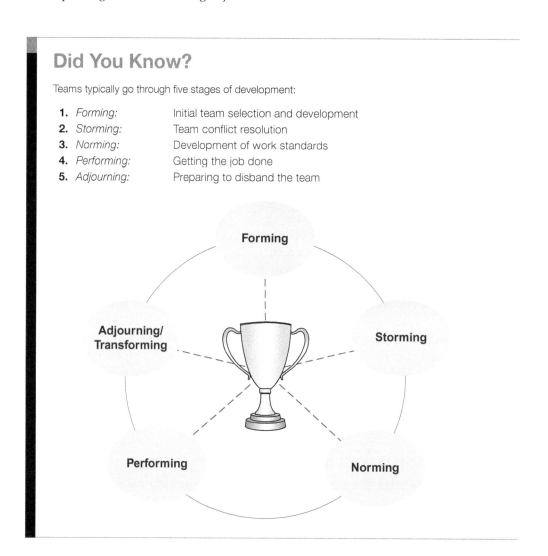

Did You Know?

Teams typically go through five stages of development:

1. *Forming:* Initial team selection and development
2. *Storming:* Team conflict resolution
3. *Norming:* Development of work standards
4. *Performing:* Getting the job done
5. *Adjourning:* Preparing to disband the team

8.4 Stages of Team Development

Teams go through different stages of development, encountering various issues along the way. Five general stages of team development have been identified, with each stage typically representing characteristic issues.

You can contribute to the team and improve its processes by recognizing what stage your team is in and understanding what issues might surface in that stage. For example, a period of conflict is not unusual after the initial formation of the team or at the midpoint of the project.

Stage 1: Forming

The fundamental issue during the forming stage is the development of team trust. Characteristics of this stage include:

- Tentative interactions or guarded discussion
- Careful behavior; trying not to offend anyone
- Mild tension; uncomfortable feelings or polite discourse
- Concern about ambiguity and what roles the members will play
- Concern about how a particular member will be accepted by the group

Stage 2: Storming

The fundamental issue during the storming stage is the resolution of team conflicts. Characteristics of this stage include:

- Individual actions that are resisted by, or incompatible with, other members
- More frequent disagreements
- Emergence of hostility
- Conflicts with roles and procedures

Stage 3: Norming

The fundamental issue during the norming stage is the development of teamwork standards. Characteristics of this stage include:

- Emerging sense of group unity and positive relationships among members
- Development of procedures, group norms, and roles
- Lower levels of anxiety
- Progress toward goals

Did You Know?

There are many factors that can cause a team not to succeed. Some of the most common contributors are:

- Inappropriate tasks
- Lack of clear purpose
- Having the wrong members
- Poor team dynamics
- Poor communication

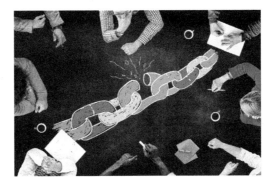

CREDIT: Rawpixel.com/Shutterstock.

Stage 4: Performing

The fundamental issue during the performing stage is the team getting the job done properly. Characteristics of this stage include:

- Good, strong decision-making
- Creative problem solving
- Mutual cooperation and buy-in
- Strong feelings of commitment to the team's success
- Goal achievement

Stage 5: Adjourning

Adjourning occurs when the project is successfully completed, deferred, or canceled. The fundamental issue during the adjourning stage is how the team is dealing with its impending breakup. Characteristics of this stage include:

- Celebration and recognition of group achievement
- Possible sense of loss about the dissolution of the team relationships
- Planning, and perhaps regret, for the change in individual work requirements
- Transforming or reforming into new or revised teams

8.5 Factors that Prevent Team Success

There are a variety of reasons why teams might not achieve their desired outcome. For example, members might not cooperate and build synergy, or the team might not have been properly constructed (i.e., team members were not the right fit for the job).

Team failure is often highly visible, especially if the team was tasked with a special project. It is crucial for process technicians to understand what blocks success so that they can watch for warning signs. If these signs are identified early and dealt with properly, problems can be resolved and the team can stay on track.

The following are some factors that can contribute to the lack of successful outcomes for teams:

Tackling tasks that are not appropriate for teams

Example:

- Situations that require fast decisions
- Tasks that require a higher degree of skill than team members have
- Painful or difficult decisions, such as deciding who gets laid off when times are bad

Lacking a clear purpose

Example:

- Absence of a clear direction and mission
- Conflicts about purpose and roles
- Lack of individual buy-in
- Taking too long to figure out a purpose

Having the wrong members

Example:

- Assignment of team members using random methods rather than by member characteristics
- Having members who appear right for the team, but turn out not to be
- Lack of a process for rectifying (correcting) poor choice of members

Having poor team dynamics

Example:

- Paying attention to the wrong issues
- Focusing too closely on tasks rather than team dynamics
- Lacking processes to handle poor team dynamics

Figure 8.3 Conflict pushes people away from each other and can lead to win-lose situations.

CREDIT: Thinglass/Shutterstock.

Resolving Conflict

Conflict can be a disruptive force or a driving force, depending on how a team handles it. If the team handles conflict poorly, team dynamics and performance can suffer. If the team can resolve a conflict and refocus it as a driving force, it can lead to improved team performance. The spirit of teamwork should be, "How can we resolve the issue, make the situation better, and bring the end goal back into focus?"

Some conflicts occur at fairly predictable times when working as a team. For example, every process has periods when stress is at a higher level because numerous tasks need to occur simultaneously or within a very small period of time. During such periods, simmering conflicts about work methods, clashing personality styles, and disagreement about goals can erupt in conflict and sidetrack success. When this happens, conflict must be dealt with swiftly and effectively. Resolution of conflict actually can create stronger teams with greater resilience in handling future conflicts.

Numerous attempts have been made to describe the elements involved in conflict and conflict resolution (Figures 8.3, 8.4, and 8.5). A useful tool for resolving conflict resolution is the **DESCC conflict resolution model**. DESCC stands for the following phases:

- Describe
- Express
- Specify
- Contract
- Consequences

DESCC conflict resolution model a model for resolving conflict, comprising the following steps: describe, express, specify, contract, and consequences.

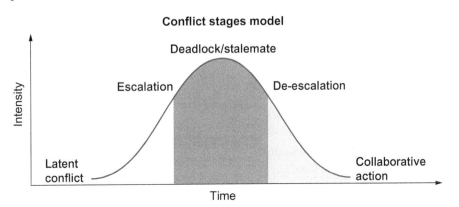

Figure 8.4 Conflict resolution stage model.

Figure 8.5 Five-step model for conflict resolution.

DESCRIBE PHASE   In this phase, team members should:
- Ask, "What is not happening that should be?"
- Ask, "What is happening that shouldn't be?"
- Describe the situation in specific behavioral terms (i.e., what someone can see or hear).
- Avoid vague descriptions such as "bad communication," "lack of commitment," and "bad attitude." For example, "Team meetings are consistently starting 10 minutes late," rather than, "People just don't care about what we're doing."
- Pay attention to the response. LISTEN.

EXPRESS PHASE   In this phase, team members should:
- Express how the situation affects them personally.
- Ask, "How does it make you feel?" For example, "When I leave what I'm doing and have to wait in an empty room for 10 minutes for the meeting to begin, my time is wasted and I feel as though other people think their time is more valuable than mine."
- Pay attention to the response. LISTEN.

SPECIFY PHASE   In this phase, team members should:
- Identify what must happen for them to be satisfied.
- Identify what the improved situation should look like. For example, "I need to know that others are going to take the meeting times seriously and show up at the agreed-upon time."
- Be very specific and use behavioral terms.
- Pay attention to the response. LISTEN.

CONTRACT PHASE   In this phase, team members should:
- Negotiate an agreement as to what will change.
- Be specific in the agreement terms. For example, "Commit to a time you know you can make; get to the meeting on time; and if you absolutely cannot make the meeting on time, text the members when you will be arriving. Also, alert people if you are away from work on the day of a meeting."
- Pay attention to the response. LISTEN.
- Achieve individual member buy-in and sign-off.

CONSEQUENCES PHASE   In this phase, team members should:
- Explain the anticipated outcome if the changes are made. For example, "We can all accomplish more and feel better about it if we keep to a focused time."
- Explain the anticipated outcome if the changes are not made. For example, "We will waste a lot of time if we start our meetings being annoyed with each other."
- Initiate consequences only when people have not lived up to the contract.
- Pay attention to the response. LISTEN.

Giving Feedback

As mentioned, teams not only have to work on tasks and projects, but they also must deal with work processes. One key issue related to work processes is feedback. When working with team members, vendors, clients, and other individuals, process technicians must learn how to give and receive feedback (Figure 8.6).

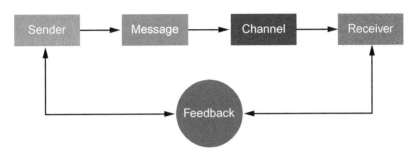

Figure 8.6 Communication model with feedback.

Often, people equate feedback with criticism. Feedback and criticism are actually quite different. **Feedback** is evaluative or corrective information provided to the originating source about a task or a process. Feedback is intended to be helpful and redirecting. **Criticism** is a serious examination and judgment of something. Although criticism can be positive (constructive) or negative (destructive), it is often negative, hurtful, or punishing.

To illustrate the concept of constructive feedback, think about driving a car. When driving, the car is seldom pointed directly at the destination. Instead, we start in a general direction and then make corrections along the way until our destination is reached. Teamwork operates under similar principles.

With teamwork, the team starts out pointed in the general direction of the goal. As the project progresses, each team member must make small corrections in individual processes to keep the team headed in the right direction. These corrections come from feedback.

Feedback can become emotionally loaded when conflicts arise. Conflict is a natural occurrence in any team project. The following tips can help process technicians and other team members give feedback that is objective, fact based, and less likely to create or increase conflict.

Feedback evaluative or corrective information provided to the originating source about a task or a process.

Criticism a serious examination and judgment of something; criticism can be positive (constructive) or negative (destructive).

When Giving Feedback	Example
Speak only for yourself; use only "I" messages.	"I don't like it when you send large print documents to the shared printer."
Critique the problem, not the person.	"When large documents are sent to the shared printer, the unit is unavailable for those who need to print items to complete their work."
Let the other person respond; do not interrupt.	Remember, you would not want to be interrupted when you are talking.
Speak from your perspective.	"I have to get printing done before my boss's 2 p.m. calls to clients, so it's hard when large jobs tie up the machine."
Offer alternatives.	"The print shop downstairs can handle large print jobs quickly. Could you take large jobs to them?"

When Receiving Feedback	Example
Listen carefully. Do not just wait for your turn to speak.	Concentrate on what the other person is saying.
Be patient.	Relax and stay focused.
Consider the other person's viewpoint.	Apply the old adage, "Walk a mile in another person's shoes." (What would it be like if you had this problem?)
Restate what the other person said, to make sure you understand it.	"So what you're saying is you don't like it when I send large print jobs to the shared printer."
Discuss ways to provide a positive resolution.	"The print shop downstairs costs too much. How about if I send it to Susan's printer instead?"
Never take feedback personally; focus on the issue to be resolved. Remember, feedback is good for the team and for you.	"I'm open to alternatives as long as they don't cost too much." Check your ego. Don't let emotions get in the way. If someone cares enough to provide feedback, you should care enough to listen.

8.6 Workforce Diversity

Diversity is the presence of a wide range of variations in qualities or attributes. In today's work environment, the term *diversity* is generically used to refer to the ways that we are different and unique from one another. *Diversity training* is a shorthand reference for anti-discrimination training.

Diversity goes beyond this, though. It also involves considering individual backgrounds, personality differences, learning styles, approaches to a task, and group dynamics.

When dealing with teams, work groups, or other people on the job, a firm understanding and acceptance of diversity is crucial.

> **Diversity** the presence of a wide range of variation in qualities or attributes; in the workplace, it also can refer to antidiscrimination training.

From the Melting Pot to the Salad Bowl

At one time, the ideal vision of the United States was a melting pot, where immigrants were expected to drop their language, customs, beliefs, traditional clothes, and other aspects of their culture in order to blend in with other Americans and become virtually "indistinguishable."

Organizations and businesses, however, have come to realize that people are most productive and creative when they feel valued and when their individual and group differences are accepted and taken into account. This new way of thinking is often referred to as the "salad bowl" concept.

The salad bowl concept uses a vision of a mixed bowl of colorful ingredients and flavors that complement each other well, yet retain their own colors, textures, and flavors. In other words, the individuality of each is respected and accepted for its unique characteristics.

In the workplace, the salad bowl approach means that individual workers must be accepting of the cultures, beliefs, races, religions, and lifestyle choices of others. In other words, they must be accepting of diversity. At the same time, however, clear communications require a common language. It is important that everyone speaks and understands a certain degree of "common language." Spelling out terms and definitions clearly can help. It can also be useful to paraphrase what the other person says: "So, what you're saying is . . . ."

Did You Know?

Diversity is an important part of the workplace. In order for employees to feel satisfied and productive, diversity must be accepted and respected.

Negative beliefs and behaviors (e.g., stereotyping, ethnocentrism, and prejudice) cause a person to feel devalued and limit a person's opportunities for success.

Terms Associated with Discussion of Diversity

Diversity is about allowing people to do their jobs without having to leave parts of themselves out of the workplace. It requires learning to be comfortable around people who are different from you and who might not necessarily want to be like you. Some individuals are not comfortable with diversity and might exhibit negative behaviors such as stereotyping, ethnocentrism, and prejudice.

Stereotyping describes beliefs about individuals or groups that are based on opinions, habits of thinking, or rumors. Stereotyping leads to generalizations about every member of a particular group.

> **Stereotyping** maintaining beliefs about individuals or groups based on opinions, habits of thinking, or rumors, which lead to generalizations about all members of a group.

- Many of our beliefs about other groups come from stereotypes.
- This way of thinking assumes that all members of a group have similar qualities.
- Some stereotypes are errors that are repeated until they seem true.
- Stereotyping greatly limits a person's opportunities.

Ethnocentrism describes a person's belief that other groups and individuals are inferior and should think and interpret things exactly the way that person does.

- Ethnocentrism generally decreases when individuals spend time outside their own culture.
- People who are ethnocentric rarely recognize it in themselves.
- A major cause of ethnocentric misunderstandings is lack of communication and lack of exposure to other groups.

Ethnocentrism belief in the superiority of one's own ethnic group; belief that others should believe and interpret things exactly the way you do.

Prejudice describes an attitude toward a group or its individual members based on stereotyped beliefs.

- Prejudice can result in selective perception (i.e., noticing only the bad qualities of a group, while not noticing the same qualities in one's own group).
- Another result of prejudice is use of "trigger" behaviors or words uttered by one group about another (e.g., a male manager calling professional women "girls"). Even if the communicator intends no harm, trigger words generate a heightened sensitivity and a feeling of hostility. Uninformed or insensitive people often mean no harm when using trigger words, and they might feel hurt or puzzled by negative reactions they receive.
- Prejudice causes a breakdown in communication.

Prejudice attitude toward a group or its individual members based on stereotyped beliefs.

Respecting Diversity

In any workplace, there will be people who are different from you. Be open-minded and sensitive to other views, opinions, and approaches. Look for common ground on projects and tasks. Seek out win-win solutions and approaches for you and your coworkers, no matter how different they are.

Your company's management will expect you not only to get along with coworkers who are different from you but also to be successful when working with them as part of a work group or a team.

If you are part of a work group whose people have difficulties dealing with diversity, the problem can compound itself when you are forced to work together almost every day. Although the situation probably has less pressure than a team environment, it still can result in conflicts that affect performance, job satisfaction, and stress.

If you are part of a team whose members have difficulties dealing with diversity, the problem can build to a boiling point because teams are often faced with intense pressure and tight deadlines.

In either situation, it is critical for everyone in the work environment to understand and appreciate diversity. Diversity goes beyond obvious issues such as gender, race, and religion. It involves all of the ways that we are unique as individuals.

Summary

A work group is a group of people organized by logical structure within a company, having a designated leader, and performing routine tasks. A team is a small group of people with complementary skills, committed to a common set of goals and tasks. Not all work groups are teams.

The process industries contain many different types of teams (e.g., audit, maintenance, safety, and operations). Each of these teams is created with a specific task or function in mind.

High-performance teams share certain characteristics. They usually consist of 10 or fewer people with complementary skills. They seldom start out with all of the skills they need. They support each other as they work toward a common goal.

Two key concepts that pertain to teams are synergy and team dynamics. Synergy occurs when team members combine their individual expertise, talents, and contributions for a common outcome. Synergy allows greater achievement than is individually possible. Team dynamics determine how team members get along with each other and function together. Both concepts are vital to team success.

As teams grow and develop, they go through several stages (forming, storming, norming, performing, and adjourning). An understanding of these stages can help teams resolve issues that often occur.

Many factors can cause a team to be unsuccessful. These factors include tackling tasks that are not appropriate for teams, lacking a clear purpose, having the wrong members, and having poor team dynamics.

Regardless of the team and its purpose, all employees should understand and respect team member diversity and differences. Negative and limiting behaviors, such as stereotyping, ethnocentrism, and prejudice, harm the work environment and reduce the industry's chances for success.

Checking Your Knowledge

1. Define the following key terms:
   a. diversity
   b. ethnocentrism
   c. feedback
   d. prejudice
   e. stereotyping
   f. team
   g. work group

2. Which of the following is a characteristic of a team?
   a. Members are answer able to each other.
   b. Team has one leader.
   c. Supervisor resolves conflict between members.
   d. Team handles only safety audits.

3. During studies of highly effective teams, researchers found:
   a. There was no disagreement.
   b. Members argued often.
   c. Members responded constructively to others' views.
   d. There were "free riders."

4. Which of the following types of teams are commonly used in the process industries? Select all that apply.
   a. Process or question improvement
   b. Operations
   c. Safety
   d. Envisioning and troubleshooting
   e. Maintenance

5. _____ is the result when the total effect of a whole is greater than the sum of its individual parts.

6. Which of the following is the best definition of team dynamics?
   a. how energetic the team members are
   b. how the team functions interpersonally
   c. how the team members cheer each other on
   d. how the team responds to criticism

7. Which of the following is NOT a stage of team development?
   a. forming
   b. storming
   c. reforming
   d. norming

8. List three reasons why teams might not be successful.

9. List the components of the DESCC conflict resolution model?
   a. D_____
   b. E_____
   c. S_____
   d. C_____
   e. C_____

10. Diversity is best described as:
    a. unique differences
    b. selective perception
    c. stereotyping
    d. ethnocentrism

11. The _____ approach states that the United States should be like a mixed container of colorful ingredients and flavors that complement each other well, while retaining their own colors, textures, and flavors.

12. Match the term to its description.

I. stereotyping	a.	a belief that other groups should think and interpret things exactly the way you do
II. ethnocentrism	b.	an attitude toward a group or its individual members
III. prejudice	c.	a vision that America has changed from "melting pot" to "salad bowl"
IV. diversity	d.	a perception that all members of a particular group have the same qualities

NOTE: Answers to Checking Your Knowledge questions are in Appendix I.

Student Activities

1. Think about a team you have joined. How and why was the team formed? What were its good qualities? What could have been improved? Did the team meet its goals? Why (or why not)?

2. List the types of teams in a process industry company. List the skill sets each team needs and discuss possible goals the team might have.

3. Think of a team that you consider high performance. List 10 characteristics that make that team high performance. Explain your choices.

4. Discuss a real world example of synergy (you can draw on your experience, sports teams, business, or other organizations).

5. Pair up with a classmate and present a demonstration of how to perform a particular task or skill (approximately 5 minutes long). Then, have your classmate provide feedback on your presentation. Next, switch roles. Finally, try your presentation again, using feedback from your classmate.

6. With your classmates, discuss all the ways you are similar. Then, describe ways that you are different. How does it make you feel when you have things in common with your classmates? How does it make you feel to be different?

Chapter 9
Safety, Health, Environment, and Security

 Objectives

Upon completion of this chapter, you will be able to:

9.1 Discuss the safety, health, and environmental hazards found in the process industries. (NAPTA SHE 1) p. 109

9.2 Explain and describe the responsibility of the following regulatory agencies:

EPA

OSHA

DOT

NRC (NAPTA SHE 9.2) p. 110

9.3 Describe the intent and application of the primary regulations affecting the process industries:

29 CFR (Code of Federal Regulations)

OSHA 1910.119—Process Safety Management (PSM)

OSHA 1910.132—Personal Protective Equipment (PPE)

OSHA 1910.1200—Hazard Communication (HAZCOM)

OSHA 1910.120—Hazardous Waste Operations and Emergency Response (HAZWOPER)

DOT CFR 49.173.1—Hazardous Materials—General Requirements for Shipments and Packaging

40 EPA CFR 239-282—Resource Conservation and Recovery Act (RCRA)

40 CFR 60-63—Clean Air Act (NAPTA SHE 9.2) p. 114

9.4 Describe the role of the process technician in protecting the safety and health of the company, employees, and community while achieving successful compliance with regulations. (NAPTA SHE 9.3) p. 117

9.5 Describe the personal attitudes and behaviors that can help to prevent workplace accidents and incidents. (NAPTA SHE 9.4) p. 117

9.6 Describe the components of the fire triangle and the fire tetrahedron. (NAPTA SHE 9.5) p. 118

9.7 Identify the consequences of noncompliance with regulations:

Legal

Moral and Ethical

Safety, Health, and Environmental

Economic (NAPTA SHE 9.6) p. 120

9.8 Explain the legal, managerial and engineering controls used in the industry to minimize hazards and maximize worker and system protection in the workplace. (NAPTA SHE 9.7-9.10) p. 121

9.9 Explain physical and cybersecurity requirements in the process industries. (NAPTA SHE 9.7) p. 126

Key Terms

Administrative controls—the implementation programs (e.g., policies and procedures) and activities to address a hazard, **p. 121.**

Air pollution—the contamination of the atmosphere, especially by industrial waste gases, fuel exhausts, smoke, or particulate matter (finely divided solids), **p. 110.**

Attitude—a state of mind or feeling with regard to some issue or event, **p. 117.**

Behavior—an observable action or reaction of a person under certain circumstances, **p. 117.**

Biological hazard—any danger that comes from a living, or once living, organism such as viruses, mosquitoes, or snakes that can cause a health problem, **p. 109.**

Chain reaction—a series of occurrences or responses in which each reaction is initiated by the energy produced in the preceding one, **p. 119.**

Chemical hazard—any danger or risk that comes from a solid, liquid, or gas element, compound, or mixture that could cause health problems or pollution, **p. 109.**

Cybersecurity—measures intended to protect information and information technology from unauthorized access or use, **p. 110.**

DOT—U.S. Department of Transportation; a U.S. government agency with a mission of developing and coordinating policies to provide an efficient and economical national transportation system, taking into account need, the environment, and national defense, **p. 113.**

Engineering controls—equipment and/or standards that use technology and engineering practices to isolate, diminish, or remove a hazard from the workplace, **p. 121.**

EPA—Environmental Protection Agency; a federal agency charged with authority to make and enforce the national environmental policy, **p. 113.**

Ergonomic hazard—any danger or risk that can create physical and psychological stresses because of forceful or repetitive work, improper work techniques, or poorly designed tools and work spaces, **p. 109.**

Fire triangle/fire tetrahedron—the elements of fuel, oxygen, and heat that are required for a fire to start and sustain itself; a fire tetrahedron adds a fourth element: a chemical chain reaction, **p. 118.**

Fuel—any material that burns; can be a solid, liquid, or gas, **p. 118.**

Hazardous agent—the substance, method, or action by which damage or destruction can happen to personnel, equipment, or the environment, **p. 109.**

Heat—added energy that causes an increase in the temperature of a material (sensible heat) or a phase change (latent heat); the energy required by the fuel source to generate enough vapors for the fuel to ignite, **p. 118.**

Information technology—the equipment, tools, processes, and methodologies (coding/programming, data storage and retrieval, systems analysis, systems design, and so on) that are used to collect, process, and present information, **p. 126.**

ISO 9000—an international standard that provides a framework for quality management by addressing the processes of producing and delivering products and services, **p. 125.**

ISO 14000—an international standard that addresses how to incorporate environmental aspects into operations and product standards, **p. 125.**

NRC—Nuclear Regulatory Commission; a U.S. government agency that protects public health and safety through regulation of nuclear power and the civilian use of nuclear materials, **p. 114.**

OSHA—Occupational Safety and Health Administration; a U.S. government agency created to establish and enforce workplace safety and health standards, conduct workplace inspections, propose penalties for noncompliance, and investigate serious workplace incidents, **p. 113.**

Personal protective equipment (PPE)—specialized gear that provides a barrier between hazards and the body and its extremities, **p. 121.**

Physical hazard—any danger or risk that comes from environmental factors such as excessive levels of noise, temperature, pressure, vibration, radiation, electricity, or mechanical hazards (Note: OSHA has its own definition of physical hazard that relates specifically to chemicals), **p. 109.**

Physical security—measures intended to protect specific assets such as production facilities, pipelines, control centers, tank farms, and other vital areas, **p. 110.**

Safety Data Sheet (SDS)—(formerly Material Safety Data Sheet or MSDS)—a document that provides key safety, health, and environmental information about a material, **p. 115.**

Soil pollution—the accidental or intentional discharge of any harmful substance into the soil, **p. 110.**

Voluntary Protection Program (VPP)—an OSHA program designed to recognize and promote effective safety and health management, **p. 125.**

Water pollution—the introduction, into a body of water or the water table, of any EPA-listed potential pollutant that affects the chemical, physical, or biological integrity of that water, **p. 110.**

Introduction

In the process industries, workers routinely work with hazardous agents or environmental factors that can cause injury, illness, or death. Some of these hazardous agents also can have impacts on the environment in the short and long term.

Government regulations are in place to protect workers' health and safety, the community, and the environment. Industries comply with regulations by using engineering controls, administrative controls, and PPE.

Companies use physical and cybersecurity measures to protect assets and workers from internal and external threats. Physical security focuses on protecting facilities and components such as pipelines, control centers, and other vital areas from damage or theft. Cybersecurity protects information assets and computing systems.

Process technicians must be trained to recognize hazardous agents and security threats. They must understand the impact on themselves, the plant or facility where they work, and the surrounding community.

Safety, Health, Environment, and Security **109**

This chapter provides an overview of various hazardous agents that process technicians might encounter in the workplace; the government agencies and regulations that address safety, health, environment, and security (SHE); controls for hazards; personal protective equipment; the cost of noncompliance; some voluntary programs that promote workplace safety, and measures to protect against physical and cybersecurity threats.

9.1 Safety, Health, and Environmental Hazards Found in the Process Industries

Hazardous agents are the substances, methods, or actions by which damage or destruction can happen to personnel, equipment, or the environment (Figure 9.1).

Different government agencies, industry groups, and individuals have created various ways of classifying and describing hazardous agents. Many companies and their safety professionals use the following classification system to categorize hazardous agents, dividing these agents into five major types: chemical, physical, ergonomic, biological or physical security, and cybersecurity.

Chemical hazard—any hazard that comes from a solid, liquid, or gas element, compound, or mixture that could cause health problems or pollution.

Physical hazard—any danger or risk that comes from environmental factors such as excessive levels of noise, temperature, pressure, vibration, radiation, electricity, or mechanical hazards. OSHA has its own description of physical hazard, namely, "a chemical for which there is scientifically valid evidence that it is a combustible liquid, a compressed gas, explosive, flammable, an organic peroxide, an oxidizer, pyrophoric, unstable (reactive), or water-reactive."

Ergonomic hazard—any danger or risk that can create physical and psychological stresses because of forceful or repetitive work, improper work techniques, or poorly designed tools and work spaces.

Biological hazard—any danger or risk that comes from a living, or once living, organism such as viruses, mosquitoes, or snakes, that can cause a health problem.

> **Hazardous agent** the substance, method, or action by which damage or destruction can happen to personnel, equipment, or the environment.
>
> **Chemical hazard** any danger or risk that comes from a solid, liquid, or gas element, compound, or mixture that could cause health problems or pollution.
>
> **Physical hazard** any danger or risk that comes from environmental factors such as excessive levels of noise, temperature, pressure, vibration, radiation, electricity, or mechanical hazards.
>
> **Ergonomic hazard** any danger or risk that can create physical and psychological stresses because of forceful or repetitive work, improper work techniques, or poorly designed tools and work spaces.
>
> **Biological hazard** any danger that comes from a living, or once living, organism such as viruses, mosquitoes, or snakes, that can cause a health problem.

Figure 9.1 Hazardous substances. **A.** Chemical hazards; **B.** High noise as a physical hazard; **C.** Ergonomic hazard; **D.** Natural or biological hazard; **E.** Security hazard.

CREDIT: **A.** Alan Poulson Photography/Shutterstock; **B.** yellomello/Fotolia; **C.** Lemurik/Shutterstock; **D.** Somboon Bunproy/Shutterstock; **E.** designer491/Fotolia.

A.

B.

C.

D.

E.

Figure 9.2 Air, water, and soil pollution. **A.** Air pollution (smog) in a city. **B.** Oil slick on the water near a cargo ship. **C.** Worker gathering samples of hazardous waste on a beach.
CREDIT: **A.** BlackCat Imaging/Shutterstock; **B.** Nightman1965 /Shutterstock; **C.** Microgen/Shutterstock.

A.

B.

C.

Security hazard—a danger, risk, or threat from a person or group seeking to do intentional harm to people, computer resources, or other vital assets.

- **Physical security**—using security measures intended to prevent physical threats from a person or group seeking to intentionally harm other people or vital assets
- **Cybersecurity**—measures intended to protect electronic assets from illegal access and sabotage.

Environmental hazards (Figure 9.2) fall into one of three broad categories: air pollution, water pollution, and soil pollution.

Air Pollution The contamination of the atmosphere by industrial waste gases, fuel exhausts, smoke, or particulate matter (finely divided solids).

Water Pollution The introduction, into a body of water or the water table, of any EPA-listed potential pollutant that affects the chemical, physical, or biological integrity of water.

Soil Pollution The accidental or intentional discharge of any harmful substance into the soil.

9.2 Regulatory Agencies and Their Responsibilities

Unfortunately, laws that protect workers and the public are often driven by catastrophic events. The following are a few disasters that prompted safety legislation and development of regulatory agencies:

1. Triangle Shirtwaist Factory fire of 1911, where 145 immigrant women burned to death because of numerous failures in the building's construction and safety code. Laws passed after this tragedy mandated fire sprinklers, fire drills, and unlocked and outward-swinging doors. Other provisions required removal of fire hazards, the use of fireproof waste receptacles, protection of gas jets, and prohibition of smoking in the factory setting.

2. Mining disasters in West Virginia and other states have led to legislation and increased regulation. In 1968, the Farmington Coal Mine explosion killed 78 coal miners and led to federal legislation that standardized coal mine health and safety practices. A 2006 disaster led to legislation requiring improved communications, supplies of oxygen in underground vaults (Figure 9.3), and quicker emergency responses. A 2006 explosion at the West Virginia Sago Mine led to the Mine Improvement and New Emergency Response (MINER) Act. MINER increased fines for mine safety violations and authorizes the government to close mines that are "pattern violators." Despite this, infractions continue. Mining deaths globally are worst where legislation is not well enforced.

Physical security measures intended to protect specific assets such as production facilities, pipelines, control centers, tank farms, and other vital areas.

Cybersecurity measures intended to protect information and information technology from unauthorized access or use.

Air pollution the contamination of the atmosphere, especially by industrial waste gases, fuel exhausts, smoke, or particulate matter (finely divided solids).

Water pollution The introduction, into a body of water or the water table, of any EPA-listed potential pollutant that affects the chemical, physical, or biological integrity of that water.

Soil pollution the accidental or intentional discharge of any harmful substance into the soil.

Figure 9.3 Underground mining vault.
CREDIT: Getty Images/borchee.

3. In December 1984, more than 40 tons of methyl isocyanate gas leaked from an internationally owned pesticide plant in Bhopal, India. It immediately killed at least 3,800 people and caused illness and premature death for thousands more. At the time there were no enforceable international standards for environmental safety, no preventative strategies to avoid such accidents, and no plans for disaster preparedness. The disaster led the Indian government to pass the Environment Protection Act and create the Ministry of Environment and Forests (MoEF). These steps strengthened India's commitment to the environment.

4. In April 1986, the Unit 4 reactor at the Chernobyl Nuclear Power Plant in northern Ukraine failed, leading to the worst global nuclear disaster in history. Core explosions and open-air fires completely destroyed Unit 4 and caused 31 (direct) deaths. The blast released 400 times the radioactivity released by the bomb dropped on Hiroshima in World War II. Radiation caused Pripyat, the factory city built to house Chernobyl workers, to be evacuated. It remains uninhabitable (Figure 9.4). Plans to encase the radioactive plant in cement proved unsuccessful. In 2016, the entire plant was covered by a massive steel arch placed over the reactor. After Chernobyl, countries adopted regulations on nuclear safety. International labor groups and agencies adopted international standards and safety guidelines. Law enforcement and inspection systems now maintain greater control over design and process operations of nuclear plants. Power plants emphasize involvement of workers in training, education, information, and participation to promote a "safety culture."

Figure 9.4 Pripiyat, the ghost city of Chernobyl.
CREDIT: © Alan Bracken | Shutterstock.com

> ### Did You Know?
>
> Many scientists believe that the Chernobyl nuclear disaster occurred because the plant was improperly designed and plant operators ignored important safety measures.
>
> As a result, large amounts of radioactive materials were emitted into the environment. This led to serious health problems or death for many of those who were exposed. The area will be quarantined indefinitely.

CREDIT: Al/Fotolia.

Figure 9.5 A. Platform oil rig in flames. **B.** Oil spill on a beach. Cleanup after an oil spill can take months or years.

CREDIT: A. © curraheeshutter/Shutterstock. **B.** © Joseph Sohm/Shutterstock

A.   B.

5. In April 2010, one of the worst U.S. environmental disasters occurred, caused by a leaking undersea pipe after the explosion and sinking of the Deepwater Horizon oil rig about 67 kilometers (42 mi) off the Louisiana coast. There were 11 fatalities. Five million barrels of oil were released into the Gulf of Mexico, creating oil slicks on the surface of the water, oil blankets on the ocean floor, and oil-coated beaches (Figure 9.5). Coral reefs, turtles, seabirds, and other sea life died from effects of the spill. Following years of cleanup, the U.S. Department of the Interior proposed that oil companies be required to use stronger "blowout preventers." The USDI also mandated use of stronger well casings to reinforce wells.

In order to protect workers, the public, and the environment from environmental and safety hazards, the U.S. government has created several different agencies. Among these agencies are the Environmental Protection Agency (EPA), the Occupational Safety and Health Administration (OSHA), the Department of Transportation (DOT), and the Nuclear Regulatory Commission (NRC).

Environmental Protection Agency (EPA)

On January 1, 1970, President Richard Nixon signed the National Environmental Policy Act (NEPA). NEPA was enacted to set national policy regarding the protection of the environment, to promote efforts to prevent or eliminate pollution of the environment, to advocate knowledge of ecological systems and natural resources, and to establish a Council on Environmental Quality (CEQ) to oversee the aforementioned policy.

It was soon clear that the CEQ, as structured, did not have the resources or the power to fulfill its mission. In Reorganization Order No. 3, issued on July 9, 1970, the President stated: "It also has become increasingly clear that only by reorganizing our federal efforts can we develop that knowledge, and effectively ensure the protection, development, and enhancement of the total environment itself."

Through Reorganization Order No. 3, the **Environmental Protection Agency (EPA)** and the National Oceanic and Atmospheric Administration (NOAA) were formed. Control of many environment-related functions was transferred from other governmental offices and agencies to the EPA and NOAA.

EPA Environmental Protection Agency; a federal agency charged with authority to make and enforce the national environmental policy.

The EPA's mission is "to protect human health and the environment." The EPA works for a cleaner, healthier environment for Americans. NOAA seeks to "observe, predict, and protect our environment."

Occupational Safety and Health Administration (OSHA)

On December 29, 1970, President Richard Nixon signed the Occupational Safety and Health Act of 1970. The purpose of the OSH Act was, and continues to be, "to assure so far as possible every working man and woman in the nation safe and healthful working conditions and to preserve our human resources."

The OSH Act established several agencies to oversee the protection of American workers. These included:

- **The Occupational Safety and Health Administration (OSHA)**—created to establish and enforce workplace safety and health standards, conduct workplace inspections, propose penalties for noncompliance, and investigate serious workplace incidents.
- **The Occupational Safety and Health Review Commission (OSHRC)**—formed to conduct hearings when employers who are cited for violation of OSHA standards contest their penalties.
- **The National Institute for Occupational Safety and Health (NIOSH)**—established to conduct research on workplace safety and health problems, specifically injuries and illnesses that may be attributed to exposure to toxic substances.

OSHA Occupational Safety and Health Administration; a U.S. government agency created to establish and enforce workplace safety and health standards, conduct workplace inspections, propose penalties for noncompliance, and investigate serious workplace incidents.

Department of Transportation (DOT)

On October 15, 1966, President Lyndon Johnson signed Public Law 89-670, which established the **Department of Transportation (DOT)**. DOT's mission is "to develop and coordinate policies that will provide an efficient and economical national transportation system, with due regard for need, the environment and the national defense."

DOT U.S. Department of Transportation; a U.S. government agency with a mission of developing and coordinating policies to provide an efficient and economical national transportation system, taking into account need, the environment, and national defense.

On September 23, 1977, Secretary of Transportation Brock Adams established the Research and Special Programs Administration (RSPA) within the Department of Transportation consolidating various diverse functions that dealt with intermodal activities.

Eventually, the RSPA came to oversee the Office of Pipeline Safety (OPS) and the Office of Hazardous Materials Safety (OHMS), two entities with considerable jurisdiction over the petrochemical industry.

Did You Know?

The Department of Transportation (DOT), the governmental institution responsible for regulating our highways, is also responsible for regulating the transportation of natural gas, petroleum, and other hazardous materials through pipelines.

CREDIT: AXpop/Shutterstock.

Nuclear Regulatory Commission (NRC)

Congress established the Atomic Energy Commission (AEC) in the Atomic Energy Act of 1946. The AEC's mission was regulation of the nuclear industry. Eight years later, Congress replaced that act with the Atomic Energy Act of 1954, which enabled the development of commercial nuclear power. The AEC's mission became twofold: encouraging the use of nuclear power and regulating its safety.

In the 1960s, critics charged that the AEC's regulations were not strict enough in several important areas, including radiation protection standards, reactor safety, plant location, and environmental protection. The AEC was disbanded in 1974 under the Energy Reorganization Act. This act created the **Nuclear Regulatory Commission (NRC)**, which started operations in 1975. Today, the NRC's regulatory activities focus on reactor safety oversight, materials safety oversight, materials licensing, and management of both high- and low-level radioactive waste. The NRC also licenses those who operate and supervise commercial nuclear reactors; i.e. power plants, and research and test reactors.

The NRC also regulates instruments in the process industries that use radioactive materials, such as testing devices (e.g., gas chromatographs) and inspection equipment (e.g., X-ray machines).

NRC Nuclear Regulatory Commission; a U.S. government agency that protects public health and safety through regulation of nuclear power and the civilian use of nuclear materials.

9.3 Regulations Affecting the Process Industries

The U.S. government has enacted numerous regulations to minimize workplace hazards. These regulations are administered through various federal agencies such as OSHA, the EPA, and the DOT. Some regulations are generic in scope and affect a variety of industries. Other regulations were created specifically to regulate a certain industry and even certain hazardous substances.

OSHA administers many of the government regulations that significantly affect the day-to-day operations of the process industries. Four of the most important regulations are described in this section: Process Safety Management (PSM), Personal Protective Equipment (PPE), Hazard Communication (HAZCOM), and Hazardous Waste Operations and Emergency Response (HAZWOPER).

Two other major regulations administered by other agencies, the EPA and DOT, are also described in this section. These regulations address hazardous materials and their shipment.

Additional regulations that are not covered in this textbook are discussed in a separate textbook, *Safety, Health, and Environment*.

29 CFR Chapter XIV Equal Employment Opportunity Act

The Equal Employment Opportunity Act was established to prevent discrimination against employees based upon such conditions as sex, religion, age, disability, or national origin. It includes regulations concerning affirmative action and privacy. This act is enforced by the Equal Employment Opportunity Commission (EEOC).

OSHA 1910.119—Process Safety Management (PSM)

The OSHA Process Safety Management of Highly Hazardous Materials (PSM)—29 CFR 1910.119 standard seeks to prevent or minimize the consequences of catastrophic releases of toxic, reactive, flammable, or explosive chemicals.

This standard establishes 14 elements aimed at improving worker safety:

- Employee involvement
- Process safety information
- Process hazard analysis

- Operating procedures
- Training
- Contractors
- Pre-startup safety review
- Mechanical integrity
- Hot work permit system
- Management of change (MOC)
- Incident investigation
- Emergency planning and response
- Compliance audits
- Trade secrets

OSHA 1910.132—Personal Protective Equipment (PPE)

The OSHA Personal Protective Equipment (PPE)—29 CFR 1910.132 standard aims to prevent worker exposure to potentially hazardous substances through the use of equipment that establishes a barrier between the hazardous substance and the individual's eyes, face, head, respiratory system, and extremities.

This standard requires employers to do the following to assess workplace hazards:

- Determine whether personal protective equipment (PPE) is necessary
- Provide required PPE to their employees
- Train employees in the proper use and care of the PPE
- Ensure that employees use the PPE appropriately.

OSHA 1910.1200—Hazard Communication (HAZCOM)

The OSHA Hazard Communication (HAZCOM)—29 CFR 1910.1200 standard seeks to ensure that the hazards of all produced or imported chemicals are evaluated and that information relating to the hazards is provided to employers and employees.

This standard requires the transmittal of information through comprehensive hazard communication programs. Information must include container labeling and other forms of warning, **safety data sheets (SDS)**, and employee training.

SDS provide key safety, health, and environmental information about a material (Figure 9.6). This information includes physical properties, proper storage and handling,

Safety Data Sheet (SDS) (formerly Material Safety Data Sheet or MSDS)—a document that provides key safety, health, and environmental information about a material.

Figure 9.6 Safety Data Sheet (SDS)
CREDIT: Travis Klein/Shutterstock.

toxicological data, established exposure limits, firefighting information, and other useful data. This information is provided in a standardized format. SDS information must be made available for any material manufactured, used, stored, or repackaged by an organization.

OSHA 1910.120—Hazardous Waste Operations and Emergency Response (HAZWOPER)

The OSHA Hazardous Waste Operations and Emergency Response (HAZWOPER)—29 CFR 1910.120 standard outlines the establishment of safety and health programs. It also describes levels of training required for employees involved in hazardous waste operations and emergency response. Employers must identify, evaluate, and control safety and health hazards in operations involving hazardous waste or emergency response.

OSHA 1910.1000 Air Contaminants

The OSHA Air Contaminants standard (29 CFR 1910.1000) establishes the permissible exposure limits (PELs) for a variety of toxic and hazardous substances. A PEL describes the amount of an airborne toxic or hazardous substance to which an employee can be exposed over a specified time period. OSHA 1910.1000 through 1910.1500 list specific toxic or hazardous substances and the assigned PEL for each.

DOT CFR 49.173.1—Hazardous Materials—General Requirements for Shipments and Packaging

The DOT Hazardous Materials—General Requirements for Shipments and Packaging—49 CFR 173.1 standard establishes requirements for preparing hazardous materials for shipment by air, highway, rail, or water, or any combination of these.

It also covers the inspection, testing, and retesting responsibilities for personnel who retest, recondition, maintain, repair, and rebuild containers used or intended for transporting hazardous materials.

EPA 40 CFR Parts 239–282—Resource Conservation and Recovery Act (RCRA)

The EPA Resource Conservation and Recovery Act (RCRA)—40 CFR 260 through 270 standards promote "cradle-to-grave" management of hazardous wastes.

These standards classify and define requirements for hazardous waste generators, transporters and treatment, storage, and disposal facilities. Additionally, it requires industries to identify, quantify, and characterize their hazardous wastes prior to disposal. It holds the generator of the hazardous waste responsible for management from the point of inception to the final disposal of materials.

EPA Clean Air and Clean Water Acts

The 1990 Clean Air Act (40 CFR 60-63) "sets limits on how much of a pollutant can be in the air anywhere in the United States." The act ensures that all Americans are covered using the same basic health and environmental protections. Each state must carry out its own implementation plan to meet the standards of the act. For example, it would be up to a state air pollution agency to grant permits to power plants or chemical facilities, fine companies for violating the air pollution limit, and so on.

In 1972, the Federal Water Pollution Control Act Amendments were enacted, reflecting growing public concern for controlling water pollution. When amended in 1977, this law became commonly known as the Clean Water Act. This act regulates the discharges of pollutants into the waters in the United States and gives the EPA the authority to implement pollution control programs (e.g., setting wastewater standards for industry). The act also sets water quality standards for all contaminants in surface waters, making it illegal to discharge any pollutant from a source into navigable waters unless a permit is obtained.

9.4 The Role of the Process Technician

Process technicians play a vital role in a company's efforts to comply with government regulations and other safety, health, and environmental policies and procedures (Table 9.1).

General safety tips that process technicians should follow regardless of where they work are shown in Table 9.2.

Table 9.1 Ways Process Technicians Must Comply with Regulations

Process Technicians Must Comply with Regulations by:	
▪ Familiarizing themselves with applicable government regulations	▪ Performing all job tasks in a timely and accurate way while following safe work practices
▪ Following all plant policies and procedures, because many of these are written to ensure compliance with regulations	▪ Maintaining a safe work environment by performing good housekeeping functions, as required by the job
▪ Attending all mandatory training to stay current with applicable regulations	▪ Having a safe attitude and exhibiting safe behavior
▪ Learning to recognize hazards and reporting and handling them appropriately	

Table 9.2 General Safety Tips

Process Technicians Should Always Follow These Guidelines:	
▪ Recognize all alarms and know the corresponding response procedures	▪ Review all safety procedures
▪ Smoke only in designated areas	▪ Understand and properly use the equipment with which you work
▪ Stay focused and alert (e.g., get adequate rest, eat properly, and refrain from abusing drugs and alcohol)	▪ Watch for hazardous conditions and report or correct them
▪ Report injuries and incidents immediately to appropriate personnel	▪ Be prepared and keep a clear head in emergency situations
▪ Obey traffic regulations in the plant and never park in fire lanes	▪ Stay in your assigned area; if you must go to another area, make sure to tell appropriate personnel
▪ Use the proper tool for the job	▪ Know how to use safety equipment and protective gear

9.5 Attitudes and Behaviors That Help to Prevent Accidents

Safety studies show that, historically, human error is a significant factor in almost every accident at a plant. Personal attitudes and behaviors toward safety can play an important role in preventing accidents or incidents.

An **attitude** is defined as a state of mind or feeling with regard to some issue or event. Process technicians who maintain a safety mindset and always think about safety tend to experience fewer accidents than process technicians who are not safety-oriented.

Attitude a state of mind or feeling with regard to some issue or event.

A **behavior** can be defined as an observable action or reaction of a person under certain circumstances. Process technicians must respond immediately and appropriately to potential hazards, do a job right the first time, and perform housekeeping duties in a timely manner.

Behavior an observable action or reaction of a person under certain circumstances.

Studies of chemical plant accidents found that factors such as insufficient knowledge, procedural errors, and operator errors contributed to the severity of the incident.

Other factors that have contributed to accidents in the petrochemical and refining industry were equipment and design failures, both operator and maintenance errors, inadequate or improper procedures, and inadequate or improper inspections.

The past decades have seen industrial accidents causing loss of human life, damage to the environment, and in some cases, damage to the cultural and societal structure of a region. Some examples are:

- Bhopal 1984—a disastrous methylisocyanate leak exposed 500,000 people to the gas and killed at least 3,800 people.
- Exxon Valdez 1989—an oil tanker ran aground in Prince William Sound, spilling 11 million to 38 million gallons of crude oil onto the reef and surrounding shoreline. This spill was one of the most devastating environmental disasters ever recorded.
- BP 2005—during startup of an isomerization unit, a column was overflowed into the blowdown system, allowing vapors to reach the atmosphere. The vapors ignited, creating an explosion that killed 17 people and injured 170 others.

These incidents and others like them created the need for better education and training of employees. The Chemical Safety Board (CSB) is an independent federal agency resulting from the Clean Air Act. It was created in 1998 to investigate and determine the root cause of major industrial accidents and to prevent similar incidents in the future. This agency works closely with the EPA and OSHA.

The National Transportation Safety Board (NTSB) is another independent federal agency. Its purpose is investigation of incidents relating to civil aviation and other modes of transportation, including railroad, highway, marine, and pipeline. The CSB and the NTSB work diligently to promote safety in the industrial workplace.

One result of the ongoing effort to promote safety of industry workers is a process called Behavior-Based Safety (BBS). This process takes a scientific approach to safety and hazard control by observing employees during routine procedures. These observations reveal what an employee does and how he or she does it. BBS then provides a scientific intervention strategy for improvement. The BBS process is used for all employees from the management level down through the plant operations group and other hourly workers.

Process technicians must understand and follow not only governmental regulations regarding safety, health, the environment, and security, but also plant policies and procedures, general safety principles, and common sense. Unsafe conditions or unsafe behaviors of other individuals must be reported to a supervisor.

Many employers use techniques during a job interview to try and determine whether a candidate will exhibit a safe attitude and behave safely on the job.

9.6 Components of the Fire Triangle and the Fire Tetrahedron

Hydrocarbons, chemicals, and many other materials used in the process industries are extremely flammable and/or combustible. Because of this, one of the greatest potential hazards to process technicians is fire and/or explosions.

Fire is a chemical reaction that starts when a substance (fuel) in the presence of air (oxygen) is heated to an ignition point (heat), resulting in combustion. Fire must have all these elements (fuel, oxygen, and heat) present to start. Removing one of these elements will extinguish a fire.

These three elements are referred to as a **fire triangle** (Figure 9.7):

1. **Fuel**—any combustible material; can be a solid, liquid or gas.
2. **Oxygen**—air is composed of 21% oxygen; generally, fire needs only 16% oxygen to ignite.
3. **Heat**—added energy that causes an increase in the temperature of a material (sensible heat) or a phase change (latent heat); the energy required by the fuel to generate enough vapors for the fuel to ignite.

Fire triangle/fire tetrahedron the elements of fuel, oxygen, and heat that are required for a fire to start and sustain itself; a fire tetrahedron adds a fourth element: a chemical chain reaction.

Fuel any material that burns; can be a solid, liquid, or gas.

Heat added energy that causes an increase in the temperature of a material (sensible heat) or a phase change (latent heat); the energy required by the fuel source to generate enough vapors for the fuel to ignite.

Figure 9.7 Fire triangle.
CREDIT: L. Cosmo/Fotolia.

Fire Tetrahedron

The fire triangle represents the elements necessary to create a fire. Once a fire has started, the fire tetrahedron represents the elements necessary to sustain combustion.

The **fire tetrahedron** consists of the components of the fire triangle and another component—the chain reaction (Figure 9.8). A **chain reaction** is a series of reactions in which each reaction is initiated by the energy produced in the preceding reaction (i.e., toppling dominoes: when the first domino is knocked over, it causes the second domino in the series to topple, which topples the third domino, and so on). This type of reaction occurs when fuel, oxygen, and heat come together in proper amounts under certain conditions. Chain reactions are what cause fires to build on themselves and spread. In order to stop a fire, one of the four components of the fire tetrahedron must be removed.

Extinguishing agents (such as dry chemical) stop a fire by preventing the chain reaction from occurring, not by removing fuel, heat, or oxygen.

Chain reaction a series of occurrences or responses in which each reaction is initiated by the energy produced in the preceding one.

Figure 9.8 Fire tetrahedron.
CREDIT: BALRedaan/Shutterstock

Classes of Fire

Fires are classified according to five groups (Figure 9.9):

- Class A—combustible materials such as wood, paper, and plastic
- Class B—flammable gases or liquids
- Class C—fire involving live electrical equipment
- Class D—combustible metals (e.g., aluminum, sodium, potassium, and magnesium)
- Class K—cooking oil, fat, grease, or other kitchen fires.

Figure 9.9 The five classes of fire.

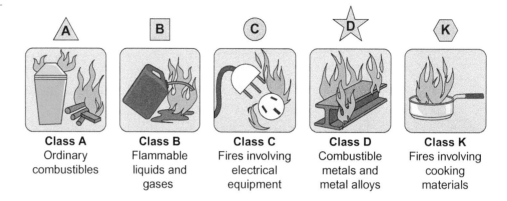

Process technicians must understand the elements required to start a fire, how to prevent fires and control them, and the combustible and flammable properties of the materials with which they are working.

9.7 Consequences of Noncompliance with Regulations

If a process technician fails to comply with regulations, it can result in legal, moral and ethical, safety, health, or environmental consequences. These consequences can be imposed as a result of a minor accident, a major accident, or an onsite inspection by a government agency representative.

Legal

Legal consequences fall into one of two major types:

- Fines and/or citations levied by federal, state, or local regulatory agencies (and possibly even criminal charges)
- Lawsuits filed by affected parties, such as injured workers or local residents

Moral and Ethical

Moral and ethical consequences can manifest as the following:

- Burden of contributing to injuries or deaths
- Responsibility for causing damage to equipment, lost production, and associated costs
- Guilt for not complying with regulations, policies, and procedures

Safety, Health, and Environmental

Numerous safety, health, and environmental consequences can result from noncompliance. These include the following:

- Exposed or injured workers
- Exposed or injured nonworkers
- Air pollution
- Water pollution
- Soil pollution

Economic

Noncompliance with government regulations as denoted by the appropriate agencies can result in the following:

- Fines, including some that could apply to managers and those directly under their supervision
- Loss of plant productivity
- Increased funds budgeted to correct violations
- Increased funds budgeted to add specific pollution abatement equipment
- Increased product cost to customers and consumers

9.8 Engineering Controls, Administrative Controls, and PPE

The process industries use three methods to minimize or eliminate worker exposure to hazards.

Engineering controls—controls that use technological and engineering improvements to isolate, diminish, or remove a hazard from the workplace. Examples of engineering controls are:

- Using a non hazardous material in a process that will work just as well as a hazardous material
- Placing a sound-reducing housing around a pump to muffle the noise it makes
- Adding guards to rotating equipment

Engineering controls equipment and/or standards that use technology and engineering practices to isolate, diminish, or remove a hazard from the workplace.

Administrative controls—if an engineering control cannot be used to address a hazard, an administrative control is used. Administrative controls involve implementing programs and activities to address a hazard.

Programs consist of written documentation such as policies and procedures. Activities involve putting a program into action.

Administrative control is also called a work practice control or managerial control. Examples of administrative controls are:

- Writing a procedure to describe the safe handling of a hazardous material
- Limiting the amount of time a worker is exposed to loud noises
- Training a worker on how to perform a potentially dangerous activity safely
- Documenting how workers should select and properly wear personal protective equipment suited to a specific task

Administrative controls the implementation programs (e.g., policies and procedures) and activities to address a hazard.

Personal protective equipment (PPE)—when engineering and administrative controls are not adequate enough to protect workers, PPE is used. PPE is specialized gear that provides a barrier between hazards and the body and its extremities. Examples of PPE are:

- Hearing protection
- Eye protection
- Hard hats
- Flame retardant clothing (FRC)
- Work gloves and safety shoes.

Personal protective equipment (PPE) specialized gear that provides a barrier between hazards and the body and its extremities.

The next section covers PPE in more detail.

Did You Know?

Noise-induced hearing loss (NIHL) can be caused by a single loud impulse noise (e.g., an explosion) or by loud, continuous noise over time (e.g., noise generated in a wood-working shop).

Hearing protection should always be worn when working in environments with sounds louder than 85 decibels for an 8-hour period (normal conversation is around 60 decibels).

Other sounds that can cause NIHL include motorcycles, firecrackers, and firearms, all of which range from 120 to 140 decibels.

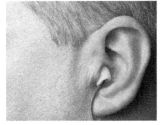

CREDIT: Rob Byron /Fotolia.

Correct Use of Personal Protective Equipment (PPE)

OSHA requires employers to use personal protective equipment (PPE) to reduce employee exposure to hazards when engineering and administrative controls are not feasible or effective. Employers are required to determine all exposures to hazards in their workplace and determine whether PPE should be used to protect their workers.

Different types of PPE are used to protect process technicians from head to toe in a variety of situations and hazards. Figure 9.10 shows some examples of PPE.

- Head protection
- Face protection
- Eye protection
- Ear protection
- Respiratory protection
- Body protection
- Hand protection
- Foot protection

The process technician also should be familiar with the location and operation of eyewashes and safety showers in the operating area. If used quickly and properly, eyewashes and safety showers can greatly reduce the severity of a chemical exposure.

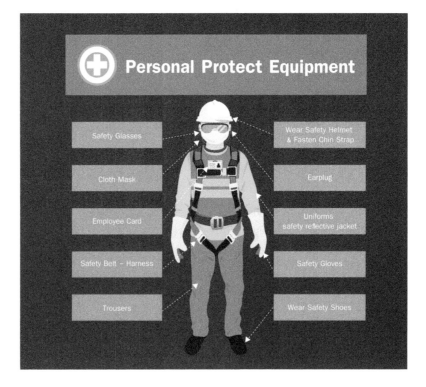

Figure 9.10 Examples of Personal Protective Equipment (PPE)

CREDIT: Ching Design/Shutterstock.

Head Protection

Head protection is used when a person's head is in danger of being bumped or struck by falling or flying objects.

Safety helmets (also referred to "hard hats") must be impact-resistant and meet the American National Standards Institute (ANSI) standard for protective headwear.

Types of PPE are illustrated in Figure 9.11.

Bump hats or caps do not meet the ANSI standard, as they are intended only to protect against bumping an obstruction.

Face Protection

Face protection is used to protect the face (and often the head and neck) against impact, chemical or hot metal splashes, heat, radiation, and other hazards.

Plastic face shields protect the face and eyes during activities such as sawing, buffing, sanding, grinding, or handling chemicals (Figure 9.12).

Acid-proof hoods protect the head, face, and neck against splashes from corrosive chemicals.

Welding helmets protect against splashes of molten metal and radiation.

Arc flash hoods protect against heat and flames associated with an electrical arc flash explosion.

Figure 9.11 Hard hat, respirator, gloves, filter, safety glasses, ear protectors.
CREDIT: photka/Fotolia.

Figure 9.12 Plastic face shield.
CREDIT: indigolotos/Fotolia.

Eye Protection

Eye protection is used to protect the eyes from flying objects, splashes of corrosive liquid or molten metals, dust, and harmful radiation.

Cover goggles are used when there is a danger of flying objects or splashing. Regular safety glasses with side shields are used during all other process-related activities (see Figure 9.11).

Ear Protection

Ear protection is used when excessive noise is present in the workplace (Figure 9.13). Proper placement is essential for effective protection. Earplugs and earmuffs can be used together when extreme noise is present to provide extra hearing protection. Use of both ear plugs and muffs can add 5 dB to the noise reduction rating of the most protective device.

Respiratory Protection

Respiratory protection is used when airborne contaminants are present.

Filter and cartridge respirators protect against nuisance dusts and hazardous chemicals (Figure 9.14). It is important to choose the correct filter or cartridge, based on the chemical exposure that is anticipated. Respirators are not suitable for high concentrations of contaminants. They do not supply oxygen so they cannot be used in oxygen-deficient atmospheres.

Figure 9.13 Ear protection devices.
CREDIT: Fotosenmeer.nl /Fotolia.

Figure 9.14 Respirator
CREDIT: Murushk I /Fotolia.

Air-supplying respirators deliver breathing air through a hose connected to the wearer's face piece. The air can be supplied through a tank attached to the back of the wearer (called a self-contained breathing apparatus, or SCBA) or from a tank that stays in one place with a long hose connecting the tank to the face piece (called a hose line respirator).

Body Protection

Body protection is used to provide additional safety for the body. This type of protection comes in various forms depending on the level of chemical protection required:

- Aprons protect the worker from chemical splashes.
- Harnesses with lifelines provide workers with fall protection.
- Arc flash suits provide protection from heat and flames associated with electrical arc flash explosions.
- Reflective clothing protects against radiant heat.
- Flame-resistant clothing protects against sparks and open flames (Figure 9.15).

Figure 9.15 Flame retardant coveralls (FRCs).
CREDIT: Andrey Popov/Shutterstock.

Hand Protection

Hand protection is used to protect the hands and fingers from cuts, scratches, bruises, chemicals, and burns.

- Extreme temperature protection protects against burns (hot or cold).
- Metal mesh protects against knives and sharp objects.
- Rubber, neoprene, and vinyl protect against different types of chemicals (see Figure 9.11).
- Leather protects against rough objects.

Foot Protection

Foot protection is used to protect the feet and toes against falling or rolling objects.

A variety of footwear can be required for use, based on the type of job or task you perform. For example, you might be required to wear safety shoes or boots (Figure 9.16), which are manufactured with a protective toe guard made of steel, alloy, or composite, rubber boots, flat-soled shoes, high-tops, etc. Some footwear is not permitted (e.g., open-toed shoes or sandals). A defined heel might be required if working off-ground, such as climbing a ladder or working on a scaffold.

The OSHA Voluntary Protection Program (VPP)

OSHA established the **Voluntary Protection Program (VPP)** to recognize and promote effective safety and health management. The VPP program requires a cooperative relationship between management, employees, and OSHA. The steps required, in order to be recognized as a VPP program participant, include:

1. Management agrees to operate an effective program that meets an established set of criteria.
2. Employees agree to participate in the program and work with management to ensure a safe and healthful workplace.
3. OSHA initially verifies that a site's program meets the VPP criteria.
4. OSHA publicly recognizes the site's exemplary program and removes the site from routine scheduled inspection lists (Note: OSHA still investigates major accidents, valid formal employee complaints, and chemical spills).

There are two OSHA VPP Ratings: Star and Merit.

- *Star*—participants meet all VPP requirements.
- *Merit*—participants have demonstrated the potential and willingness to achieve Star program status and are implementing planned steps to fully meet all Star requirements.

Periodically (every three years for the Star program and every year for the Merit program), OSHA reassesses the site to confirm that it continues to meet VPP criteria.

Figure 9.16 Safety boots.
CREDIT: Duncan Andison /Shutterstock.

Voluntary Protection Program (VPP) an OSHA program designed to recognize and promote effective safety and health management.

The ISO 14000 Standard

The International Organization for Standardization (ISO), headquartered in Geneva, Switzerland, consists of a network of national standards institutes from more than 140 countries.

ISO has published more than 13,700 international standards. ISO standards are voluntary, because the organization is nongovernmental and has no legal authority to enforce the standards. The standards that affect the process industries most are ISO 9000 and ISO 14000.

ISO 9000 provides a framework for quality management by addressing the processes of producing and delivering products and services. The chapter on *Quality* covers ISO 9000.

ISO 14000 addresses how to incorporate environmental aspects into operations and product standards. It requires a site to implement an environmental management system (EMS) using defined, internationally recognized standards as described in the ISO 14000 specification.

ISO 14000 establishes criteria for the following:

- Establishing an environmental policy
- Determining environmental aspects and impacts of products, activities, and services
- Planning environmental objectives and measurable targets
- Implementing and operating programs to meet objectives and targets
- Checking against the standard and making corrective actions
- Performing management review

ISO 14001, one of the subclassifications of ISO 14000, addresses the following requirements:

- Sites must document and make available to the public an environmental policy.
- Procedures must be established for ongoing review of environmental aspects and impacts of products, activities, and services.
- Environmental goals and objectives must be established that are consistent with the environmental policy, and programs must be set in place to implement goals and objectives.

ISO 9000 an international standard that provides a framework for quality management by addressing the processes of producing and delivering products and services.

ISO 14000 an international standard that addresses how to incorporate environmental aspects into operations and product standards.

- Internal audits of the EMS must be conducted routinely to ensure that instances of noncompliance in the system are identified and addressed.
- Management review must ensure top management involvement in the assessment of the EMS and, as necessary, address the need for change.

The *environmental management system (EMS)* document is the central document that describes the interaction of the core elements of the system.

The environmental policy and the environmental aspects and impacts sections of the EMS provide:

- Analysis, including legal and other requirements
- Direction for the environmental program by influencing the selection of specific and measurable environmental goals, objectives, and targets
- Recommendations for specific programs and/or projects that must be developed to achieve environmental goals, objectives, and targets
- Ongoing management review of the EMS and its elements to help ensure continuing suitability, adequacy, and effectiveness of the program

9.9 Physical Security and Cybersecurity

Physical security is intended to protect specific assets such as pipelines, control centers, tank farms, and other vital areas. Cybersecurity is the means of protecting information and information technology against unauthorized access and use (Figure 9.17). **Information technology** encompasses the equipment, tools, processes, and methodologies (coding/programming, data storage and retrieval, systems analysis, systems design, and son on) that is used to collect, process, and present information.

Physical security and cybersecurity threats can come from these sources:

- Terrorist organizations and hostile nation-states
- Insiders
- Criminal elements

Process technicians should recognize threats to physical security and cybersecurity. They include the following:

- Terrorist threats and acts
- Workplace violence
- Criminal acts
- Industrial espionage

Information technology the equipment, tools, processes, and methodologies (coding/programming, data storage and retrieval, systems analysis, systems design, and so on) that are used to collect, process, and present information.

Figure 9.17 Security and cybersecurity are important to process industries.

CREDIT: designer491/Fotolia; aijohn784/Fotolia.

To reduce the hazards of such threats, companies can create threat response and emergency action plans. In these plans, companies analyze their critical resources and operations, determine vulnerabilities and threats and identify what level of risk exists. The plan then establishes the processes and procedures required to lessen or eliminate threats.

Physical security aims to protect a company's critical assets from unauthorized access, thereby preventing them from being damaged or stolen. It involves the use of measures to protect and monitor assets and resources. Protective measures include physical access barriers (e.g., fences and doors), monitoring devices (e.g., cameras and motion detectors), and security patrols or guard stations.

Along with access and perimeter security, physical security also includes operations planning, communications planning, personnel background checks, and more.

Cybersecurity involves protecting information assets and capabilities from unauthorized access, modification, or destruction. It is a way of securing a company's information and information technology (IT) infrastructure, to ensure that sensitive data and computer networks are protected.

Cybersecurity includes the following:

- Preventing unauthorized computer access (e.g., through password protection and computer firewalls)
- Making sure those with access are trustworthy, and monitoring computer access and usage
- Preventing physical threats (e.g., direct access to sensitive information and computer networks)
- Having communication safeguards (e.g., cell phone and PDA protection)
- Preventing groups or individuals from using sensitive information and computer networks to carry out physical threats (e.g., hacking into a pipeline monitoring network and taking control of the system)

Summary

The process industries have many safety, health, and environmental hazards. These hazards can be chemical, physical, ergonomic, or biological.

In order to prevent or minimize many of these hazards, the U.S. government has created several agencies. The Environmental Protection Agency (EPA) protects human health and the environment. The Occupational Safety and Health Administration (OSHA) establishes workplace safety and health standards, conducts workplace inspections, and proposes penalties for noncompliance. The Department of Transportation (DOT) develops and coordinates policies for an efficient and economical national transportation system. The Nuclear Regulatory Commission (NRC) protects public health and safety through regulation of nuclear power and nuclear materials.

In an attempt to protect workers and the environment, OSHA, the DOT, and the EPA have created many regulations. These regulations help minimize the consequences of catastrophic releases of toxic, reactive, flammable, or explosive chemicals; prevent worker exposure to potentially hazardous substances; ensure that the hazards of all produced or imported chemicals are evaluated, and that information relating to the hazards is provided to employers and employees; establish emergency response operations for the releases of hazardous substances; and set requirements for handling hazardous materials.

In the event of a major plant or process incident, two independent federal agencies are charged with investigating to find the root cause and to help prevent future problems. The Chemical Safety Board (CSB) is charged with investigation of industrial accidents. The National Transportation Safety Board (NTSB) is charged with investigation of railway, highway, marine, and pipeline accidents.

Process technicians play a vital role when it comes to safety and health. Technicians must always maintain a safety-conscious attitude and behave in a safe, responsible, and appropriate manner. By following plant policies and procedures, attending training, learning to manage hazards appropriately, and more, process technicians facilitate compliance and create a safer and healthier workplace.

A fire triangle is a set of the three essential elements required for combustion: fuel, oxygen, and heat. If any one of these elements is removed, combustion cannot occur. A fire tetrahedron adds a fourth element, a chain reaction, as part of the combustion process.

Failure to comply with regulations can have legal, moral, ethical, safety, health, environmental, and economic consequences. Process industries often employ engineering controls, administrative controls, and personal protective equipment to make a safer workplace.

Physical security and cybersecurity measures are used to protect assets from internal and external threats.

Checking Your Knowledge

1. Define the following key terms:
   a. administrative controls
   b. biological hazard
   c. chain reaction
   d. chemical hazard
   e. cybersecurity
   f. engineering controls
   g. ergonomic hazard
   h. fire triangle
   i. hazardous agent
   j. physical hazard
   k. physical security

2. Which type of hazard involves noise and radiation?

3. Which type of pollution occurs when there is an emission of a potentially harmful substance or pollutant into the atmosphere?

4. Which government regulation deals with establishing emergency response operations for releases of, or substantial threats of releases of, hazardous substances?

5. Which of the following is addressed by OSHA regulation 1910.119?
   a. process hazard analysis
   b. process waste analysis
   c. process communications analysis
   d. process safety analysis

6. Personal attitudes toward safety can play a significant part in preventing accidents or incidents. What is the definition of an attitude?

7. What are the three parts of a fire triangle?
   a. air, wood, and a match
   b. fuel, combustion, and air
   c. earth, wind, and fire
   d. fuel, oxygen, and heat

8. What type of eye protection should be worn to guard against flying objects or splashing?

9. The intent of OSHA's Voluntary Protection Program is to:
   a. Regulate all environmental activities of the process industries.
   b. Recognize and promote effective safety and health management.
   c. Gauge the impact that the process industries have on the safety and health of process technicians.
   d. Help companies implement operation programs that are within EPA guidelines.

10. What is the central document that describes the interaction of the core elements of ISO 14001?

11. Which of these relates to physical, not cyber, security?
    a. Protecting intellectual property
    b. Protecting information from destruction
    c. Protecting remote-control cameras from damage
    d. Protecting process drawing from unauthorized access

NOTE: Answers to Checking Your Knowledge questions are in Appendix I.

Student Activities

1. Select one of the following government regulations and write a one-page report. Use any available resources. Describe the regulation, its history, what area it impacts most in the process industries (safety, health, or environmental), and why it is an important regulation. Be sure to include a list of all your reference sources.

   - OSHA Process Safety Management PSM (29 CFR 1910.119)
   - OSHA Personal Protective Equipment PPE (29 CFR 1910.132)
   - OSHA Hazard Communication HAZCOM (29 CFR 1910.1200)
   - OSHA Hazardous Waste Operations and Emergency Response HAZWOPER (29 CFR 1910.120)
   - DOT Hazardous Materials (49 CFR 173.1)
   - EPA Resource and Conservation Recovery Act RCRA (40 CFR 264.16)

2. Keep a journal for a week, describing how you use a safe attitude and behaviors on a daily basis (e.g., checking my blind spot before changing lanes, or being alert to my surroundings).

3. For each item in the table below, indicate which category it falls into (i.e., engineering control, administrative control, or personal protective equipment).

Item	Engineering Control	Administrative Control	PPE
a. Adding a machine guard to a place of equipment			
b. Writing a procedure for dealing with hazardous materials			
c. Training workers on hearing protection			
d. Providing hearing protection			
e. Documenting how to perform a hazardous task safely			
f. Wearing a face shield			
g. Improving ventilation in a work area			
h. Setting up shifts to limit exposure to hearing hazards			
i. Writing up fall protection procedures			
j. Training workers on fire extinguishers			
k. Making respirators available			
l. Training in how to use respirators			
m. Adding soundproofing around a pump			
n. Wearing a chemical protective suit			
o. Creating procedures for handling spills			

Chapter 10
Quality

Objectives

Upon completion of this chapter, you will be able to:

10.1 Identify responses in the process industries to quality issues. (NAPTA 1) p. 132

10.2 Describe the role each of the following played in quality implementation:

E. Deming

Joseph Juran

Philip Crosby (NAPTA 2) p. 132

10.3 Describe the four components of total quality management (TQM) and how it is applied in today's workplace. (NAPTA 3) p. 135

10.4 Describe the application of the International Standardization Organization, ISO 9000 series, as it relates to the petrochemical and petroleum industry. (NAPTA 4) p. 136

10.5 Describe the use of statistical process control (SPC) in the workplace. (NAPTA 5) p. 137

10.6 Describe the roles and responsibilities of the process technician in supporting quality improvement within the workplace. (NAPTA 6) p. 144

Key Terms

Assignable variation—statement that, when a product's variation goes beyond the limits of natural variation, it is the result of a worker-related cause, **p. 133.**

Attributes—also called discrete data; data that can be counted and plotted as distinct or unconnected events (such as percentage of late shipments or number of mistakes made during a process), **p. 138.**

Cause and effect diagrams—graphics that show the relationship between an event or incident (effect) and the potential causes that created it; sometimes called a fishbone diagram, **p. 139.**

Control charts—documents that establish operating limits for the amount of variation in a process, **p. 138.**

Fishbone diagrams—cause and effect diagrams (sometimes called Ishikawa diagrams, after Kaoru Ishikawa); used to help identify possible causes of an event or incident, **p. 139**.

Flowcharts—documents that represent a sequence of operations schematically (or visually), **p. 138**.

Histograms—bar graphs of a frequency distribution in which the widths of the bars are proportional to the classes into which the variable has been divided, and the heights of the bars are proportional to the class frequencies, **p. 140**.

ISO—taken from the Greek word *isos*, which means equal, ISO is the International Standardization Organization, which consists of a network of national standards institutes from more than 140 countries, **p. 136**.

Overall equipment effectiveness (OEE)—a concept that assesses availability (time), performance (speed), and quality (yield) to evaluate manufacturing productivity, **p. 143**.

Pareto charts—graphics that rank causes from most significant to least significant; they represent the 80-20 rule described by Juran, stating that most undesired effects come from relatively few causes, **p. 139**.

Pareto principle—a quality principle, also called the 80-20 rule, that states 80% of problems come from 20% of the causes, **p. 134**.

PPM—predictive/preventive maintenance, a program to identify potential issues with equipment and use preventive maintenance before the equipment fails, **p. 143**.

Scatter plots—graphs drawn using dots or a similar symbol to represent data, **p. 140**.

Six Sigma—an advanced quality management method that increases output by minimizing variability and defects in manufacturing processes, **p. 141**.

SPC—statistical process control; uses mathematical laws dealing with probability to gather data (numbers) and study the characteristics of processes, and then uses the data to make the processes behave the way they should, **p. 137**.

TPM—total productive maintenance; an equipment maintenance program that emphasizes a company-wide effort to involve all levels of staff in various aspects of equipment maintenance, **p. 143**.

TQM—total quality management; a collection of philosophies, concepts, methods, and tools used to manage quality; TQM consists of four parts: customer focus, continuous improvement, managing by data and facts, and employee empowerment, **p. 136**.

Variables—also called continuous data; pieces of information that can be measured and plotted on a constant scale (such as flow through a pipeline or liquid in a tank), **p. 138**.

Zero defects—the goal of a quality practice with the objective of reducing defects, thus increasing profits, **p. 135**.

Introduction

Quality is an important part of the process industries. Without quality measures, products and services become deficient or unsatisfactory. Unsatisfactory products lead to unhappy customers, increased waste, inefficiencies, increased costs, reduced profits, and an inability to maintain a competitive edge.

To maintain a competitive edge, many companies have adopted theories and philosophies from famous pioneers in the study of quality. By incorporating these philosophies and acting in ways that support them, companies and process technicians can improve processes, reduce waste, increase efficiency, reduce costs, produce superior products, and maintain a competitive edge in a global marketplace.

10.1 What is Quality?

The term *quality* has different meanings to different people, but generally, quality has two major definitions in regard to the process industries:

- A product or service free of deficiencies
- The characteristics of a product or service that bear on its ability to satisfy stated or implied needs

This chapter will explore these definitions, as well as other meanings that have been given to the term *quality*.

Industry Response to Quality Issues and Trends

Why are organizations concerned with quality? Because producing quality products and services is critical to the success of organizations, process industries implement a variety of methods to:

- Satisfy and retain customers
- Maintain a competitive advantage and respond to rapidly changing markets
- Capture a leading position in the global marketplace
- Improve profitability
- Manage change more effectively
- Maintain or bolster the organization's reputation

Companies in the process industries also implement quality processes to address such additional business needs as:

- Offering standardized products on a consistent basis
- Improving efficiency of operations and maintenance
- Reducing waste of resources such as utilities and feedstocks
- Decreasing downtime of people and equipment
- Ensuring certifications necessary to trade internationally
- Tapping new technologies and methods

10.2 The Quality Movement And Its Pioneers

Today, most companies take quality seriously. They show great concern for providing quality products and services to their customers. However, this was not the case when competition was almost nonexistent and customers did not have a large number of choices. As more competitors emerged both locally and internationally, and as economic conditions changed, companies have made greater efforts to improve quality.

During the early part of the twentieth century, F.W. Taylor pioneered the concept of "scientific management," which sought to improve manufacturing by using engineers to develop plans that supervisors and workers executed. Taylor's system successfully raised productivity, but human relations were affected negatively. This approach assumed workers did not really know what they were doing. It kept knowledge about how to produce a quality product in the hands of engineers and management. It used people simply as tools for the job. In process units with this form of management, worker involvement was sometimes referred to as "working from the neck down."

Following Taylor, Dr. Walter Shewhart developed the theory of statistical quality control in the 1920s. While working for Bell Telephone Labs, Shewhart recognized that not all

> **Did You Know?**
>
> Shewhart developed his ideas about quality when he was asked to develop a radio headset for use during World War I.
>
> After measuring people's heads to determine a standard size for the headset, he used charts to show a pattern in the measurements.

products created during work processes were exactly alike. Shewhart stated that every process produces variation in its products because of random natural causes.

According to Shewhart, product variation that exceeds the limits of a natural variation is the result of a worker-related cause. He called this an **assignable variation**, meaning it could be attributed to a specific worker's action. Shewhart developed control charts that were used to identify assignable causes of variation. By identifying and correcting the variation, the company could improve the quality of its products. (His approach did not address such causes as defective raw product or interplant issues.)

Assignable variation states that when a product's variation goes beyond the limits of a natural variation, it is the result of a worker-related cause.

Taylor and Shewhart laid the groundwork for the quality movement during the early part of the twentieth century. Since the middle of the century, three individuals have had a significant impact on quality with their ideas, approaches, and tools:

- Dr. W. Edwards Deming
- Joseph Juran
- Philip Crosby

Many organizations have adopted some or all of these pioneers' definitions of quality into their corporate culture.

Dr. W. Edwards Deming

Deming studied statistical process control under Shewhart in 1927 and adopted his theories for quality control. During World War II, Deming worked with the War Department, applying Shewhart's statistical quality control principles to the production of *materiel* (French term for military supplies and materials). He assisted the war effort by recommending that engineers be trained in the basics of applied statistics.

After the war, Deming began working in Japan to help rebuild its war-torn industries. Deming convinced Japanese companies to apply statistical methods to help improve the quality of products and services. During the decade that followed, Japanese companies implemented his principles and established themselves as leaders in quality manufacturing.

> **Did You Know?**
>
> The Union of Japanese Scientists and Engineers awards the Deming Prize for quality to companies that meet or exceed customers' needs based on Deming's 14 points. It is the highest quality award in Japan.

CREDIT: Armita/Shutterstock.

Deming developed a theory on quality control, referred to as the Deming Cycle (Figure 10.1). The Deming Cycle states that every task or every job is part of a process. Specifically, a system is a group of interrelated components that work toward optimization of the system, even if the result does not benefit the individual component.

His 14-point theory of management emphasized the need to build customer awareness, reduce variation, and foster constant change and improvement.

He defined quality as meeting and exceeding the customer's needs and expectations, then continuing to improve. He also urged conformity to specifications. Deming associated

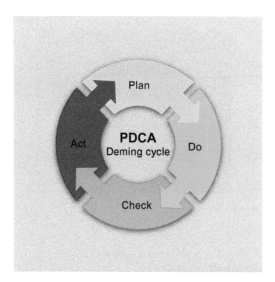

Figure 10.1 Deming Cycle.
CREDIT: Alena Kubikov/Shutterstock.

quality with management, a view unknown and not taught in the foremost business schools in America at that time.

American car manufacturing had taken a back seat to the Japanese when current management at the company called on Deming for help. Deming's management principles and philosophies are largely credited with the turnaround of Ford Motor Company in the mid-1980s.

Joseph M. Juran

Juran worked with the government during World War II to develop supply processes for the allied nations under the Lend Lease Pact. Like Deming, Juran also worked with Japanese companies after the war and was instrumental in assisting the Japanese to rebuild their economy.

In the mid-1950s, the Union of Japanese Scientists and Engineers group invited Juran to conduct quality control courses for middle and top management. These courses extended the philosophy of quality control to every aspect of an organization's activities. Juran emphasized that everything should be viewed as a process. He promoted the idea of using quality control as a management tool. In the mid-1960s, he predicted that the Japanese would lead the world in quality.

Juran's best-known quality teachings emphasized the **Pareto principle**, an idea of separating the vital few factors from the many trivial ones. The 80-20 rule is an example of the Pareto principle. It states that 80% of problems come from 20% of the causes.

Pareto principle a quality principle, also called the 80-20 rule, that states 80% of problems come from 20% of the causes.

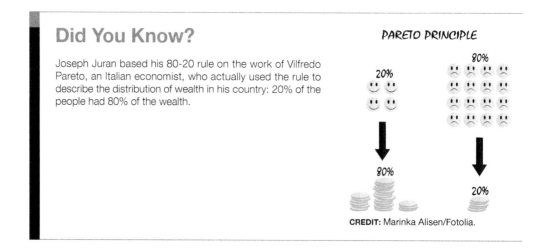

Did You Know?

Joseph Juran based his 80-20 rule on the work of Vilfredo Pareto, an Italian economist, who actually used the rule to describe the distribution of wealth in his country: 20% of the people had 80% of the wealth.

CREDIT: Marinka Alisen/Fotolia.

Juran's other contributions to the quality movement involved top management and stressed the need for widespread training in quality. The main components of the Juran Trilogy of managerial processes are:

- Quality planning
- Quality control
- Quality improvement

He defined quality as fitness for use as perceived by customers.

Philip B. Crosby

Crosby is best known for creating the concept of zero defects in the early 1960s (Figure 10.2). **Zero defects** is a quality practice with the objective of reducing defects, thereby increasing profits.

Zero defects the goal of a quality practice with the objective of reducing defects, thus increasing profits.

Figure 10.2 Philip B. Crosby developed the concept of zero defects.

In the 1970s, Crosby became a leader in the quality movement. He served as vice president for ITT (International Telephone and Telegraph) before forming his own management-consulting firm.

He published a book titled *Quality Is Free*. The single word Crosby uses to sum up quality is prevention. He stated that quality involves doing "it right the first time." He advocated a proactive approach to quality: Fix the process so errors will not occur.

Crosby's quality improvement program promotes 14 points covering management commitment, zero defects, training, goals, teams, corrective action, and removal of error causes.

He defined quality as conformance to requirements and the elimination of variation.

The Japanese Influence

Kaoru Ishikawa, considered a pioneer of the Japanese quality movement, studied statistical quality control with Deming and quality management with Juran. He played a major role in the growth and concept of the quality circle, which demonstrates causes and effects using diagrams.

In the 1980s, in response to Japan's quality movement, many American companies instituted quality programs or revamped their existing programs. Dr. Genichi Taguchi, a Japanese engineer and scientist, began working with the Ford Motor Company to provide seminars to its managers. Dr. Taguchi contributed his expertise in the field of industrial research to the quality movement, focusing on the design of experiments.

10.3 Quality Initiatives

Quality initiatives that process industries have adopted include:

- Total quality management (TQM)
- ISO 9000 series
- Statistical process control (SPC)

- Six Sigma
- Self-directed or self-managed work teams
- Malcolm Baldrige Criteria for Performance Excellence
- Maintenance programs such as total productive maintenance and predictive/preventive maintenance

Total Quality Management (TQM)

TQM total quality management; a collection of philosophies, concepts, methods, and tools used to manage quality; TQM consists of four parts: customer focus, continuous improvement, managing by data and facts, and employee empowerment.

The quality principles from Deming, Crosby, Juran, and other experts in the field have evolved into a concept referred to as total quality management (TQM). **TQM** is not a specific, well-defined program, but a collection of philosophies, concepts, methods, and tools used to manage quality. Consequently, not every company practices TQM the same way or even calls it by the same name.

TQM, whether referred to as customer satisfaction, reengineering, or some other name, encompasses four major components:

- Customer focus—the customer viewpoint determines what quality is. Companies use customer satisfaction with its products and services as a measure of quality.
- Continuous improvement—companies improve the activities used to produce products and services, resulting in higher quality products and services. The philosophy is to create a process that makes it easy to do things the right way and difficult to do them the wrong way.
- Management by data and facts—companies gather information to understand how their processes work, what can be produced, and where improvements can be made.
- Employee empowerment—top management must demonstrate its buy-in, commitment, and involvement to quality improvements. Every employee is free to question, challenge, and help change the way that products and services are produced.

10.4 ISO

ISO taken from the Greek word *isos*, which means equal, ISO is the International Standardization Organization, which consists of a network of national standards institutes from more than 140 countries.

ISO, the International Organization for Standardization, is headquartered in Geneva, Switzerland, and consists of a network of national standards institutes from more than 140 countries. The American National Standards Institute (ANSI) is a member of ISO.

Between its founding after World War II and the present day, ISO has published more than 13,700 international standards. These standards address everything from screw sizes to symbols, computers to shipping containers, and more.

ISO standards are voluntary because the organization is nongovernmental and has no legal authority to enforce the standards. However, without the consent of both government and private sectors to use ISO standards, many vital economic segments around the world would be affected, including manufacturing, trade, science, technology, and many others.

The standards with the greatest impact on the process industries are ISO 9000 and ISO 14000. ISO 9000 provides a framework for quality management by addressing the processes of producing and delivering products and services. The ISO 14000 standard addresses environmental management systems, helping organizations improve their environmental performance.

This section addresses ISO 9000 in more detail. The Safety, Health, Environment, and Security chapter describes ISO 14000 in more detail.

ISO 9000 provides global standards of product and service quality. These standards form a quality management system that ensures the correction of quality system defects and makes products and services conform to stated standards.

More than half a million organizations in more than 60 countries have implemented or will implement ISO 9000. Most companies in the process industries have ISO 9000 certification. Their ISO 9000 certification efforts usually overlap with TQM programs.

Did You Know?

The term *ISO* is derived from the Greek word *Isos*, meaning equal; whatever the country or language, the short form of the organization's name is always ISO.

CREDIT: Bankrx/Shutterstock.

Although TQM policies and practices are comparable in some ways to the ISO 9000 model, TQM is defined in many different ways, whereas ISO standards are consistent and have been adopted by companies around the world.

Because ISO certification is recognized internationally, organizations can participate in the global marketplace and be assured of conformity. Both manufacturers and suppliers know that quality standards are being met when dealing with ISO certified organizations.

Companies seek ISO 9000 certification for a variety of reasons:

- Contractual—many companies require certification from suppliers.
- Liability—certification can result in improved product liability procedures and documentation.
- Cost savings—companies with certification report a significant increase in operational efficiency, which results in improved profitability.
- Competition—certification allows companies to compete on a global basis and levels the playing field between small and large companies.
- Customer satisfaction—customer confidence is increased when dealing with companies that are certified.

ISO 9000 provides companies with a common approach for documenting and maintaining a quality system. An accredited auditor reviews of a company's quality system to determine whether that system complies with the ISO 9000 standard. In other words, ISO 9000 certification targets the quality system itself, not products or services.

ISO 9000 certification guarantees that a plan for continuous improvement is in place. Recertification relies on continuous adherence to the standards and is based on regular audits.

To obtain ISO 9000 certification, a company must have a quality program in place that meets documentation and operational criteria.

As a process technician, your role is to make sure that all processes and documentation are carried out in accordance with your company's quality program.

10.5 Statistical Process Control and Other Analysis Tools

Walter Shewhart developed the concept of Statistical Process Control (SPC) during World War I, with Dr. W.E. Deming improving and expanding it. Their work influenced the quality initiatives of many companies that integrated the approach of SPC into their quality programs.

SPC uses mathematical laws dealing with probability, or how often certain events could occur under certain conditions. Data play a crucial role in SPC, allowing companies to monitor, control, correct, and improve their processes. Companies use SPC to gather data and study the characteristics of processes, then use the data to make the processes behave the way they should.

Many companies use SPC or statistical thinking to make decisions, basing them on meaningful data and statistics. It basically applies a scientific method to decision making. Decisions are not made unless some type of proof (data or statistics) is available to substantiate the changes.

SPC statistical process control; uses mathematical laws dealing with probability to gather data (numbers) and study the characteristics of processes, then uses the data to make the processes behave the way they should.

Attributes also called discrete data, or data that can be counted and plotted as distinct or unconnected events (such as percentage of late shipments or number of mistakes made during a process).

Variables also called continuous data; pieces of information that can be measured and plotted on a constant scale (such as flow through a pipeline or liquid in a tank).

Specifically, SPC is a way of determining whether a process is producing predictable results. SPC is also a basic tool for identifying both immediate systemic problems and opportunities for improvement.

Data derived through SPC can predict how a process will function in the future, making it possible to avoid off-spec products and unnecessary process changes. Data can be placed into one of two categories:

- **Attributes**—also called discrete data, or data that can be counted and plotted as distinct or unconnected events (such as percentage of late shipments or number of mistakes made during a process).
- **Variables**—also called continuous data, or data that can be measured and plotted on a constant scale (such as flow through a pipeline or liquid in a tank).

Data are critical to the operation and maintenance of a plant and help process technicians understand how the processes in a plant are performing. So, taking accurate samples and readings is vital to quality.

Data observed, collected, and recorded can describe processes and products, imply what might be happening to them, and predict what can be done to improve them. One of a process technician's daily tasks is collecting data to monitor, improve, control, and correct processes and products.

For example, technicians must track crucial data called process variables, such as temperature, pressure, flow, and level. They also must ensure that the data is accurate, thorough, and timely.

SPC uses specific tools to analyze data. These include:

- Control charts
- Flowcharts
- Cause and effects diagrams
- Fishbone diagrams
- Pareto charts
- Histograms
- Scatter plots

Control charts establish limits for the amount of variation in a process. Figure 10.3 shows an example of a control chart.

Control charts documents that establish operating limits for the amount of variation in a process.

Figure 10.3 Control chart example.

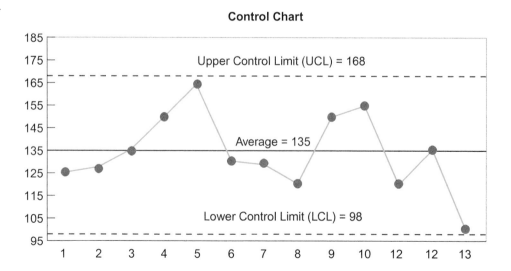

Flowcharts documents that represent a sequence of operations schematically (or visually).

Flowcharts are charts that represent a sequence of operations schematically (or visually). Figure 10.4 shows an example of a flowchart.

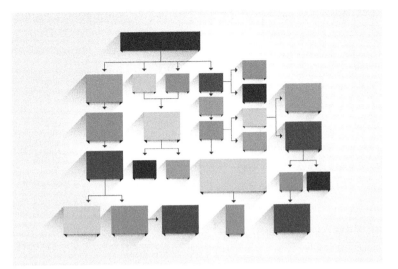

Figure 10.4 Flowchart example.
CREDIT: kubko/Fotolia.

Cause and effect diagrams or **fishbone diagrams** are graphics that show the relationship between a cause and effect. On a cause and effect diagram, activities (causes) are connected to attributes (effects) by arrows. These diagrams (also called Ishikawa diagrams, after Kaoru Ishikawa) are used to help identify possible sources of a problem. Figure 10.5 shows an example of a cause and effect diagram.

Cause and effect diagrams graphics that show the relationship between an event or incident (effect) and the potential causes that created that result it; sometimes called a fishbone diagram.

Fishbone diagrams cause and effect diagrams (sometimes called Ishikawa diagrams, after Kaoru Ishikawa); used to help identify possible causes of an event or incident.

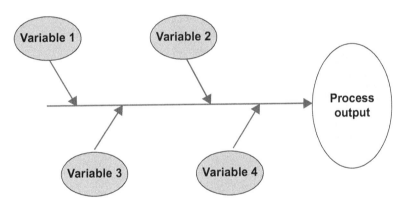

Figure 10.5 Cause and effect diagram example. Each variable in the process acts as a cause, and the cumulative action of all the variables creates the final effect (process output).

Pareto charts are graphics that rank causes from most significant to least significant. They represent the 80-20 rule described by Juran, stating that most effects come from relatively few causes. Figure 10.6 shows an example of a Pareto chart.

Pareto charts graphics that rank causes from most significant to least significant; they represent the 80-20 rule described by Juran, stating that most undesired effects come from relatively few causes.

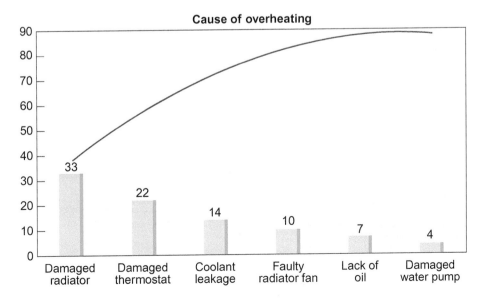

Figure 10.6 Pareto chart example.

Histograms bar graphs of a frequency distribution in which the widths of the bars are proportional to the classes into which the variable has been divided, and the heights of the bars are proportional to the class frequencies.

Histograms are bar graphs of a frequency distribution in which the widths of the bars are proportional to the classes into which the variable has been divided, and the heights of the bars are proportional to the class frequencies. Figure 10.7 shows an example of a histogram.

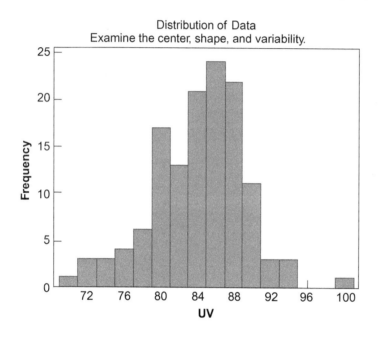

Figure 10.7 Histogram example.
CREDIT: Courtesy of Eastman Chemical Company.

Scatter plots graphs drawn using dots or a similar symbol to represent data.

Scatter plots, also called scatter graphs or scattergrams, are graphs drawn using dots or a similar symbol to represent data. Figure 10.8 shows an example of a scatter plot.

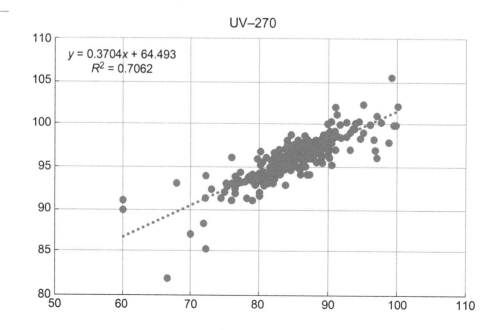

Figure 10.8 Scatter plot example.
CREDIT: Courtesy of Eastman Chemical Company.

Six Sigma

Compared to other quality initiatives, **Six Sigma** is a newer TQM approach used in the process industries. The Six Sigma quality initiative was first used in the Motorola Company in 1985. Bill Smith, a senior quality engineer with the company, is credited with invention of Six Sigma. Six Sigma is considered advanced quality management using a data-driven approach and methodology to eliminate defects. Six Sigma concentrates on measuring product quality and improving process engineering.

Six Sigma was developed in the 1980s using the concept of zero defects promoted by Philip Crosby. Six Sigma aims for a measure of quality that is near perfection. Sigma (σ), a Greek letter, represents standard deviation (or variation) in a process. Companies using Six Sigma aim to reduce process variation so that no more than 3.4 defects per 1 million opportunities (DPMO) result (99.99966% with no variation).

The system requires that all processes are inspected, errors (and where they occur) are identified and then corrected, and measures are implemented to control the processes. Along with TQM efforts, Six Sigma focuses on strategy and leadership development.

Six Sigma uses two sub-methodologies for process improvement:

- DMAIC—define, measure, analyze, improve, control; an improvement system used for existing processes that fall below specifications and will be improved incrementally.
- DMADV—define, measure, analyze, design, verify; an improvement system used to develop new processes or products at quality levels that meet Six Sigma. This system also can be used for existing processes that require more than incremental improvement.

Six Sigma an advanced quality management method that increases output by minimizing variability and defects in manufacturing processes.

Did You Know?

Leaders in Six Sigma efforts are sometimes referred to using karate belt-like designations, such as green belt or black belt.

CREDIT: cristovao31/Fotolia.

Self-Directed or Self-Managed Work Teams

Self-directed and self-managed teams are not quality-specific initiatives, but they can involve such efforts because teams are a key component of various quality systems (like Crosby's quality improvement program).

Companies often refer to the use of such teams as "empowerment." Teams work together to perform a specific function, such as to create a product or provide a service, while also managing that work (i.e., performing tasks such as scheduling and recognition). This frees up managers and supervisors from tasks such as directing and controlling, allowing them to take on a facilitation role by coaching, developing, and teaching.

Work becomes restructured around the whole process of providing the product or service. Team members are involved in all aspects of the process, from design to development to deployment. This approach integrates the needs of the team members with the work to be done.

There are differences between self-directed teams and self-managed teams. A self-directed team involves a group of people working together toward a set of common goals. The team defines these goals; it also determines compensation and discipline for team members. Self-directed teams determine their own future and act like a profit center within the company. For example, if the team holds down costs, improves quality, and increases profitability, the team is rewarded for its efforts.

A self-managed team is different from a self-directed team. It consists of a group of individuals working in their own way to achieve a set of common goals defined outside of the team. In a self-managed team, the team receives goals from management and then handles its own processes, training, scheduling, and rewards to achieve them.

Malcolm Baldrige National Quality Award

Created by President Ronald Reagan in 1987, the Malcolm Baldrige National Quality Improvement Act established an annual U.S. National Quality Award.

Figure 10.9 Malcolm Baldrige.

Named after the 26th secretary of commerce (Figure 10.9), the award honors Baldrige (who died in 1987) for his managerial excellence, which contributed to long-term improvement in efficiency and effectiveness of government. The award, managed by the National Institute of Standards and Technology (NIST), aims to establish a standard of excellence that can help U.S. organizations achieve world-class quality.

The U.S. president presents this award to manufacturing and service businesses of all sizes (as well as educational and healthcare organizations) for their achievements in quality and performance.

Baldrige Award recipients must be judged outstanding in seven areas:

1. Leadership
2. Strategic planning
3. Customer and market focus
4. Information and analysis
5. Human resources focus
6. Process management
7. Business results

> **Did You Know?**
>
> Malcolm Baldrige, who died while competing in a rodeo in 1987, believed that quality management was a key to the United States' prosperity and long-term strength.
>
>
>
> **CREDIT:** puckillustrations/Fotolia.

The award focuses on performance excellence for the entire organization, not just for products or services. Organizations applying for the award must undergo an intense self-assessment process, identifying and tracking organization results in customer products and services, financials, human resources, and organizational effectiveness.

This award has proven very important to U.S. businesses as a model of high standards against which they can evaluate their own organizations. Companies can use the criteria from the seven areas to measure their own efforts, even though they may not apply for the award.

Maintenance Programs

U.S. plants spend billions of dollars in maintenance costs annually. A large percentage of those costs are wasted because of inefficient maintenance programs (e.g., repairing equipment after it fails instead of preventing the failure).

There are four types of maintenance program approaches:

- Corrective—waiting for a failure to occur and then fixing the equipment as quickly as possible to restore production.
- Preventive—conducting maintenance tasks on a set schedule to keep failures from happening.
- Predictive—monitoring the condition of equipment and interpreting the data to identify possible failures and prevent them from occurring.
- Detective—checking equipment such as alarms and detectors on a regular basis to make sure they work and to prevent failures.

Two practical maintenance programs that can reduce costs and improve quality are total productive maintenance and predictive/preventive maintenance.

Total Productive Maintenance (TPM)

Total productive maintenance (**TPM**) employs a series of methods to ensure every machine in a process is always able to perform so that production is never interrupted. It emphasizes a company-wide equipment management program that involves all levels of staff in various aspects of equipment maintenance.

TPM also uses continuous improvement techniques, one of the four components of total quality management.

Small teams perform activities aimed at maximizing equipment effectiveness, such as prioritizing problems, applying problem solving, evaluating processes (in order to simplify them), and measuring data. Each team establishes a thorough system of preventive maintenance for the life span of the equipment.

TPM uses a concept of **overall equipment effectiveness (OEE)**, a concept that assesses availability (time), performance (speed), and quality (yield) to evaluate manufacturing productivity. It identifies the percentage of manufacturing time that is truly productive. OEE is considered a best practice in manufacturing. OEE take into account six big losses:

1. Equipment downtime
2. Setup and adjustment
3. Minor stoppages
4. Slow-running processes
5. Time spent making nonconforming products during normal operation
6. Waste during start-up or warm-up

The elements of availability (time), performance (speed), and quality (yield) are described in the following equation:

$$OEE = Time \times Speed \times Quality$$

The TPM program sets goals for OEE and measures variations. The team then seeks to eliminate problems and improve performance.

Predictive/Preventive Maintenance (PPM)

Predictive/preventive maintenance (**PPM**) bases equipment maintenance on conditions, not schedules. Some maintenance programs are time driven. Maintenance tasks are based on elapsed running hours of equipment and historical/statistical data.

Instead of relying on manufacturer's recommended maintenance schedules or plant-established maintenance schedules, PPM uses actual operating conditions and direct monitoring of the equipment to determine maintenance tasks and frequency.

TPM total productive maintenance; an equipment maintenance program that emphasizes a company-wide effort to involve all levels of staff in various aspects of equipment maintenance.

Overall equipment effectiveness (OEE) a concept that assesses availability (time), performance (speed), and quality (yield) to evaluate manufacturing productivity.

PPM predictive/preventive maintenance, a program to identify potential issues with equipment and use preventive maintenance before the equipment fails.

The philosophy of PPM is to improve productivity and efficiency, optimize plant operations, and improve quality by using operating conditions to predict potential problems before they occur. Preventive maintenance then is performed on an as-needed basis.

PPM allows plants to do the following:

- Prevent equipment deterioration
- Reduce potential equipment failures
- Decrease the number of equipment breakdowns
- Increase the life span of equipment

A PPM approach uses direct monitoring of equipment, checking operating conditions (such as heat or vibration), efficiency, and other indicators to predict when failures or loss of efficiency might occur. Data and facts are gathered by direct observation or with tools such as thermal monitors and vibration sensors.

After monitoring conditions and gathering data, the information is analyzed. Potential problems are detected before they become serious, and appropriate preventive maintenance is planned and scheduled. Preventive maintenance tasks include:

- Inspection
- Lubrication
- Cleaning
- Replacement of worn parts
- Minor adjustments
- Repairs
- Overhauls

It is the process technician's responsibility to maintain the health of the equipment being used. The observations and readings taken by a process technician are recorded in a data base. These observations and readings are used to determine maintenance needs for specific pieces of equipment. Many pieces of operating equipment have regular maintenance schedules as part of the TPM/PPM criteria. The process technician is responsible to ensure the equipment is safe for a maintenance crew to open and complete the needed work.

10.6 Process Technicians and Quality Improvement

In process industries, the company determines the quality terms, methods, processes, and tools that will be used as part of the process technician's everyday tasks.

A process technician's responsibilities typically include the following:

- Becoming familiar with and understanding the quality program
- Practicing good quality habits, using the company's quality manual as a basis
- Providing good customer service (even if a technician does not deal with customers directly, there are always internal customers to satisfy)
- Gathering data for use with statistical quality control tools (e.g., flowcharts and cause and effect diagrams) and using those tools
- Following documented procedures, such as operating procedures and work instructions
- Monitoring and controlling processes and operations
- Assisting with equipment maintenance tasks such as TPM or PPM
- Identifying and troubleshooting problems with a goal of continuous improvement
- Communicating effectively

- Keeping clear and complete records
- Working with teams to meet quality goals
- Participating in quality-oriented training sessions
- Using skills in time management, organization, planning, and prioritization.

Summary

Quality is an important part of the process industries. It helps minimize waste, inefficiencies, increased costs, and unhappy customers. It helps companies maintain a competitive edge.

To compete more effectively, many companies have adopted theories and philosophies from famous quality pioneers like Dr. W.E. Deming, Joseph Juran, and Philip B. Crosby.

Deming created a theory called the Deming Cycle, which states that every task or job is part of a process and that all interrelated components work toward optimization of the system, even if the result does not benefit the individual component. Deming emphasized the need to build customer awareness, reduce variation, and foster constant change and improvement. Quality, according to Deming, is meeting and exceeding the customer's needs and expectations, then continuing to improve.

Juran created the Pareto principle (the 80-20 rule), an idea that distinguishes the vital few factors from the many trivial ones. He defined quality as fitness for use as perceived by customers.

Crosby developed the concept of zero defects. Crosby advocated a proactive approach to quality, "doing it right the first time," and fixing processes proactively so errors will not occur. Crosby defined quality as conformance to requirements.

Some of the many different quality initiatives that have been adopted over the years are total quality management (TQM), the ISO 9000 series, and statistical process control (SPC). Six Sigma, invented by Bill Smith, is an advanced quality management program implemented in the mid 1980s.

Total quality management (TQM) can be described in different ways but has four main components: (1) customer focus, (2) continuous improvement, (3) managing by data and facts, and (4) employee empowerment.

The International Organization for Standardization (ISO) is an international standards organization. ISO certification is recognized internationally, so organizations can participate in the global marketplace and be assured of conformity. Both manufacturers and suppliers know that quality standards are being met when dealing with ISO certified organizations.

Statistical process control (SPC) uses the mathematical laws of probability to determine how often certain events could occur under certain conditions. Data play a crucial role in SPC. Through data collection, companies can monitor, control, correct, and improve their processes. Data derived through SPC can predict how a process will function in the future, making it possible to avoid off-specification products and unnecessary process changes.

Process technicians should always be familiar with their company's quality program and work in ways that support the goals of that program.

Checking Your Knowledge

1. Define the following key terms:
   a. assignable variation
   b. ISO
   c. Pareto principle
   d. PPM
   e. Six Sigma
   f. SPC
   g. TPM
   h. TQM
   i. zero defects
2. List five ways that quality affects process industries.
3. The ___ cycle states that every task or every job is part of a process.
4. Explain the 80-20 rule.
5. Who discovered and determined the concept of zero defects?
   a. Dr. W. Edwards Deming
   b. Joseph M. Juran
   c. Philip B. Crosby
   d. Kaoru Ishikawa
6. What do the initials TQM stand for?

7. Which of the following is one of the four major components of TQM?
   a. the Pareto principle
   b. the idea that every task or job is part of a process
   c. ISO certification
   d. customer focus

8. What is the purpose of ISO 9000 certification?

9. Define statistical process control.

10. Which of the following tools are used for statistical process control? (Select all that apply)
    a. control charts
    b. cause and effect diagrams
    c. ISO reports
    d. histograms

11. Which of the following is an aim of total productive maintenance (TPM)?
    a. to maximize equipment effectiveness (overall effectiveness)
    b. to use control charts for quality control
    c. to achieve ISO 9000 certification
    d. to involve only management in TPM

12. PPM involves:
    a. using SPC for TQM to achieve ISO
    b. creating a quality tree for all staff members to follow
    c. promoting PPM through TQM
    d. preventing potential equipment failures

NOTE: Answers to Checking Your Knowledge questions are in Appendix I.

Student Activities

1. Write a one-page paper describing whichever one of the following three individuals you think had the most significant impact on process industries. Discuss why you selected that person.
   - Dr. W. Edwards Deming
   - Joseph Juran
   - Philip Crosby

2. Discuss how ISO 9000 and TQM are similar and how they are different.

3. For statistical process control, list the items in the table below in the correct category (attribute or variable).

Data	Attribute	Variable
a. Pounds per square inch (PSI)		
b. Gallons per minute		
c. Temperature reading		
d. Bad batches per shift		
e. Flow through a pipe		
f. Downtime for equipment		
g. Days lost to accidents		
h. Level in a tank		
i. Amount of electricity used		
j. Percent of delayed deliveries		

4. Draw examples of the following types of charts and provide descriptions of each:
   a. control chart
   b. flowchart
   c. cause and effect diagram
   d. fishbone diagram
   e. Pareto chart
   f. histogram
   g. scatter plot

5. You are a process technician working in a baking plant. Your boss has asked you to review XYZ process that produces chocolate brownies. Brainstorm and create a list of how a quality initiative can affect the product. Then, outline a brief plan with three tasks that can be done to improve quality for that product.

PART 3 Basic Knowledge for Process Technicians

Chapter 11
Basic Physics

 Objectives

Upon completion of this chapter, you will be able to:

11.1 Define the application of physics in the process industries. (NAPTA Physics 1) p. 150

11.2 Describe matter and the states of matter (liquid, gas, and solid). (NAPTA Physics 2, 3) p. 150

11.3 Define and provide examples of the following terms:

mass
density
elasticity
viscosity
buoyancy
specific gravity
flow
evaporation
temperature
British thermal unit (BTU)
calorie
velocity
friction
electricity
pressure
force
leverage (NAPTA Physics 4, 11) p. 151

11.4 Convert between scales using mass flow and temperature, which are commonly used in the process industry. (NAPTA Physics 12) p. 154

11.5 Describe the three methods of BTU heat transfer:

convection
conduction
radiation (NAPTA Physics 5) p. 158

11.6 Describe how Boyle's law explains the relationship between pressure and volume of gases. (NAPTA Physics 6) p. 161

11.7 Describe how Charles' law explains the relationship between temperature and volume of gases. (NAPTA Physics 7) p. 161

11.8 Describe how Dalton's law explains the relationship between total and partial pressure of a gas. (NAPTA Physics 8) p. 162

11.9 Describe how the General (or Combined) Gas Law explains the relationships among temperature, pressure, and volume of gas. (NAPTA Physics 9) p. 162

11.10 Describe how Bernoulli's law explains the flow of liquids and gases. (NAPTA Physics 10) p. 163

Key Terms

Absolute pressure (psia)—gauge pressure plus atmospheric pressure; pressure referenced to a total vacuum (zero psia), **p. 160.**

API gravity—the American Petroleum Institute (API) standard used to measure the density of hydrocarbons, **p. 154.**

Atmospheric pressure—the pressure at the surface of the earth (14.7 psia at sea level), **p. 159.**

Baume gravity—the industrial manufacturing measurement standard used to measure the gravity of nonhydrocarbon materials, **p. 154.**

Bernoulli's law—a physics principle stating that as the speed of a fluid in a constricted space increases, the pressure inside the fluid, or exerted by it, decreases, **p. 163.**

Boiling point—the temperature at which liquid physically changes to a gas at a given pressure, **p. 160.**

Boyle's law—a physics principle stating that at a constant temperature, as the pressure of a gas increases, the volume of the gas decreases, **p. 161.**

British thermal unit (BTU)—the amount of heat energy required to raise the temperature of 1 pound of water 1 degree Fahrenheit, **p. 157.**

Buoyancy—the principle that a solid object will float if its density is less than the fluid in which it is suspended; the upward force exerted by the fluid on the submerged or floating solid is equal to the weight of the fluid displaced by the solid object, **p. 154.**

Calorie—the amount of heat energy required to raise the temperature of 1 gram of water by 1 degree Celsius (centigrade), **p. 157.**

Charles' law—a physics principle stating that, at constant pressure, the volume of a gas increases as the temperature of the gas increases, **p. 161.**

Conduction—the transfer of heat through matter via vibrational motion; exchange media must be touching, **p. 158.**

Convection—the transfer of heat through the circulation or movement of a liquid or a gas, **p. 158.**

Dalton's law—a physics principle stating that the total pressure of a mixture of gases is equal to the sum of the individual partial pressures, **p. 162.**

Density—the ratio of an object's mass to its volume, **p. 152.**

Elasticity—an object's tendency to return to its original shape after it has been stretched or compressed, **p. 153.**

Evaporation—conversion of a liquid into a vapor, **p. 160.**

Flow—the movement of fluids, **p. 154.**

Fluid—substances, usually liquids or vapors, that can be made to flow, **p. 153.**

Force—energy that causes a change in the motion of an object, involving strength or direction of push or pull, **p. 158.**
Friction—the resistance encountered when one material slides against another, **p. 156.**
Gases—substances with a definite mass but no definite shape, whose molecules move freely in any direction and completely fill any container they occupy, and which can be compressed to fit into a smaller container (vapor), **p. 150.**
Gauge pressure (psig)—pressure measured with respect to Earth's surface at sea level (zero psig), **p. 160.**
General (or Combined) Gas Law—relationships among pressure, volume, and temperature in a closed container; pressure and temperature must be in absolute scale ($P1V1/T1 = P2V2/T2$), **p. 162.**
Heat—the transfer of energy from one object to another as a result of a temperature difference between the two objects, **p. 157.**
Heat tracing—a coil of heated wire or tubing that is adhered to or wrapped around a pipe and piping components in order to increase the temperature of the process fluid, reduce fluid viscosity, and facilitate flow, **p. 155.**
Hydrometer—an instrument designed to measure the specific gravity of a liquid, **p. 153.**
Latent heat—heat that does not result in a temperature change but causes a phase change, **p. 157.**
Latent heat of condensation—the amount of heat energy given off when a vapor is converted to a liquid without a change in temperature, **p. 157.**
Latent heat of fusion—the amount of heat energy required to change a solid to a liquid without a change in temperature, **p. 157.**
Latent heat of vaporization—the amount of heat energy required to change a liquid to a vapor without a change in temperature, **p. 157.**
Leverage—an assisted advantage, usually gained through the use of a tool (such as lever and fulcrum), **p. 158.**
Liquids—substances with a definite volume but no fixed shape that demonstrate a readiness to flow with little or no tendency to disperse, and are limited in the amount in which they can be compressed, **p. 150.**
Lubrication—a friction-reducing film placed between moving surfaces in order to reduce drag and wear, **p. 156.**
Mass—the amount of matter in a body or object measured by its resistance to a change in motion, **p. 151.**
Mass flow rate—amount of mass passing through a plane per unit of time, **p. 154.**
Matter—anything that takes up space and has inertia and mass, **p. 150.**
Phase change—transition of a substance from one physical state to another, such as when ice melts to form water, **p. 151.**
Plasma—a gas that contains positive and negative ions, **p. 150.**
Pressure—the amount of force a substance or object exerts over a particular area, **p. 159.**
Radiation—the transfer of heat energy through electromagnetic waves, **p. 158.**
Sensible heat—heat transfer that results in a temperature change, **p. 157.**
Solids—substances with a definite volume and a fixed shape that are neither liquid nor gas, and that maintain their shape independent of the shape of the container, **p. 150.**
Specific gravity—the ratio of the density of a liquid or solid to the density of pure water, or the density of a gas to the density of air at standard temperature and pressure (STP), **p. 153.**
Specific heat—the amount of heat required to raise a unit of mass by one degree (e.g., the temperature of 1 gram of a substance 1 degree Celsius), **p. 157.**
Temperature—the measure of the thermal energy of a substance, **p. 156.**
Vacuum pressure (psiv)—any pressure below atmospheric pressure, **p. 160.**
Vapor pressure—a measure of a substance's volatility and its tendency to form a vapor, **p. 160.**

Velocity—the distance traveled over time or change in position over time, **p. 155**.
Viscosity—the measure of a fluid's resistance to flow, **p. 153**.
Weight—a measure of the force of gravity on an object, **p. 151**.

Introduction

The concepts of basic physics are important to process technicians. By learning these concepts, process technicians increase their ability to visualize what is occurring in a process, improve their ability to troubleshoot, and reduce the likelihood they will make costly or catastrophic errors.

11.1 Applications of Physics in the Process Industries

During normal, everyday operations, process technicians are required to open valves, check pressures, monitor fluid and gas flows, monitor furnace operations, and much more. In order to perform these tasks effectively, a technician must have a firm understanding of physics.

Consider the task of rerouting fluid flow using valves such as the ones shown in Figure 11.1. If a technician does not have a firm understanding about fluid flow principles, the process technician might not realize that one must open a new path (valve A) before closing off the original path (valve B). Failure to perform these steps in this order could cause an increase in backflow pressure that could damage or "deadhead" the pump.

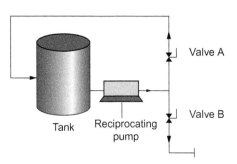

Figure 11.1 Technicians should never "deadhead" a positive displacement pump.

Matter anything that takes up space and has inertia and mass.

Solids substances with a definite volume and a fixed shape that are neither liquid nor gas, and that maintain their shape independent of the shape of the container.

Liquids substances with a definite volume but no fixed shape that demonstrate a readiness to flow with little or no tendency to disperse, and are limited in the amount in which they can be compressed.

Gases substances with a definite mass but no definite shape, whose molecules move freely in any direction and completely fill any container they occupy, and which can be compressed to fit into a smaller container (vapor).

Plasma a gas that contains positive and negative ions.

11.2 States of Matter

The term *matter* is used frequently in physics. **Matter** is anything that takes up space and has inertia and mass (described below). In other words, matter is the substance of which things are made. An atom is the smallest indivisible unit of matter.

Matter is found in four states: solid, liquid, gas, and plasma (Figure 11.2). Each of these states has its own unique properties and characteristics.

- **Solids** are substances with a definite volume and a fixed shape. They are neither liquid nor gas, and they maintain their shape independent of the shape of the container.
- **Liquids** are substances with a definite volume but no fixed shape. They demonstrate a readiness to flow with little or no tendency to disperse and are limited in the degree to which they can be compressed (unlike gases, which are highly compressible).
- **Gases** are substances with definite mass but no definite shape. The molecules in a gas move freely in any direction, completely fill any container they occupy, and can be compressed to fit into a smaller container. Gases can mix freely with each other and can be liquefied through compression or temperature reduction.
- **Plasma** is a gas that contains positive and negative ions. Because plasma is not commonly found in the process industries, it will not be discussed in great detail in this chapter.

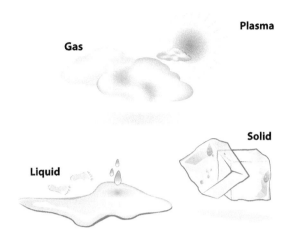

Figure 11.2 Matter can be solid, liquid, gas, or plasma.
CREDIT: designua/Fotolia.

As the states of matter (specifically, solids, liquids, and gases) are exposed to different environmental conditions, they can be induced to change form. This process is called a **phase change**. During a phase change, the substance (solid, liquid, or gas) maintains the same chemical composition, but physical properties change. (For example, ice is a solid that melts to form water, a liquid, yet both have the same chemical formulation of H_2O.)

Process technicians need to understand the concept of phase changes, since many of the substances with which they work will undergo these changes. Failure to understand the process and potential impacts of phase changes could cause process problems or pose serious safety risks. Table 11.1 lists and describes the six phase changes, as well as potential problems or safety risks associated with each.

Phase change transition of a substance from one physical state to another, such as when ice melts to form water.

Table 11.1 Six Phase Changes

Description	Term	Movement of Heat	Example of Change	Examples of Problems or Risks
Solid to liquid	Melting	Heat is absorbed by the solid as it melts.	Ice melting to form a puddle of water	Melting ice could form a puddle of liquid that poses a slipping hazard.
Liquid to solid	Freezing	Heat leaves the liquid as it freezes.	Water freezing to form ice	Leaving an unheated vessel (e.g., a tank or pipe) full of water during freezing temperatures could cause equipment damage or vessel rupture as ice builds up and expands inside.
Liquid to vapor	Vaporization (includes boiling and evaporation)	Heat is absorbed by the liquid as it vaporizes.	Water boiling to make steam or a puddle evaporating after a rain shower	Liquids exposed to extreme heat form a vapor that could overpressurize an enclosed vessel or pose a burn risk.
Vapor to liquid	Condensation	Heat leaves the vapor as it condenses.	Water vapor (humidity) in the air condensing into water droplets on a cold drinking glass	Sensitive electronic equipment left outside overnight could be damaged by humidity or condensation.
Solid to vapor	Sublimation	Heat is absorbed by the solid as it sublimates.	"Dry ice" (compressed carbon dioxide) changing from a solid to a fog like vapor	Carbon dioxide (CO_2) "fog" displaces oxygen; in a confined space this could cause suffocation.
Vapor to solid	Deposition	Heat leaves the vapor as it solidifies.	Water vapor condensing to form frost on the surface of a pipe	Exposing uncovered skin to frost could result in frostbite, a skin injury that requires medical attention.

11.3 Key Concepts in Physics for Process Industries

Mass

Mass is the amount of matter in a body or object measured by its resistance to a change in motion. It is also a measure of an object's resistance to acceleration.

Mass should not be confused with weight. **Weight** is a measure of the force of gravity on an object. To illustrate this, think about an apple. An apple would weigh more on Jupiter

Mass the amount of matter in a body or object measured by its resistance to a change in motion.

Weight a measure of the force of gravity on an object.

than it does on Earth because Jupiter's gravity is stronger. However, the mass is the same, regardless of where the apple is located (Figure 11.3).

Figure 11.3 Gravity has a direct impact on weight.

CREDIT: (*Earth*) robert/Fotolia; (*Jupiter*) mode/Fotolia; (*apples*) ZoneCreative/Fotolia.

Apple mass = Apple mass
Apple weight = 2 oz (57 gm)   Apple weight = 4.6 oz (130 gm)

METRIC MEASUREMENTS Metric measurements are in common use in the process industries. Process technicians must become familiar with conversion tables or conversion applications to use when measuring supplies or processes. The following is an example of equivalencies that exist in the oil industry:

1 U.S. barrel of oil = 5.6 cubic feet (ft^3) = 0.16 cubic meters (m^3) = 42 U.S. gallons

In an increasingly international market, process technicians must be alert to which systems of measurement are being used in equipment and processes. An incorrect unit of measure can have a huge impact. For example, a conversion error occurred with NASA's Mars probe Climate Orbiter in October of 1999. Instead of entering Mars' orbit smoothly, Climate Orbiter crashed into the Martian atmosphere and was disabled. The error lost the probe and cost NASA $125 million. This failure occurred because engineers neglected to make the conversion from English units to metric units.

Density

Density the ratio of an object's mass to its volume.

Density is a scientific way to determine or compare the "heaviness" of an object or objects. To determine density, one must compare the ratio of the object's mass to its volume.

Consider a canister of marshmallows such as the one shown in Figure 11.4.

In this example, a canister of marshmallows at normal atmospheric pressure weighs 16 ounces (454 grams). If you were to compress or compact all of those marshmallows so they filled only half the container, they would still weigh 16 ounces, but they would occupy only half the volume. In other words, the weight of the marshmallows remains the same, but the density increases. If you add 16 more ounces of compressed marshmallows to the container, the volume will again be the same as the original, noncompacted container. However, the mass will increase from 16 ounces to 32 ounces (454 to 908 grams) and the marshmallows will be more compact (dense). This is the concept of density.

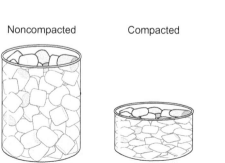

Figure 11.4 Marshmallows become denser when air is removed.

Increasing density allows more mass in the same volume of space.

Elasticity

Elasticity is an object's tendency to return to its original shape after it has been stretched or compressed. The *elastic limit* is the greatest amount of stress a material is able to sustain without being permanently deformed.

To illustrate the concept of elasticity, think about a balloon. If you stretch a balloon without blowing it up, it becomes thinner and longer. After you release it, it returns to its original shape and size. However, if you blow up the balloon, stretching it outward as well as long, its elastic limit is surpassed, and it is permanently deformed. The balloon will not return to its original size. This is elasticity.

Elasticity an object's tendency to return to its original shape after it has been stretched or compressed.

Viscosity

The term **viscosity** refers to the measure of resistance to flow of a **fluid**. To demonstrate this concept imagine a spoon full of water and a spoon full of honey (Figure 11.5).

If you fill a spoon with water and tilt it, the water will drip rapidly off the spoon. This is because water is a low-viscosity liquid (not viscous). If you try the same experiment with honey, the honey drips off the spoon much more slowly. This is because honey is more viscous than water and so is more resistant to flow.

Viscosity the measure of a fluid's resistance to flow.

Fluid substances, usually liquids or vapors, that can be made to flow.

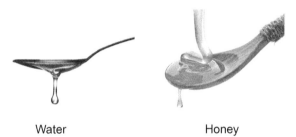

Figure 11.5 Honey is more viscous than water.

CREDIT: (*left*) BalancePhoto / Shutterstock (*right*) MaraZe /Shutterstock.

Water   Honey

Specific Gravity

Specific gravity describes the ratio of the density of a liquid or solid to the density of pure water. It also describes the ratio of the density of a gas to the density of air at standard temperature and pressure (STP). In other words, it is the "heaviness" of a substance.

Pure water has a density of 1 gram per cubic centimeter (g/cc). Materials that are lighter than water (specific gravity less than 1.0) will float. Those that are heavier than water (specific gravity greater than 1.0) will sink.

Hydrocarbons are organic compounds found in petroleum, natural gas, and coal; they contain only carbon and hydrogen. Process technicians in the chemical and refining industries know that most hydrocarbons have a specific gravity less than 1.0. A good example is oil, which tends to float on water.

Knowing the specific gravity of a substance is important. By knowing the specific gravity, a technician can better understand the characteristics of a substance and how it will react in the presence of other substances with differing specific gravities. This can be very important in both day-to-day operations and emergency situations.

For example, light, flammable liquids such as gasoline can spread. If ignited, they will burn on the surface of a body of water. If a remediation or firefighting crew is unaware of this, they might take actions that make the situation worse instead of better. (They might douse the fire with water, causing it to spread even more.)

In nonemergency situations, process technicians need to understand the concept of specific gravity because this measurement is often used when testing samples.

As an example, in a heat exchanger, many process fluids are routed through a set of tubes surrounded by a shell full of flowing cool water. To make sure the fluid is moving properly and to verify that the tubes are not leaking, process technicians must regularly sample the cooling water and test its specific gravity. They use a **hydrometer** (an instrument

Specific gravity the ratio of the density of a liquid or solid to the density of pure water, or the density of a gas to the density of air at standard temperature and pressure (STP).

Hydrometer an instrument designed to measure the specific gravity of a liquid.

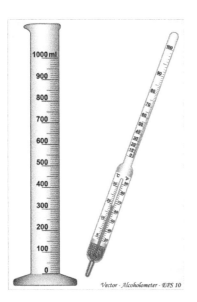

Figure 11.6 Hydrometer example.
CREDIT: Fouad A. Saad / Shutterstock.

Baume gravity the industrial manufacturing measurement standard used to measure the gravity of nonhydrocarbon materials.

API gravity the American Petroleum Institute (API) standard used to measure the density of hydrocarbons.

Buoyancy the principle that a solid object will float if its density is less than the fluid in which it is suspended.

designed especially for measuring the specific gravity of a liquid). If the hydrometer reading is normal, the technician knows the system is working properly. However, if the specific gravity indicates the presence of hydrocarbons, the technician will know there is a leak and can take corrective action to isolate the exchanger.

Hydrometers also are used in the food and beverage industries to test the specific gravity of beer, wine, and other food products. Figure 11.6 shows an example of a hydrometer.

In the process industries, the two most common gravity measurement types are Baume gravity and API gravity.

Baume gravity is the industrial manufacturing measurement standard used to measure nonhydrocarbon heaviness. In contrast, **API gravity** is the American Petroleum Institute standard used to measure the density of hydrocarbons. In order to determine API gravity, technicians must use a specially designed hydrometer marked in API units (the higher the API reading, the lower the fluid's specific gravity).

Buoyancy

Buoyancy refers to the principle that a solid object will float if its density is less than the fluid in which it is suspended, and if the upward force exerted by the fluid on the submerged or floating solid is equal to the weight of the fluid displaced by the solid object.

The higher the specific gravity of the fluid, the more buoyant the object is. For example, distilled water has a specific gravity of 1.000, while natural seawater has a specific gravity somewhere around 1.025. For this reason, objects and people (e.g., boats and scuba divers) float better in salt water than they do in fresh water.

Did You Know?

People are more buoyant in salt water than in fresh water because salt water has a higher specific gravity. In the Dead Sea, swimmers float high on the water. Scuba divers would have to wear more weight when diving there to maintain their neutral buoyancy (a position in water where you go neither up nor down).

CREDIT: ra66 / fotolia.

11.4 Flow

Flow the movement of fluids.

Flow is the movement of fluids. When we use the term *flow*, we refer to any uninterrupted movement of a liquid or gas.

Many factors can restrict or enhance the flow of a substance. These factors include mechanical restriction (e.g., valves), environmental factors (e.g., extreme heat or extreme cold), or the physical characteristics of the substance itself (e.g., viscosity).

Mass flow rate amount of mass passing through a plane per unit of time.

In process industry, the total flow is the amount of flow past a point, usually a metered (mass flow) point. Total flow is generally described as a quantity in a 24-hour period. **Mass flow rate** is the amount of mass passing through a plane per unit of time.

The calculation formula for mass flow rate is:

$$\text{Mass flow rate} = \text{density} \times \text{velocity} \times \text{flow area}$$

or

$$\text{Mass flow rate} = \text{volume flow rate} \times \text{density}.$$

Process technicians must understand the concept of flow, especially as it applies to process monitoring. A large part of a process technician's job involves monitoring process variables, troubleshooting, and opening and closing valves.

Valves, which are discussed more in later chapters, have a definite impact on flow rates of fluids. Depending on the type of valve used, process technicians can regulate the flow of a substance so it is completely on, completely off, or somewhere in between.

Temperature has a direct impact on flow because there is a direct relationship between temperature and molecular movement. As temperature decreases, so does molecular movement. (Remember the old saying about moving "as slow as molasses in January.") If a process occurs in extremely cold temperatures, whether because of weather or an endothermic (heat-absorbing) chemical reaction, the flow rate might be affected. If the impact of this decreased flow is significant enough, then plant technicians might have to apply **heat tracing**. Heat tracing is a coil of heated wire or tubing that adheres to or wraps around a pipe to increase the temperature of process fluid, reduce fluid viscosity, and facilitate flow (Figure 11.7).

Other factors can affect flow besides the standard factors of valve position and temperature. Many of these factors are more directly related to malfunctions that would be detected during troubleshooting. For example, a plugged tube might stop the flow of a substance, but a ruptured tube might only change the flow rate. With a rupture, the rate change would depend on the size of the ruptured area, the pressure of the flow, and the meter location.

Heat tracing a coil of heated wire or tubing that is adhered to or wrapped around a pipe or piping components in order to increase the temperature of the process fluid, reduce fluid viscosity, and facilitate flow.

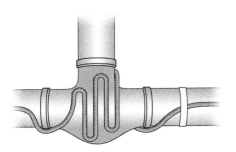

Figure 11.7 Heat tracing applied to process piping. Note, most heat tracing on process pipes is hard-wired into a circuit.

Velocity

Velocity is the distance traveled over time or change in position over time.

In the process industries, the concept of velocity is most often used in relationship to equipment such as pumps and compressors, which are discussed in later chapters. As velocity increases, so does the amount of flow.

The simplest relationship between weight flow rate and fluid velocity is as follows:

$$w = \rho \times v \times A$$

Where:

- w = weight flow rate of fluid in lb/sec (or kg/sec)
- ρ = fluid density in lb/ft^3 (or kg/m^3)
- v = average fluid velocity in ft/sec (or m/sec)
- A = the cross-section area of the pipe carrying the fluid.

This equation shows that, for a constant weight flow rate, the velocity of fluid will increase if the cross-sectional area or diameter (d) of the pipe decreases. Because area is proportional to d^2, velocity increases rapidly as pipe diameter decreases.

In relation to process flow, increases in temperature can affect flow rate by decreasing viscosity, thereby decreasing density.

Velocity the distance traveled over time or change in position over time.

Friction

Friction the resistance encountered when one material slides against another.

Lubrication a friction-reducing film placed between moving surfaces in order to reduce drag and wear.

Friction is the resistance encountered when one material slides against another. Friction can be both desirable and undesirable. The amount of friction generated depends on how smooth the contact surfaces are and how much force is applied when the two surfaces are pressed together. Friction can produce heat and potential wear to two sliding surfaces.

In order to reduce friction, lubrication is often employed. **Lubrication** is the introduction of a friction-reducing film between moving surfaces in order to reduce drag and wear. Lubricants can be fluid, solid, or plastic substances.

It is important to understand that *sliding* an object back and forth over a surface produces more friction, requires more energy, and causes more wear than *rolling* an object (e.g., a tire or ball bearing). Because of this, lubrication, ball bearings, and other friction-reducing techniques often are used in rotating equipment such as pumps and compressors. Failure to maintain proper lubrication during operations can cause excessive wear or heat that can damage or ruin equipment.

Although friction is most often thought to result from the movement of two solid objects against each other, it may also occur with the movement of fluids. Fluid friction occurs when the molecules of a gas or liquid are in motion. Unlike solid friction, fluid friction varies with velocity and area. It is important to understand the concept of fluid friction, because some chemical substances will ignite or explode as a result of friction that builds up when flow changes too rapidly.

Temperature

Temperature the measure of the thermal energy of a substance.

Temperature is a measure of the thermal energy of a substance (hotness or coldness). It can be determined using a thermometer. Process technicians use a variety of temperature scales to measure temperature. These include Kelvin, Celsius (formerly called Centigrade), Fahrenheit, and Rankine. Figure 11.8 compares and contrasts each of these four temperature measurement scales.

It is worth noting that both degrees K and degrees R are known as *absolute temperature scales* based on a true zero temperature in which molecular motion is believed to cease. When dealing with ideal gases later in this chapter, we will have to use absolute temperatures to calculate changes in vapor pressures and temperatures.

Looking at Figure 11.8, you can see that there is a wide range among the four temperature measuring scales. For example, the boiling point of water in Celsius is 100 degrees. In Rankine, however, it is 672 degrees. Because of this variability, process technicians need to

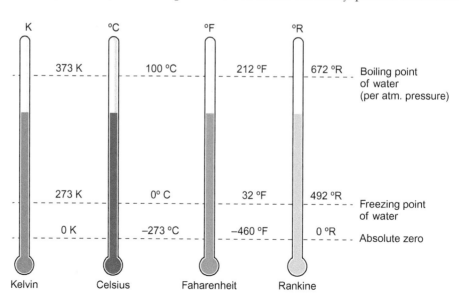

Figure 11.8 Kelvin, Celsius, Fahrenheit, and Rankine temperature scales.

report not just their measurements but also the scale they are using when measurements are taken.

Table 11.2 lists several formulas process technicians will find useful when performing temperature conversions.

Table 11.2 Temperature Conversion Formulas

Conversion Type	Formula
Celsius to Kelvin (approximate)	K = °C + 273
Celsius to Fahrenheit	°F = 1.8 × °C + 32
Fahrenheit to Rankine (approximate)	°R = °F + 460
Fahrenheit to Celsius	°C = (°F − 32)/1.8

When studying temperature, it is also important to know that temperature and heat are not the same thing. As previously discussed, temperature is a measure of the thermal energy of a substance (its hotness or coldness), determined by using a thermometer. **Heat** is the transfer of energy from one object to another as a result of a difference in temperature between the two objects.

According to the principles of physics, heat energy always moves from hot to cold. It cannot be created or destroyed. It can only be transferred from one object or substance to another.

Heat is typically measured in British thermal units or in calories. A **British thermal unit (BTU)** is the amount of heat energy required to raise the temperature of 1 pound (0.9 kg) of water by 1 degree Fahrenheit. A **calorie** is the amount of heat energy required to raise the temperature of 1 gram of water by 1 degree Celsius. One BTU is equivalent to 252 calories.

Heat comes in many different forms: specific heat, sensible heat, latent heat, latent heat of fusion, latent heat of crystallization, latent heat of vaporization, and latent heat of condensation. Table 11.3 lists and describes each of these different forms of heat.

Figure 11.9 illustrates the difference between sensible heat and latent heat when an ice cube at 0 degrees F is converted to steam at 300 degrees F (−18 degrees to 149 degrees C).

Heat the transfer of energy from one object to another as a result of a temperature difference between the two objects.

British thermal unit (BTU) the amount of heat energy required to raise the temperature of 1 pound of water 1 degree Fahrenheit.

Calorie the amount of heat energy required to raise the temperature of 1 gram of water by 1 degree Celsius (centigrade).

Did You Know?

The term Calorie (with a capital "C"), found on food labels and in diet books, is really 1,000 of the calories referred to in this chapter.

CREDIT: anaumenko / Fotolia.

Specific heat the amount of heat required to raise the temperature of 1 unit of mass by 1 degree (e.g., 1 gram of a substance by 1 degree Celsius).

Sensible heat heat transfer that results in a temperature change.

Latent heat heat that does not result in a temperature change but causes a phase change.

Latent heat of fusion the amount of heat energy required to change a solid to a liquid without a change in temperature.

Latent heat of vaporization the amount of heat energy required to change a liquid to a vapor without a change in temperature.

Latent heat of condensation the amount of heat energy given off when a vapor is converted to a liquid without a change in temperature.

Table 11.3 Types of Heat

Type of Heat	Description
Specific heat	The amount of heat required to raise the temperature of 1 gram of a substance by 1 degree Celsius, or 1 pound (0.9 kg) of a substance by 1 degree Fahrenheit. *Example: The specific heat of water is 1 cal/g °C or 1 BTU/lb °F. The specific heat of copper is 0.10 cal/g °C or 0.10 BTU/lb °F.*
Sensible heat	Heat transfer that results in a temperature change. *Example: The heat required to bring a pot of water up to boiling.*
Latent heat	Heat that does not result in a temperature change but causes a phase change. *Example: The heat required to keep a pot of water boiling with the temperature remaining constant.*
Latent heat of fusion	The amount of heat energy required to change a solid to a liquid without a change in temperature. *Example: The heat required to change ice to water at a constant temperature.*
Latent heat of vaporization	The amount of heat energy required to change a liquid to a vapor without a change in temperature. *Example: The heat required to change water into steam at a constant temperature.*
Latent heat of condensation	The amount of heat given off when a vapor is converted to a liquid without a change in temperature. *Example: The heat removed to change steam into water at a constant temperature.*

Figure 11.9 Comparison of sensible heat and latent heat. Ascending lines represent sensible heat; horizontal lines represent latent heat.

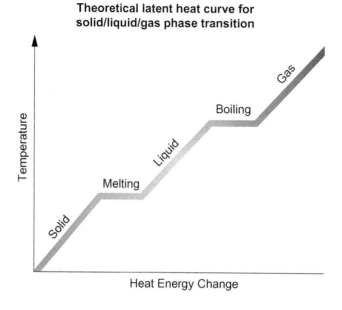

Convection the transfer of heat through the circulation or movement of a liquid or a gas.

Conduction the transfer of heat through matter via vibrational motion, exchange media must be touching.

Radiation the transfer of heat energy through electromagnetic waves.

11.5 Heat Transfer

Heat energy is commonly transferred by three methods: conduction, convection, and radiation (Figure 11.10). **Conduction** is the transfer of heat through matter by vibrational motion (e.g., transferring heat energy from a frying pan to an egg or from heat tracing around a pipe to a process fluid). **Convection** is the transfer of heat through the circulation or movement of a liquid or gas (e.g., warm air being circulated in a furnace). **Radiation** is the transfer of heat energy through electromagnetic waves (e.g., warmth emitted from the sun or an open flame).

Figure 11.10 Heat can be transferred through conduction, convection, or radiation.

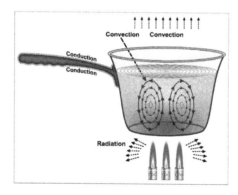

Electricity

Electricity is a phenomenon associated with stationary or moving electric charges. Electrons are the charge carriers that flow from the negative to positive terminals around a circuit. This movement is called an electric current. Electricity is discussed in depth in Chapter 19, Electricity and Motors.

Force and Leverage

Force energy that causes a change in the motion of an object, involving strength or direction of push or pull.

Leverage an assisted advantage, usually gained through the use of a tool (such as lever and fulcrum).

Force and leverage are two principles that are often at work in process industries. **Force** is energy that causes a change in the motion of an object. A common example is the force of gravity, which causes an apple released from a tree to drop to the ground.

Leverage is an assisted advantage, usually gained through the use of a tool. Leverage utilizes tools such as a lever and fulcrum to move materials efficiently and with less expenditure of energy. In process industries, the principles of leverage are often used in installation, relocation, or demolition of equipment. Another example is the forklift, a common tool used in the process industries to move pallets of product bags, cylinders, and many other items located on plant sites. In a forklift, the counterweight on the machine acts as a counterbalance or lever, the front wheel is the pivot point or fulcrum, and the fork is the location of the load. Whether motorized and man-powered or operated by hand, each version utilizes the principles of hydraulics and can lift and carry many more pounds than a human being alone could do. Pascal's Law is put to work here: an incompressible liquid is used to multiply the effects of the force applied to lift something very large or heavy. A larger surface area requires a smaller amount of force to do the same amount of work (lift).

Pressure

Pressure is the amount of force a substance or object exerts over a particular area. The formula for pressure is defined as force per unit area and is usually expressed in pounds per square inch (psi) or metrically as kilopascals (kpa).

It is important to understand the concept of pressure, especially with regard to fluids, because pressure is a fundamental part of process operations.

Within a plant, a technician can encounter pressure in components such as towers (columns), compressors, pumps, hydraulic lines, storage tanks, fluid lines, and more. By understanding the concept of pressure, technicians improve their ability to troubleshoot and reduce the likelihood of equipment damage or serious injury.

Pressure the amount of force a substance or object exerts over a particular area.

ATMOSPHERIC PRESSURE Air pushes on every surface it touches. Air pressure is the weight of the air from the earth's atmosphere down to the surface. At sea level, air exerts a pressure of about 14.7 pounds per square inch (psi). This is called **atmospheric pressure** or psi absolute (psia). One atmosphere = 14.7 psi absolute = 760 millimeters of mercury (mmHg) = 29.92 inches of mercury (in. Hg) absolute = 101 kilopascals (kpa, the metric equivalent).

To illustrate the concept of atmospheric pressure, think about a piece of paper that is 10 inches square. In a 10-inch by 10-inch piece of paper, there are 100 square inches. So, at sea level, the atmosphere is actually exerting 1470 pounds (10,100 kilopascals) of pressure on the paper. At 10,000 feet (3050 meters), because of the fewer layers of atmosphere at that altitude, the same piece of paper would only have 1010 pounds (6,953 kpa) of pressure on it.

Atmospheric pressure the pressure at the surface of the earth (14.7 psia at sea level).

The abbreviations psi and psig are used interchangeably. A good example of a tool that uses psi is a tire gauge. The psig obtains the *difference* between the measured environment (the inside of the car tire) and the ambient pressure, whether the car is below sea level or at high altitudes.

Instruments that measure pressure must obviously take atmospheric pressure into account. Process technicians may need to verify that pressure-reading instruments have been properly calibrated, especially when working with newly installed equipment.

Vacuum is any pressure less than atmospheric pressure. Vacuum, discussed later, is an important part of many industrial processes.

Equivalencies among different methods of measuring atmospheric pressure are shown in Table 11.4.

Table 11.4 Equivalency of Units Measuring Atmospheric Pressure

Unit of Measure	Equivalent Measurements
PSIA	14.7
Millimeters mercury (mmHg)	760
Inches of mercury (in. Hg)	29.92
Kilopascals (kpa)	101
Inches of H_2O	407.17

PRESSURE GAUGE MEASUREMENTS In order to determine the amount of pressure that is present in a particular process, process technicians must be able to read and interpret various pressure gauges and measurements. The three main types of pressure measurements a technician will encounter are gauge, absolute, and vacuum (Figure 11.11).

Figure 11.11 Pressure gauges can present gauge, absolute, or vacuum pressure readings.

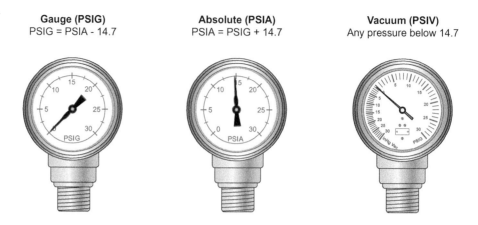

Gauge pressure (psig) pressure measured with respect to Earth's surface at sea level (zero psig).

Absolute pressure (psia) gauge pressure plus atmospheric pressure; pressure referenced to a total vacuum (zero psia).

Vacuum pressure (psiv) any pressure below atmospheric pressure.

Vapor pressure a measure of a substance's volatility and its tendency to form a vapor.

Boiling point the temperature at which liquid physically changes to a gas at a given pressure.

- **Gauge pressure (psig)** is pressure measured with respect to Earth's surface at sea level (zero psig).
- **Absolute pressure (psia)** is gauge pressure plus atmospheric pressure (pressure referenced to a total vacuum, which is zero psia).
- **Vacuum pressure (psiv)** is any pressure below atmospheric pressure. Vacuum is typically measured in inches of mercury (in. Hg) or millimeters of mercury (mmHg).

VAPOR PRESSURE AND BOILING POINT **Vapor pressure** is a measure of a substance's volatility and its tendency to form a vapor. Vapor pressure can be directly linked to the strength of the molecular bonds of a substance.

Molecular bond strength and vapor pressure have an indirect relationship, meaning that the stronger the molecular bond, the lower the vapor pressure. The weaker the molecular bond, the higher the vapor pressure. As heat is added to a liquid, molecular motion and vapor pressure increase.

Vapor pressure and boiling point also have an indirect relationship. If a substance has a low vapor pressure, then its boiling point will be high, and vice versa.

Boiling point is the temperature at which liquid physically changes to a gas at a given pressure (Figure 11.12).

Figure 11.12 Vapor pressure and boiling point.

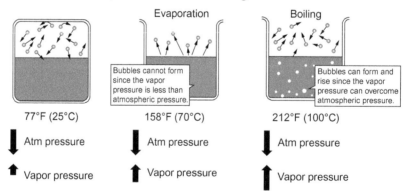

Evaporation conversion of a liquid into a vapor.

When the boiling point of a liquid is reached, the vapor pressure of the liquid becomes slightly greater than the pressure being exerted on the liquid by the surrounding atmosphere. As a result, bubbles form inside the liquid and rapid vaporization begins.

The boiling point of pure water at atmospheric pressure (14.7 psi) is 100 degrees C or 212 degrees F. However, liquids do not have to reach their boiling point in order for the process of **evaporation** (conversion of a liquid into a vapor) to occur. Think of a puddle after a rain shower. Over time, the puddle will evaporate and disappear, especially if the sun comes out, warms the air, and increases the molecular activity of the water.

IMPACT OF PRESSURE ON BOILING POINT System pressure has a direct impact on the boiling point of a substance. As pressure increases, the boiling point increases because the molecules are forced closer together. In addition, increases in pressure also can cause vapors to be forced back into solution.

As pressures decrease, as in the case of a vacuum, boiling points also decrease. By lowering the boiling point using a vacuum, plants can reduce both energy costs and heat-induced molecular damage or product degradation.

Gas Laws

Gas laws are scientific principles that describe the properties of gases and how they will react under different circumstances. The gas laws that process technicians encounter most are Boyle's law, Charles' law, Dalton's law, the General (also called the Combined) Gas Law, and Bernoulli's law (principle).

It is important to remember that gas law calculations require the use of absolute temperature and pressure units.

11.6 Boyle's Law (Pressure-Volume Law)

In the late 1600s, the English scientist Robert Boyle discovered that there is an inverse (opposite) relationship between the pressure of a gas and its volume, if the amount of the sample and temperature are held constant. During his experiments, Boyle learned that if he took a gas sample and doubled the absolute pressure, the volume decreased by 50%. If he decreased the pressure by 50%, the volume doubled. Based on his observations, he created Boyle's law. **Boyle's law** states that the pressure of an ideal gas is inversely proportional to its volume, if the temperature and amount of gas are held constant.

In layman's terms, this means that there is an opposite relationship between pressure and volume. As the pressure on a quantity of gas increases, the volume of the gas decreases. (If you double the pressure, you decrease the volume by half.)

Boyle's law a physics principle stating that, at a constant temperature, as the pressure of a gas increases, the volume of the gas decreases.

Boyle's Law

The pressure of an ideal gas is inversely proportional to its volume, if the temperature is held constant.

The formula for Boyle's law is:

$$P_1 V_1 = P_2 V_2$$

(P = Pressure, V = Volume)

NOTE: One atmosphere is equal to 14.7 psi.

11.7 Charles' Law (Temperature–Volume Law)

Jacques Alexander Cesar Charles was a French physicist. In the early 1800s (when there was much excitement over hot-air balloons), he became interested in the effect of temperature on a volume of gas. To study the effects of temperature, Charles created an apparatus that allowed him to maintain a fixed mass of gas at a constant pressure. He determined from these experiments that as temperature increased, the volume of gas also increased.

Charles' Law

Charles' law states that volume of a gas varies directly with the temperature when the pressure is held constant.

The formula for **Charles' law** is:

$$V_1/T_1 = V_2/T_2$$

(V = Volume, T = Temperature in Kelvin)

Charles' law a physics principle stating that, at constant pressure, the volume of a gas increases as the temperature of the gas increases.

11.8 Dalton's Law (Law of Partial Pressures)

In the early 1800s, John Dalton, an English scientist, discovered the relationship between gases when two or more were mixed together. Dalton determined that the total pressure of a mixture of gases is equal to the sum of the individual partial pressures.

Partial pressure is the pressure that a particular gas, contained in a mixture of gases, would have if it were alone in the container. In other words, it is the pressure that a particular gas would exert if all of the other gases were removed.

The subscripts in Dalton's formula (e.g., P_a) are used here symbolically to represent different gases. In normal applications, the symbolic subscript would be replaced with the actual formula for the gas being referenced. For example, the partial pressure of oxygen (O_2) would be written as P_{O_2} or PO_2.

Dalton's law a physics principle stating that the total pressure of a mixture of gases is equal to the sum of the individual partial pressures.

Dalton's Law

Dalton's law states that the total pressure of a mixture of gases is equal to the sum of the individual partial pressures.

The formula for is:

$$P_{total} = P_a + P_b + P_c \ldots P_n$$

(P = Pressure)

11.9 General (or Combined) Gas Law

Earlier in this chapter, we explained that Boyle's law applies to situations in which the temperature remains constant, and that Charles' law pertains to situations in which pressure remains constant. However, in many settings it is not possible to control pressure or temperature, especially if the gas storage containers (e.g., tanks or bottles) are exposed to harsh environmental conditions such as extreme heat and extreme cold. In order to compensate for pressure and temperature variances, Boyle's law and Charles' law have been combined to form the **General (or Combined) Gas Law**.

General (or Combined) Gas Law relationships among pressure, volume, and temperature in a closed container; pressure and temperature must be in absolute scale ($P1V1/T1 = P2V2/T2$).

The formula for the General (or Combined) Gas Law is: $\dfrac{P_1 V_1}{T_1} = \dfrac{P_2 V_2}{T_2}$

Where:

P = Pressure

V = Volume

T = Temperature in Kelvin

The General (or Combined) Gas Law allows you to calculate what will happen if you change conditions (i.e., pressure, temperature, or volume) given an existing set of conditions. For example, imagine a gas is stored in a container (a fixed volume, that cannot change) at a given temperature (T_1) and pressure (P_1). If you increase the temperature (T_2) on the container, you calculate how much the pressure will change (P_2) using the formula above. V_1 and V_2 would be the same because gas cylinders or storage tanks are rigid and the gas would not be able to change volume.

These principles are important when maintaining control of pressures in plant equipment.

General (or Combined) Gas Law

This law combines the principles of Boyle's law and Charles' law. The formula for the General (or Combined) Gas Law is:

$$\frac{P_1 V_1}{T_1} = \frac{P_2 V_2}{T_2}$$

(P = Pressure, V = Volume, T = Temperature in Kelvin)

11.10 Bernoulli's Law (Bernoulli's Principle)

Daniel Bernoulli was a Dutch-born Swiss mathematician. In the early 1700s, he began formulating theories of fluid flow and aerodynamics (in scientific terms, gases and liquids are both considered fluids). **Bernoulli's law**—the physics principle Bernoulli created through his experiments—states that as the speed of a fluid in a constricted space increases, the pressure inside or exerted by the fluid decreases. To better illustrate this concepts, refer to Figure 11.13. Note the pressure indicated at different points along the pipe.

Bernoulli's law a physics principle stating that as the speed of a fluid in a constricted space increases, the pressure inside the fluid, or exerted by it, decreases.

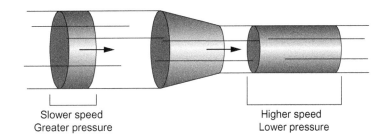

Figure 11.13
As speed increases, pressure decreases. Pressure gauges show the changes in pressure.

Bernoulli's Law

In a flowing liquid or gas, the pressure is least where the speed is greatest. Bernoulli's law (see Figure 11.13) is used in calculating process flows within lines using a restriction such as an orifice plate, venture tube, or other similar device. It is also used in calculating suction flows for centrifugal compressors using these same restrictions.

Figure 11.14 shows water traveling through a section of pipe that narrows gradually. Water enters at the left and flows toward a restriction in the center. The water flow is forced to speed up in the restricted area, causing an exchange of pressure for velocity at the center point. It returns to slower speed and higher pressure after the restriction.

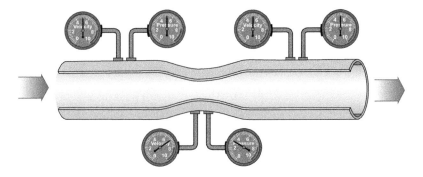

Figure 11.14
Illustration of the speed-pressure relationship.

To understand this principle, think about what happens to your shower curtain when you first turn on the water. The water-air velocity inside the curtain increases, and the pressure inside the shower *decreases* relative to the still air outside the shower. This pressure differential is what causes the shower curtain to be sucked into the tub.

Another practical example of Bernoulli's principle applies to airplanes and the air flow over a foil (or airplane wing). High wind velocity over the top of the wing causes a lower pressure above and a higher pressure on the underside of the wing. This differential creates "lift" of the wing.

Summary

Process technicians encounter physics concepts every day as they open and close valves; check process variables such as pressure, temperature, and flow; and work with different states of matter such as solids, liquids, and gases.

Each state of matter (solid, liquid, or gas) has its own unique characteristics and the potential to undergo a phase change from one form to another (such as from ice to water).

Other concepts with which process technicians need to be familiar include mass and weight, density, elasticity, viscosity, specific gravity, buoyancy, flow, velocity, friction, temperature, and pressure. These measurements are used in almost every piece of equipment or process operations function.

Heat is the transfer of energy from one object to another as a result of the temperature difference between the two objects. Heat can be measured in British thermal units (BTUs) or calories. The different types of heat include specific heat, sensible heat, latent heat, latent heat of fusion, latent heat of vaporization, and latent heat of condensation.

Process technicians also must be familiar with the concept of pressure. Pressure is the amount of force a substance or object exerts over a particular area, and it is usually measured in pounds per square inch (psi).

There is a direct relationship between pressure and boiling point. As pressure increases, so does boiling point, the temperature at which liquid physically changes to a gas at a given pressure.

When working with temperature and pressure, it is important to understand the various laws that describe the characteristics of gases and how they will react under different circumstances. These laws include Boyle's law (pressure–volume law), Charles' law (temperature–volume law), Dalton's law (law of partial pressures), the General (or Combined) Gas Law (a combination of Boyle's law and Charles' law), and Bernoulli's law (the relationship between speed and pressure).

Failure to understand the laws and the principles of physics can result in process problems and safety issues, which could lead to costly or catastrophic events.

Checking Your Knowledge

1. Define the following terms:
    a. buoyancy
    b. density
    c. elasticity
    d. flow
    e. friction
    f. mass
    g. pressure
    h. specific gravity
    i. temperature
    j. velocity
    k. viscosity

2. The phase change in which a solid (e.g., dry ice) changes to a vapor is:
    a. condensation
    b. sublimation
    c. transformation
    d. vaporization

3. The transfer of heat from heat tracing to a pipe is an example of:
    a. conduction
    b. convection
    c. radiation

4. *(True or False)* A hydrocarbon with a specific gravity less than the specific gravity of water will sink to the bottom if the two substances are mixed together.

5. State the freezing point of water for each of the listed temperature scales.
   a. Kelvin
   b. Celsius
   c. Fahrenheit
   d. Rankine

6. Using the formulas found in Table 11.2, make the following temperature conversions (round to the nearest tenth if necessary).
   a. 79 degrees C is equal to _____ K
   b. 68 degrees C is equal to _____ degrees F
   c. 200 degrees F is equal to _____ degrees R
   d. 182 degrees F is equal to _____ degrees C
   e. 102 degrees F is equal to _____ K

7. At what temperature are the Kelvin and Rankine readings the same?

8. The amount of heat required to raise 1 pound of water 1 degree Fahrenheit is a:
   a. calorie
   b. British thermal unit (BTU)
   c. Rankin

9. Atmospheric pressure is 14.7 pounds per square inch (psi). What do we call any pressure lower than atmospheric pressure?
   a. vapor pressure
   b. gauge pressure
   c. vacuum pressure

10. As atmospheric pressure increases, boiling point:
    a. increases
    b. decreases
    c. stays the same

11. List the four states of matter.

12. Describe what is meant by the term *phase change*.

13. Define specific heat, sensible heat, and latent heat.

14. Define convection, conduction, and radiation.

15. According to Boyle's law, what is the relationship between the absolute pressure of a gas and its volume?

16. According to Charles' law, what is the relationship between the absolute temperature of a gas and its volume?

17. According to Dalton's law, what does the total pressure of a mixed gas equal?

18. Boyle's law and Charles' law both pertain to situations in which temperature and pressure remain constant. What gas law compensates for both pressure and temperature differences in settings where it is not possible to keep these variables constant?

19. According to Bernoulli's law, what is the relationship between the speed of a fluid and the pressure it exerts?

20. According to Bernoulli's law, at which point on the diagram below will fluid have greater pressure, point A or point B?

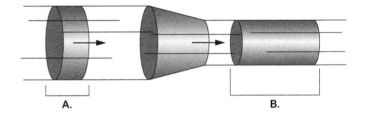

NOTE: Answers to Checking Your Knowledge questions are in Appendix I.

Student Activities

1. Investigate how adding salt to a glass of water can increase the buoyancy of an egg by doing the following:
   a. Obtain a glass of water, a raw egg, a container of table salt, a teaspoon measure, and a stirring spoon.
   b. Form a hypothesis as to how much salt you think it will take to float an egg in a glass of water.
   c. Place the egg in the glass of water (no salt added) and observe whether it floats (If the egg floats, go to Step e. If the egg does not float, go to Step d).
   d. Remove the egg from the glass, add a teaspoon of salt to the water, stir until the salt is dissolved, and then place the egg back in the water to see whether it floats (If the egg floats, go to Step e. If the egg does not float, repeat Step d).
   e. Record your results (i.e., record how much water was in the cup, how many teaspoons of salt it took to make the egg float, and whether your hypothesis was correct).

2. Complete the following phase change experiment and record your results. Wear the appropriate personal protective equipment or safety glasses and gloves to avoid burns.
   a. Fill a small plastic container with 1 cup of water. Place the container in the freezer and then monitor it to see how long it takes for the water to freeze solid. Record your results.
   b. After the water has frozen solid, remove the ice from the cup. (Note: you might have to run the outside of the cup under warm water for a few seconds to release the ice.) Place the ice in a glass bowl at room temperature, and then monitor how long it takes for the ice to melt completely. Record your results.
   c. After the ice has melted, pour the water from the ice into a saucepan. Place the saucepan on the stove and turn the heat on high. Monitor the water to see how long it takes for the water to start forming steam. Record your results.
   Safety Note: After the steam has begun to form, turn the fire off. Do NOT allow an empty pan to continue heating on the stove, as this poses a safety risk and will damage the pan.

3. In an open area or outdoors, divide into small groups. Each group will need 1 empty 20-oz or half-liter plastic water bottle with cap. Students will need to wear safety glasses.
   a. Cap the water bottle. Twist the bottom half of the bottle to compress the system.
   b. Pointing the mouth of the bottle away from students and objects, unscrew the bottle cap.
   c. Discuss the results in your small group, Determine which law was illustrated by the experiment and discuss how it worked.

4. Demonstrate the effects of force and leverage. Materials needed are 1 full copy paper box (or box of equivalent weight and 1 furniture-moving dolly. Have students test the weight of the box. Caution students not to lift the box directly but to tilt it so they can feel its weight. Have them decide how many people they think would be needed to lift and move the box safely across the room. Then insert the dolly foot under the box and carefully tilt it backwards to the balance point. Allow students to move the file cabinet and feel the lighter weight they experience with the box at the balance point. Discuss as a group why this occurs and what are the implications in process work.

5. Divide into four small groups and solve the following equations. Then have each group report on one equation, providing steps they took to calculate the correct answer:
   a. Use the formula for Boyle's law to determine the answer to the following gas problem: If 100 milliliters (mL) of oxygen gas is compressed from 10 atmospheres (atm) of pressure to 50 atm of pressure at a constant temperature, what is the new volume of the oxygen gas?
   b. Use the formula for Charles' law to determine the answer to the following gas problem: If 2 cubic feet of gas at 100 degrees F are heated to 450 degrees F, what is the volume?
   c. Use the formula for Dalton's law to determine the answer to the following gas problem: The total pressure of a sample is 760 millimeters of mercury (mmHg). In clean, dry air at sea level and 0 degrees C, the partial pressure of nitrogen (N_2) is 601 mmHg. If oxygen (O_2) is the only other component, what is the partial pressure of oxygen?
   d. Use the formula for the General Gas Law to determine the answer to the following gas problem: Five cubic feet of a gas at 90 psig and 75 degrees F is compressed to a volume of 2 cubic feet and then heated to a temperature of 250 degrees F. What is the new pressure?

Chapter 12
Basic Chemistry

 Objectives

Upon completion of this chapter, you will be able to:

12.1 Define the application of chemistry in the process industries. (NAPTA Chemistry 1) p. 169

12.2 Define the difference between organic and inorganic chemistry. (NAPTA Chemistry 3) p. 169

12.3 Describe the relationships among molecules, atoms, protons, neutrons, and electrons. (NAPTA Chemistry 2) p. 169

12.4 Explain the difference between chemical properties and physical properties. (NAPTA Chemistry 4) p. 173

12.5 Define and provide examples of the following terms:
hydrocarbon
boiling point
chemical reaction
oxidation/reduction
acid
alkaline
exothermic
endothermic
compounds
miscible
immiscible
mixtures
solutions
homogenous
equilibrium
catalyst (NAPTA Chemistry 5) p. 173

12.6 Describe acidity and alkalinity (caustic), including pH measurement. (NAPTA Chemistry 6, 7) p. 175

Key Terms

Acid—a substance with a pH less than 7 that releases hydrogen (H+) ions when mixed with water, **p. 175**.

Alkaline—having to do with a base (i.e., a substance with a pH greater than 7), **p. 176**.

Atomic number—the number of protons found in the nucleus of an atom, **p. 171**.

Atomic weight—the sum of protons and neutrons in the nucleus of an atom, **p. 171**.

Base—a substance with a pH greater than 7 that releases hydroxyl (OH−) anions when dissolved in water, **p. 176**.

Catalyst—a substance used to facilitate the rate of a chemical reaction without being consumed in the reaction, **p. 174**.

Catalytic cracking—the process of adding heat plus a catalyst to facilitate a chemical reaction, **p. 174**.

Caustic—capable of destroying or eating away human tissue or other materials by chemical action; also a process industries term that refers to a strong base, **p. 176**.

Chemical change—a reaction in which the molecular bonds between atoms of a substance are altered and a new substance is produced, **p. 173**.

Chemical formula—a shorthand symbolic expression that represents the elements in a substance and the number of atoms present in each molecule (e.g., water, H_2O, is two hydrogen atoms and one oxygen atom bonded together), **p. 174**.

Chemical properties—characteristics of elements or compounds that are associated with chemical reactions, **p. 173**.

Chemical reaction—a chemical change or rearrangement of chemical bonds to form a new product, **p. 174**.

Chemical symbol—one- or two-letter abbreviations for elements in the periodic table, **p. 171**.

Chemistry—the science that describes matter, its chemical and physical properties, the chemical and physical changes it undergoes, and the energy changes that accompany those processes, **p. 169**.

Compound—a pure and homogeneous substance that contains atoms of different elements in definite proportions and that usually has properties unlike those of its individual elements, **p. 172**.

Electrons—negatively charged particles that orbit the nucleus of an atom, **p. 171**.

Elements—substances composed of like atoms that cannot be broken down further without changing their properties, **p. 170**.

Endothermic—having to do with a chemical reaction that requires the addition or absorption of energy, **p. 174**.

Equilibrium—a point in a chemical reaction at which the rate of the products forming from reactants is equal to the rate of reactants forming from the products, **p. 174**.

Exothermic—relating to a chemical reaction that releases energy in the form of heat, **p. 174**.

Heterogeneous—having matter with properties that are not the same throughout, **p. 175**.

Homogeneous—relating to matter that is evenly distributed or consisting of similar parts or elements, **p. 172**.

Hydrocarbons—compounds that contain only carbon and hydrogen, **p. 169**.

Immiscible—having to do with liquids that do not form a homogeneous mixture when put together, **p. 175**.

Inorganic chemistry—the study of substances that do not contain carbon, **p. 169**.

Insoluble—describing a substance that does not dissolve in a solvent, **p. 175**.

Miscible—having to do with liquids that form a homogeneous mixture when put together, **p. 175**.

Mixture—two or more substances that are combined together but do not react chemically, **p. 175**.
Molecule—a set of two or more atoms held together by chemical bonds, **p. 171**.
Neutrons—particles without electrical charge, found in the nucleus of an atom, **p. 171**.
Organic chemistry—the study of carbon-containing compounds, **p. 169**.
Periodic table—a chart of all known elements listed in order of increasing atomic number and grouped by similar characteristics, **p. 170**.
pH—a measure of the quantity of hydrogen ions in a solution that can react and indicate whether a substance is an acid or a base, **p. 176**.
Physical change—an event in which the physical properties of a substance (e.g., how it looks, smells, or feels) can be altered, but the change is reversible and a new substance is not produced, **p. 173**.
Physical properties—aspects of an element or compound that are observable and do not pertain to chemical reactions, **p. 173**.
Products—substances that are produced during a chemical reaction, **p. 174**.
Protons—positively charged particles found in the nucleus of an atom, **p. 171**.
Reactants—the starting substances in a chemical reaction, **p. 174**.
Soluble—having to do with a substance that will dissolve in a solvent, **p. 175**.
Solute—the substance that is dissolved in a solvent, **p. 175**.
Solution—a homogeneous mixture of two or more substances, **p. 175**.
Solvent—the substance that is present in a solution in the largest amount, **p. 175**.

12.1 Introduction

Chemistry is a very important part of the process industries. Through chemistry, scientists and process technicians are able to understand various elements and compounds, their proportions, and how they interact with one another in the presence of heat, cold, catalysts, and other variables. By understanding these principles, the process industries are able to produce better products and safer processes.

12.2 Applications of Chemistry in the Process Industries

Chemistry is the science that describes matter, its chemical and physical properties, the chemical and physical changes it undergoes, and the energy changes that accompany those processes. This includes the study of elements, the compounds they form, and the reactions they undergo. It is the study of substances, what they are made of, and how they react.

The field of chemistry is divided into many branches including physical chemistry, organic chemistry, and inorganic chemistry.

Physical chemistry is the branch of chemistry that studies the relationships among the physical properties of substances and their chemical compositions and transformations. **Organic chemistry** is the study of carbon-containing compounds. **Inorganic chemistry** is the study of substances that do not contain carbon. Process technicians focus primarily on organic or inorganic chemistry.

Technicians in the process industries work with a variety of substances, including organic compounds such as **hydrocarbons**—compounds that contain only carbon and hydrogen. Hydrocarbons are naturally found in petroleum, natural gas, and coal.

Chemistry the science that describes matter, its chemical and physical properties, the chemical and physical changes it undergoes, and the energy changes that accompany those processes.

Organic chemistry the study of carbon-containing compounds.

Inorganic chemistry the study of substances that do not contain carbon.

Hydrocarbons compounds that contain only carbon and hydrogen.

12.3 Elements and Compounds

Matter is anything that has mass and takes up space. All matter is composed of building blocks called elements, which can exist individually or in combination.

Elements substances composed of like atoms that cannot be broken down further without changing their properties.

Periodic table a chart of all known elements listed in order of increasing atomic number and grouped by similar characteristics.

Elements are substances composed of like atoms that cannot be broken down further without changing their properties. There are currently 118 known elements, 92 of which occur naturally. Hydrogen, oxygen, nitrogen, and phosphorus are all examples of naturally occurring elements.

Each element has physical and chemical characteristics that distinguish it from other elements. For example, oxygen is a colorless, odorless gas at room temperature, whereas carbon is a black solid at room temperature.

The Periodic Table

To map out all of the elements and their properties, scientists have created a reference chart called the periodic table. The **periodic table** contains all known elements listed in order of increasing atomic number, and grouped by similar characteristics (e.g., how they react with other elements, chemical properties, and physical properties). This table allows elements to be classified into categories of metals, metalloids, and nonmetals (Figure 12.1).

Each row in a periodic table is referred to as a period. The elements are also divided into groups with elements containing similar properties. Each group is given a number.

For example, column 18 (in some sytems called Group 8A) on the far right of the periodic table contains elements known as "noble gases." Noble gases, also called *inert gases*, include helium, neon, argon, krypton, xenon, and radon. All of these are rare gases that exhibit great chemical stability and extremely low reaction rates. By knowing the major properties of a chemical family, it is possible for process technicians to predict the behavior of other elements in that family or group.

Within each group or "family" of elements, each element behaves similarly to all of the other elements in the group, although the level or intensity of the behavior can vary. In the transition from one side of the periodic table to the other, there is a noticeable change in chemical behavior and physical properties. For example, electronegativity increases from left to right, while atomic radius decreases from left to right within a period. To read a periodic

Figure 12.1 The periodic table.
CREDIT: vchalup / Fotolia.

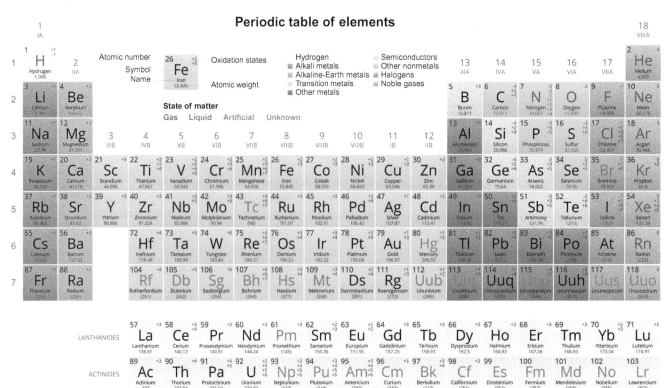

Basic Chemistry 171

table properly, a process technician must be familiar with the designated components in each element of the table. The diagram in Figure 12.2 demonstrates each of these components.

At the top of the table, you will see a column number and under it the Roman numeral group number. The column number makes it easier to find a particular set of elements. In all, there are 18 columns on the periodic table. There are 8 groups (Roman numeral I through VIII, A and B).

Each group represents a unique set of physical and/or chemical properties that each member of that group possesses. Beneath the column and group numbers is an atomic number. The **atomic number** is the number of protons found in the nucleus of an atom (protons will be explained in more detail later in this chapter).

After the atomic number is the name of the element. Beneath the element's name is a one- or two-letter abbreviation called a **chemical symbol**. This abbreviation is a shorthand way to refer to the element when writing chemical equations.

Finally, beneath the element's symbol is the atomic weight of the element. The **atomic weight** is the approximate sum of the number of protons and neutrons found in the nucleus of an atom (also to be explained in more detail later in this chapter).

Characteristics of Atoms

All elements are composed of atoms. Atoms are the smallest particles of an element that still retain the properties and characteristics of that element.

A **molecule** is a set of two or more atoms held together by chemical bonds. A molecule is the smallest unit of a compound that displays the properties of the compound.

Because atoms are too small to see with the naked eye, scientists use drawings and models to represent atoms and their components. Figure 12.3 is a two-dimensional model of a zinc atom. Figure 12.4 is a three-dimensional representation of a carbon atom.

In these drawings, you will see that an atom has a nucleus at its center. Inside the nucleus are positively charged particles called **protons** and uncharged particles called **neutrons**. Surrounding the nucleus are shells that contain negatively charged particles called **electrons**. Every atom has an equal number of protons and electrons, which allows the atom to remain balanced.

Atomic number the number of protons found in the nucleus of an atom.

Chemical symbol one- or two-letter abbreviations for elements in the periodic table.

Atomic weight the sum of protons and neutrons in the nucleus of an atom.

Molecule a set of two or more atoms held together by chemical bonds.

Protons positively charged particles found in the nucleus of an atom.

Neutrons particles without electrical charge, found in the nucleus of an atom.

Electrons negatively charged particles that orbit the nucleus of an atom.

Did You Know?

Chemical symbols containing two letters have only the first letter capitalized.

CREDIT: Steve Young / Fotolia.

Figure 12.2 Sample element (zinc) from a periodic table.
CREDIT: alexlmx / Fotolia.

Figure 12.3 Two-dimensional diagram of a zinc (Zn) atom.
CREDIT: BlueRingMedia/Shutterstock.

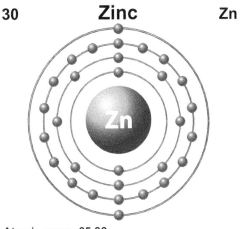

Atomic mass: 65.38
Electron configuration: 2, 8, 18, 2

Figure 12.4 Three-dimensional representation of a carbon atom.

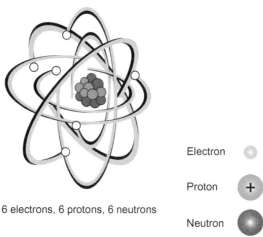

Carbon atomic structure

6 electrons, 6 protons, 6 neutrons

Electron
Proton
Neutron

Within an atom, the positive electrical charge of a proton is equal to the negative charge of an electron. However, the mass of these particles is not equal. A proton has about the same mass as a neutron, but electrons are much lighter. It takes about 1,840 electrons to equal the mass of one proton. Note: For the carbon atom, the atomic number is 6 and the approximate atomic weight is 6 + 6 or 12.

Characteristics of Compounds

With only 118 known elements, how can there be so many different substances? The reason is simple: elements combine together chemically to form compounds.

A **compound** is a pure substance that is **homogeneous** (consists of similar parts of elements). It contains atoms of different elements in definite proportions. It usually has properties unlike those of its constituent elements. Compounds can be broken down into two or more elements by chemical means.

Scientists use chemical formulas to show the kinds of atoms in a compound and their proportions. Table 12.1 lists some examples of common chemical formulas and their constituents.

If you look closely at the formulas in Table 12.1, you will see that several of the substances contain the same components. For example, ethanol, acetic acid, and glucose all contain carbon, hydrogen, and oxygen. What differs are the quantities of each element and how they are attached.

To better illustrate this concept, consider water and hydrogen peroxide.

Water contains two hydrogen atoms and one oxygen atom. Hydrogen peroxide contains two hydrogen atoms and two oxygen atoms. This extra oxygen atom makes hydrogen peroxide a good oxidizer, an effective antiseptic, and a good bleaching agent. Figure 12.5 shows some common chemical substances.

Compound a pure and homogeneous substance that contains atoms of different elements in definite proportions and that usually has properties unlike those of its individual elements.

Homogeneous relating to matter that is evenly distributed or consisting of similar parts or elements.

Table 12.1 Common Chemical Formulas

Substance	Chemical Formula	Constituent Elements of the Compound
Table salt	NaCl	Sodium (Na) and Chlorine (Cl)
Water	H_2O	Hydrogen (H) and Oxygen (O)
Hydrogen peroxide	H_2O_2	Hydrogen (H) and Oxygen (O)
Oxygen gas	O_2	Oxygen (O)
Carbon dioxide gas	CO_2	Carbon (C) and Oxygen (O)
Ethanol	C_2H_5OH	Carbon (C), Hydrogen (H), and Oxygen (O)
Acetic acid	$HC_2H_3O_2$	Carbon (C), Hydrogen (H), and Oxygen (O)
Glucose	$C_6H_{12}O_6$	Carbon (C), Hydrogen (H), and Oxygen (O)

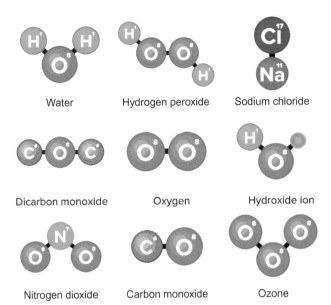

Figure 12.5 Chemical structure of common substances.

CREDIT: AlexOakenman / Fotolia.

12.4 Chemical Versus Physical Properties

In order to understand chemical reactions and the properties of matter, process technicians must be able to determine whether the characteristics of a substance and any changes associated with it are physical or chemical.

Chemical properties are characteristics of elements or compounds that are associated with a chemical reaction (pH, reactivity with water, toxicity, flammability, etc.). **Physical properties** are the aspects of elements or compounds that are observable and do not pertain to a chemical reaction (color, boiling point, density, viscosity, etc.).

During a **physical change**, the physical properties of a substance, such as how it looks, smells, or feels, can change. The change can be reversible, and a new substance is *not* produced. In a **chemical change**, however, the chemical properties of the substance produced will change, the change is not readily reversible, and a new substance *is* produced. Chemical changes often can be identified by changes in color or odor, production of gas bubbles, release or absorption of heat, or creation of a new substance with different properties.

Table 12.2 provides some examples of physical and chemical changes.

One physical property that is important to process technicians in the refining industry is boiling point. Boiling point is the temperature at which a liquid physically changes to a gas at a given pressure. Because gasoline, kerosene, and other organic compounds have different boiling points, scientists are able to separate these substances using heat, pressure, and special equipment.

12.5 Chemical Reactions

Elements react with one another to form compounds. This change, or a rearrangement of bonds to form a new product, is

Chemical properties characteristics of elements or compounds that are associated with chemical reactions.

Physical properties aspects of an element or compound that are observable and do not pertain to chemical reactions.

Physical change an event in which the physical properties of a substance (e.g., how it looks, smells, or feels) can be altered, but the change is reversible and a new substance is not produced.

Chemical change a reaction in which the molecular bonds between atoms of a substance are altered and a new substance is produced.

Did You Know?

Compounds can be broken down into two or more substances by chemical means. Elements cannot be broken down further.

Compound   Element

CREDIT: *(left)* Sebastian Studio/Fotolia. *(right)* jonnysek / Fotolia.

Table 12.2 Examples of Physical and Chemical Changes

Physical Change	Chemical Change
Cutting firewood	Burning firewood to make carbon and heat
Leaves falling from a tree	Composting leaves to form soil
Mining bauxite from the ground	Making aluminum from bauxite

Chemical reaction a chemical change or rearrangement of chemical bonds to form a new product.

Chemical formula a shorthand symbolic expression that represents the elements in a substance and the number of atoms present in each molecule (e.g., water, H_2O, is two hydrogen atoms and one oxygen atom bonded together).

Reactants the starting substances in a chemical reaction.

Products substances that are produced during a chemical reaction.

called a **chemical reaction**. Compounds can be represented symbolically by a chemical formula.

A **chemical formula** is a shorthand, symbolic expression that represents the elements in a substance and the number of atoms present in each molecule.

A chemical reaction has two main components: reactants and products.

Reactants are the starting substances in a chemical reaction. **Products** are the substances that are produced during a chemical reaction. To better illustrate these components, look at the following reaction:

$$\underbrace{\text{Sodium Metal (Na)} + \text{Chlorine Gas (Cl)}}_{\text{Reactants}} \longrightarrow \underbrace{\text{Sodium Chloride (NaCl)}}_{\text{Products}}$$

On the left side of the equation, you will see the reactants. On the right side of the equation, you will see the products.

In this particular example, the reactants include one molecule of sodium (Na) and one molecule of chlorine (Cl). These molecules react to form one molecule of sodium chloride (NaCl), which is table salt.

This process is complex and outside the scope of this textbook. These types of reactions are discussed in more detail in general chemistry courses that are required for most process technology degree programs.

Process products can be affected by changes in conditions or parameters, such as heat, pressure, and component mixture. These all relate to the chemical formula of feedstock (reactants). Conditions must be right for a chemical reaction to occur. During a chemical reaction, the chemical bonds in reactants are broken and reformed so that the end product is different from the original reactants. When the chemical reaction reaches a point at which the rate of the products forming from reactants is equal to the rate of reactants forming from the products, the reaction is in a state of dynamic **equilibrium**. The reaction is taking place, but is unseen other than as the resulting product.

Equilibrium a point in a chemical reaction at which the rate of the products forming from reactants is equal to the rate of reactants forming from the products.

Catalyst a substance used to facilitate the rate of a chemical reaction without being consumed in the reaction.

Catalytic cracking the process of adding heat plus a catalyst to facilitate a chemical reaction.

CATALYSTS In some chemical reactions, a substance called a **catalyst** is introduced to speed up the reaction. When a catalyst is added, the speed of the chemical reaction is increased, but the catalyst itself is not consumed in the reaction.

A catalyst can be in the form of a solid, liquid, or a gas. In the refining industry, catalysts and heat are used to break down large hydrocarbon molecules into smaller molecules. This process of adding heat and catalyst to facilitate a chemical reaction is called **catalytic cracking**. Because the catalyst flows through the process with the process fluids and is not incorporated into the reaction, it can be reclaimed at the end of the cracking process, "regenerated," and used again (as long as some surface area still remains on the catalyst particles themselves).

Exothermic relating to a chemical reaction that releases energy in the form of heat.

Endothermic having to do with a chemical reaction that requires the addition or absorption of energy.

ENDOTHERMIC VERSUS EXOTHERMIC REACTIONS Another characteristic of chemical reactions is their tendency to release energy or require the addition of energy. Chemical reactions that release energy (generate heat) are called **exothermic** reactions. Reactions that require the addition or absorption of energy (often as heat) are called **endothermic** reactions.

OXIDATION/REDUCTION REACTIONS Oxidation/reduction is a reaction process in which oxidation and reduction occur simultaneously. One does not occur without the other. The oxidation process involves loss of electrons; the reduction process involves gain of electrons. Examples of oxidation/reduction in industry are:

- Smelting of ore to get the metal (i.e., gold, zinc, etc.)
- Making galvanized metal—coating metal with zinc
- Electroplating
- Making fertilizers
- Producing compact discs

Mixtures and Solutions

The previous section discussed chemical reactions and how reactants are combined together to form a product that is chemically different from the original reactants. But what is produced if the combined substances do not react chemically?

When substances are combined but do not react chemically, the result is considered to be a **mixture**. If you have ever put sand and water together in a container and mixed them, you are familiar with the concept of mixtures. Sand and water are **heterogeneous**—having matter with properties that are not the same throughout.

In a mixture, each substance retains its own identity. The components of a mixture can be identified easily as two separate substances and can be separated by physical means (e.g., filtering through a screen).

A **solution** is a homogeneous mixture of two or more substances. The individual particles are uniformly distributed throughout the substance.

Solutions have two parts: a solute and a solvent. A **solvent** is the substance present in a solution in the largest amount. A **solute** is the substance being dissolved in the solvent. An example of a solute and a solvent would be salt and water. If you dissolve salt (the solute) in water (the solvent), you form a solution of salt and water. However, during this process the salt and the water do not actually react chemically. Thus, if the water were to evaporate or boil away, the salt would remain intact and chemically unchanged. Figure 12.6 shows an example of how solutes and solvents interact in a solution.

Mixture two or more substances that are combined together but do not react chemically.

Heterogeneous having matter with properties that are not the same throughout.

Solution a homogeneous mixture of two or more substances.

Solvent the substance that is present in a solution in the largest amount.

Solute the substance that is dissolved in a solvent.

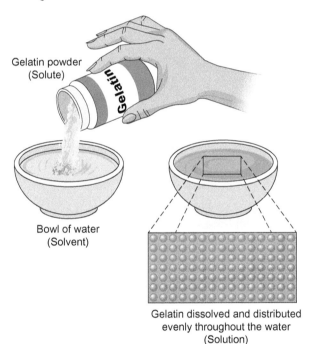

Figure 12.6 Solute molecules evenly distributed among solvent molecules in a solution.

If a solute will dissolve in a solvent, it is considered **soluble**. If a solute will not dissolve in a solvent, it is considered **insoluble**. Oil, for example, is insoluble in water. A solution is considered **miscible** if the solute and solvent remain homogeneous upon standing. **Immiscible** solutions separate upon standing. Examples of immiscible solutions are oil and vinegar or gasoline and water.

Another way homogeneous liquid mixtures can be separated is through the process of distillation. In distillation, the components of the liquid mixture are separated by boiling point. The component or "fraction" with the lowest boiling point is collected first.

Soluble having to do with a substance that will dissolve in a solvent.

Insoluble describing a substance that does not dissolve in a solvent.

Miscible having to do with liquids that form a homogeneous mixture when put together.

Immiscible having to do with liquids that do not form a homogeneous mixture when put together.

12.6 Acids and Bases

Acids and bases are two common compounds that react in water. **Acids** are corrosive substances with a pH less than 7 that release hydrogen (H+) ions when mixed with water.

Acid a substance with a pH less than 7 that releases hydrogen (H+) ions when mixed with water.

Base a substance with a pH greater than 7 that releases hydroxyl (OH−) anions when dissolved in water.

Alkaline having to do with a base (i.e., a substance with a pH greater than 7).

Caustic capable of destroying or eating away human tissue or other materials by chemical action; also a process industries term that refers to a strong base.

pH a measure of the quantity of hydrogen ions in a solution that can react and indicate whether a substance is an acid or a base.

Acids, which turn litmus paper red, neutralize bases and conduct electricity. The term *acidic* is used to refer to materials that are acids. Consumable acids, such as lemon or grapefruit juice, have a sour taste.

Bases are corrosive substances with a pH greater than 7 that usually release hydroxyl (OH−) anions when dissolved in water. Bases, which turn litmus paper blue, react with acids to form salts and water. A strong base is called **alkaline** or **caustic**. The term *basic* is used to refer to materials that are bases. Bath soap is an example of a base.

pH is a reference to the density of hydrogen ions in a solution. The scale for measuring pH is from 0 to 14, with 0 being a strong acid, 14 being a strong base, and 7 being neutral (neither acid nor base). Figure 12.7 shows an example of a pH scale.

It is important for process technicians to understand that the pH scale is based on powers of 10. This means that a substance with a pH of 3 is ten times more acidic than a substance with a pH of 4, and a thousand times more acidic that a substance with a pH of 6.

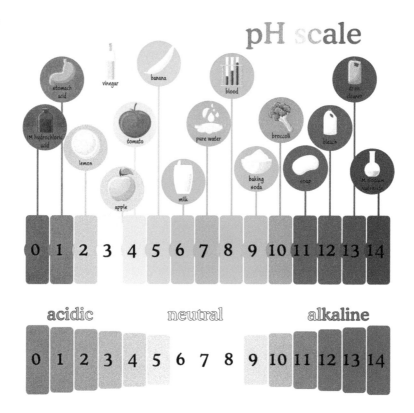

Figure 12.7 Example of a pH scale.
CREDIT: Inna Bigun / Shutterstock.

Summary

Chemistry is an important part of the process industries. It is the study of elements, the compounds they form, and the reactions they undergo. A person who studies chemistry is called a chemist.

The field of chemistry is divided into many fields, including organic and inorganic chemistry. Organic chemistry is the study of carbon-containing compounds called hydrocarbons. Inorganic chemistry is the study of compounds that do not contain carbons.

Chemistry helps process technicians understand various elements and compounds and how they interact with one another. By understanding these principles, the process industries are able to produce better products and safer processes.

All matter is composed of building blocks called elements. Elements are recorded on a reference chart called a periodic table.

Atoms are the smallest particles of an element that still retain the properties and characteristics of that element. Each atom contains positively charged particles called protons, uncharged particles called neutrons, and negatively charged particles called electrons.

A molecule is a set of two or more atoms held together by chemical bonds. A molecule is the smallest unit of a compound that still retains the properties of the compound.

Compounds are homogeneous substances that can be broken down chemically into two or more substances. Scientists use chemical formulas to represent these substances.

In any chemical reaction, there is a reactant (the beginning substance) and a product (the ending substance). Catalysts can be used to facilitate or speed up a chemical reaction, but they are not chemically incorporated into the reaction.

When a chemical reaction reaches a point where the rate of the products forming from reactants is equal to the rate of reactants forming from the products, the reaction is in a state of dynamic equilibrium.

Chemical reactions that release energy (often in the form of heat) are called exothermic reactions. Reactions that absorb energy are called endothermic reactions.

Mixtures occur when two substances combine but do not react chemically. Solutions are a class of mixtures in which the molecules are uniformly distributed (homogeneous). In any given solution, there is a solute (the substance being dissolved) and a solvent (an agent in which the solute is being dissolved). If a solute will not dissolve in a solvent (e.g., oil and water), it is said to be insoluble.

Substances can be an acid, base, or neutral. The scale we use to determine this is called the pH scale. The pH scale ranges from 0 to 14 with 0 being a strong acid, 14 being a strong base, and 7 being neutral.

Checking Your Knowledge

1. Define the following terms:
   a. acid
   b. alkaline
   c. catalyst
   d. chemical reaction
   e. compound
   f. electron
   g. endothermic
   h. equilibrium
   i. exothermic
   j. homogenous
   k. mixture
   l. molecule
   m. neutron
   n. pH
   o. proton
   p. solution

2. Explain the difference between organic and inorganic chemistry.

3. Explain the difference between chemical properties and physical properties.

4. Tell whether each of the following changes is physical or chemical. (Remember: in a chemical change, a new substance is formed.)
   a. ice melting
   b. gasoline burning
   c. iron nails rusting
   d. sugar dissolving into a pitcher of lemonade
   e. vinegar and baking soda combining with each other to form a foamy substance
   f. water boiling
   g. butter melting
   h. aluminum foil being cut in half

5. If a mixture is uniformly distributed, it is said to be:
   a. homogeneous
   b. heterogeneous

6. Which of the following is the smallest unit of an element?
   a. atom
   b. molecule
   c. compound
   d. mixture

7. Every element has a set number of protons, neutrons, and electrons. If the element silicon has 14 protons, how many electrons will it have? Where will these electrons be located?

8. Explain why a scientist or a process technician might use the process of distillation.

9. Using the periodic table shown in Figure 12.1, locate the information for calcium, lead, and chlorine and then use that information to complete the chart.

	Calcium	Lead	Chlorine
Column number (e.g., 1, 2, 3)			
Group number (e.g., IA, IIA)			
Atomic number			
Symbol			
Atomic weight			

10. If each dot represents a molecule, which of the following diagrams best represents a chemical change? (Remember: in a chemical change, a new substance is formed.)

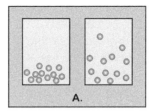
A.

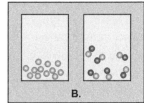
B.

11. On the following equation, name the reactants and the products.

$$NH_3 \; + \; HCl \; \longrightarrow \; NH_4Cl$$
(Ammonia) (Hydrochloric acid) (Ammonium chloride)

12. If you have a packet of hot chocolate mix and you pour it into a cup of boiling water, which item is the solute?

NOTE: Answers to Checking Your Knowledge questions are in Appendix I.

Student Activities

Note: Safety glasses and rubber or light plastic (laboratory) gloves should be worn for all student activities.

1. Using several pH test strips or litmus paper, determine the pH of the following household items and tell whether each one is an acid, base, or neutral.
   a. bleach
   b. hand lotion
   c. carbonated beverage or soda
   d. window cleaner with ammonia
   e. vinegar
   f. tap water
   g. bath soap
   h. rubbing alcohol
   i. mouthwash or toothpaste
   j. shampoo

2. To study the effect of an acid on metal, perform the following steps and identify the change that occurs.
   a. Obtain a paper towel, a glass dish, a shiny penny, and a small bottle of vinegar.
   b. Place the paper towel in the dish and then pour vinegar over it until it is completely saturated.
   c. Place the shiny penny on the paper towel and leave it for 24 hours.
   d. Describe the change(s) that occurred and identify whether the change(s) are chemical or physical.

3. To study the effects of water and vinegar on baking soda, perform the following steps and identify the change that occurs.
   a. Obtain 2 small glass dishes, 2 tsp. baking soda, 1 Tbsp. water, and 1 Tbsp. vinegar.
   b. Place 1 tsp. of baking soda in each dish.
   c. Pour 1 Tbsp. of water into dish #1 and observe what happens.
   d. Pour 1 Tbsp. of vinegar into dish #2 and observe what happens.
   e. Record your results and identify whether the reactions that occurred were physical or chemical.

Chapter 13
Process Drawings

 Objectives

Upon completion of this chapter, you will be able to:

13.1 Describe the purpose or function of process systems drawings. (NAPTA Process Print 1) p. 180

13.2 Identify the common components and information within process systems drawings. (NAPTA Process Print 2) p. 181

13.3 Identify the different drawing types and their uses:

   block flow diagrams (BFDs)
   process flow diagrams (PFDs)
   piping and instrumentation diagrams (P&IDs)
   utility flow diagrams (UFDs)
   plot plan drawing
   engineering flow drawing (EFD)
   electrical diagrams: wiring, schematic
   isometrics (NAPTA Process Print 3) p. 183

Key Terms

Application block—the main part of a drawing that contains symbols and defines elements such as relative position, types of materials, descriptions, and functions, p. 183.

Block flow diagram (BFD)—a very simple drawing that shows a general overview of a process, indicating the parts of a process and their relationships, p. 184.

CAD (or CADD)—computer aided design (and drafting), a software technology that replaces manual drafting with an automated process for design and design documentation, p. 180.

Cutaway—drawing that shows internal elements and structures, p. 190.

Electrical diagram—a drawing that shows electrical components and their relationships, p. 187.

Elevation diagram—a drawing that represents the relationship of equipment to ground level and other structures, p. 191.

Engineering flow diagram—a high-level drawing that represents the overall process, its flow, and unit equipment, and their relationships to each other; similar to a PFD, p. 186.

Equipment location diagram—a drawing that shows the relationship of units and equipment to a facility's boundaries, **p. 191.**

Isometric diagrams—drawings that show objects as they would be seen by the viewer (like a 3-D drawing, the object has depth and height). Isometrics are drawn on graph paper with equipment shown in relation to compass points and line relationships to the equipment, **p. 190.**

Legend—a section of a drawing that explains or defines the information or symbols contained within the drawing (like a legend on a map), **p. 181.**

Loop diagram—a drawing that shows all components and connections between instrumentation and a control room, **p. 191.**

Piping and instrumentation diagram (P&ID)—also called a *process and instrumentation drawing*. A drawing that shows the equipment, piping, and instrumentation of a process in the facility, along with more complex details than a process flow diagram, **p. 184.**

Plot plan diagram—illustration drawn to scale, showing the layout and dimensions of equipment, units, and buildings; also called an *equipment location drawing*, **p. 186.**

Process flow diagram (PFD)—a basic drawing that shows the primary flow of product through a process, using equipment, piping, and flow direction arrows, **p. 184.**

Schematic—a drawing that shows the direction of current flow in a circuit, typically beginning at the power source, **p. 190.**

Simulation—realistic three-dimensional representation, **p. 191.**

Symbol—figures used to designate types of equipment and instrumentation, **p. 181.**

Title block—a section of a drawing (typically located in the bottom right corner) that contains information such as drawing title, drawing number, revision number, sheet number, and approval signatures, **p. 182.**

Utility flow diagrams (UFD)—drawings that show the piping and instrumentation for the utilities in a process, **p. 186.**

Wiring diagram—a drawing that shows electrical components in their relative position in the circuit and all connections in between, **p. 190.**

Introduction

Process drawings are as critical to a process technician as a topographical map (showing hills, streams, and trails) is to a hiker in the deep woods. Just as the hiker must be able to read a map, a process technician must be able to read process systems drawings to understand the process flow and equipment of a process facility.

Process facilities use process drawings to assist with operations, modifications, and maintenance. These drawings visually explain the components of the facility and how they relate to each other in a process unit or the process system. Once drawn by hand, drawings now are often produced using **CAD**. CAD or **CADD**, computer-aided design (and drafting) is a software technology for design and design documentation.

CAD (or CADD) computer aided design (and drafting) is a software technology that replaces manual drafting with an automated process for design and design documentation.

Process technicians are exposed to different types of industrial drawings. The two most commonly used types of drawings are process flow diagrams (PFDs) and piping and instrumentation diagrams (P&IDs). Process drawings that technicians may also use include block flow diagrams (BFDs), utility flow diagrams (UFDs), isometric diagrams, electrical diagrams, and others.

13.1 Purpose of Process Drawings

Process drawings provide a process technician with a visual description and explanation of the processes, equipment, and all auxiliary components in a facility. Each drawing type represents different aspects of the process and levels of detail. Looking at combinations of these drawings gives the process technician a more complete picture of the processes at a facility.

Process drawings allow a process technician to:

- Become familiar with the process(es) in a safe environment
- Understand or explain a process and how it relates to other processes
- Prepare equipment for maintenance

All drawings have three common purposes:

1. They simplify, using common symbols to make complicated processes easy to understand.
2. They provide a visual representation of a concept, describing how all of the parts or components of a system work together. Drawings can quickly and clearly show the details of a system that otherwise might take many written pages to explain.
3. They standardize, using a common set of lines and symbols to represent components. Knowledge of these symbols allows the technician to interpret drawings at any facility.

13.2 Common Components and Process Drawings Information

Process drawings must meet several requirements to be considered proper industrial drawings. These requirements include specific, universal rules about:

- How lines are drawn
- How proportions are used
- What measurements are used
- What components are included
- What the drawing's industrial application is

All process drawings have common components that contain a great deal of useful information. These components include symbols, a legend, a title block, and an application block. Figure 13.1 shows each of these drawing components and where they are located.

Symbols

Symbols are figures used to designate types of equipment and instrumentation. A set of common symbols has been developed to represent actual equipment, piping, instrumentation, and other components. Although some symbols may differ from plant to plant, many are universal with only subtle differences.

The International Society of Automation (ISA) and the American National Standards Institute (ANSI) are the organizations that set the benchmark for drawing symbols, including line, equipment, and instrumentation (and its labeling structure). Figure 13.2 shows a few common equipment symbols.

It is critical that process technicians recognize and understand these symbols. Each equipment chapter in this textbook includes the symbols that represent that equipment on a drawing.

Symbol figures used to designate types of equipment and instrumentation.

Legend

A **legend** is a section of a drawing that explains or defines the information or symbols contained within the drawing (like a legend on a map). Legends include information such as abbreviations, numbers, symbols, and tolerances. Figure 13.3 shows an example of a legend.

Legend a section of a drawing that explains or defines the information or symbols contained within the drawing (like a legend on a map).

182 Chapter 13

Figure 13.1 Labeled components of a process and instrumentation drawing (P&ID). Drawings may have a Notes section with either general notes that apply to the entire drawing or specific (local) notes that apply to a part of the process.

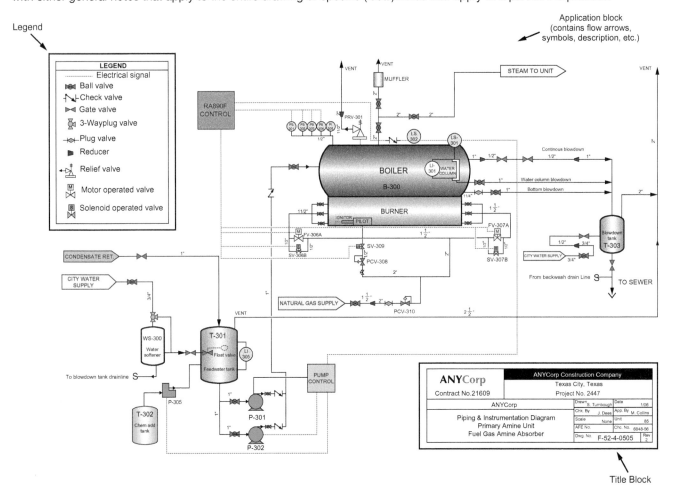

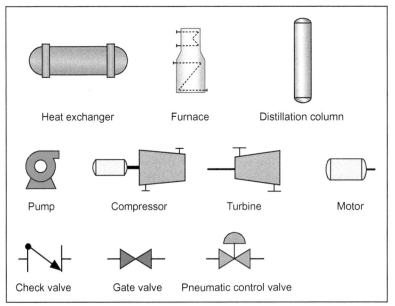

Figure 13.2 Examples of common symbols.

Title block a section of a drawing (typically located in the bottom right corner) that contains information such as drawing title, drawing number, revision number, sheet number, and approval signatures.

Title Block

The **title block** is a section of a drawing (typically located in the bottom right corner) that provides information such as drawing title, company name, drawing number, revision number, sheet number, and approval signatures. Figure 13.4 shows an example of a title block.

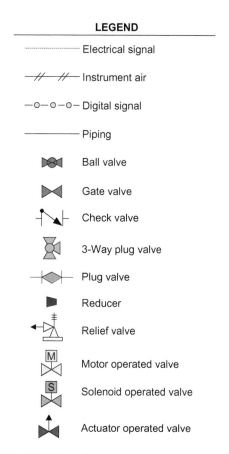

Figure 13.3 Legend example. Symbols can vary slightly from one facility to another.

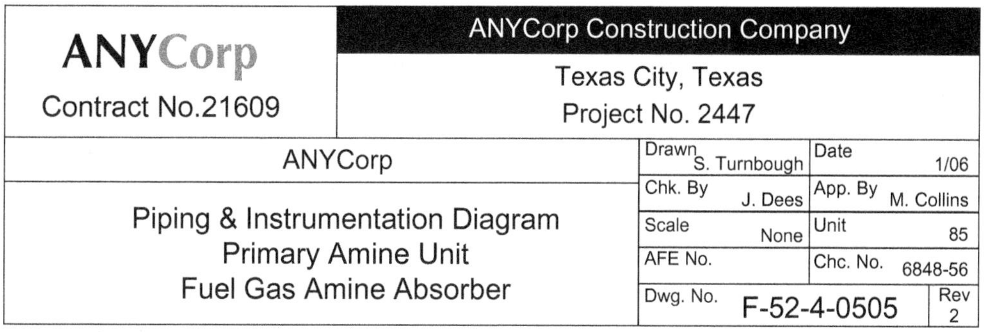

Figure 13.4 Title example.

Application Block

An **application block** is the main part of a drawing; it contains symbols and defines elements such as relative position, types of materials, descriptions, and functions. See Figure 13.1 for an example of an application block.

Application block the main part of a drawing that contains symbols and defines elements such as relative position, types of materials, descriptions, and functions.

13.3 Types of Process Drawings and Their Uses

Process technicians must recognize a wide variety of drawings and understand how to use them.

The most commonly encountered drawings are:

- Block flow diagrams (BFDs)
- Process flow diagrams (PFDs)
- Piping and instrumentation diagrams (P&IDs)
- Utility flow diagrams (UFDs)
- Electrical diagrams
- Isometric drawings

Block Flow Diagrams (BFDs)

Block flow diagram (BFD) a very simple drawing that shows a general overview of a process, indicating the parts of a process and their relationships.

Block flow diagrams are the simplest drawings used in the process industries. They provide a general overview of the process and contain few specifics. BFDs represent sections of a process (drawn as blocks) and use flow arrows to show the order and relationship of each component. Process technicians find BFDs useful for getting a high-level ("big picture") understanding of a process (see Figure 13.5).

Figure 13.5 Sample block flow diagram.

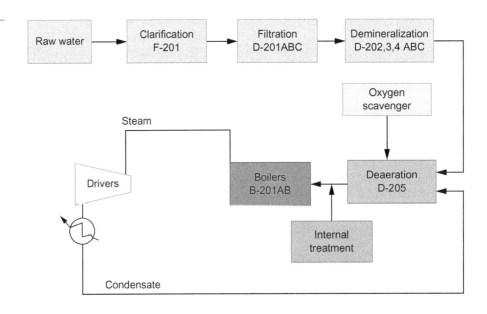

Process Flow Diagrams (PFDs)

Process flow diagram (PFD) a basic drawing that shows the primary flow of product through a process, using equipment, piping, and flow direction arrows.

Process flow diagrams allow process technicians to trace the step-by-step flow of a process. PFDs contain symbols that represent the major pieces of equipment and piping used in a process. Directional arrows show the path of the process.

The process flow is typically drawn from left to right, starting with feedstock or raw materials on the left and ending with finished products on the right. Variables such as temperature and pressure also might be shown at critical points (see Figure 13.6).

Piping and Instrumentation Diagrams (P&IDs)

Piping and instrumentation diagram (P&ID) also called a *process and instrumentation drawing*. A drawing that shows the equipment, piping, and instrumentation of a process in the facility, along with more complex details than a process flow diagram.

Piping and instrumentation diagrams, sometimes referred to as *process and instrumentation drawings*, are similar to process flow diagrams but provide more complex process information.

Typical P&IDs contain equipment, piping and flow arrows, and additional details such as equipment numbers and operating conditions, piping specifications, and instrumentation. Because a series of P&IDs generally compile a plant's equipment and instrumentation, the drawings usually include continuation arrows at the edges of the page on both sides. Utility service continuation arrows might be located at the specific piece of equipment that it is serving. These arrows are a way of communicating where the line comes from and where it goes on the previous or next drawing in the series (see Figure 13.7). For a process technician, a vital part of a P&ID is the instrumentation data. Using a P&ID, a process technician can understand how a product flows through the process and how it can be monitored and controlled. P&IDs are also critical during maintenance tasks, modifications, upgrades, and, most importantly, during the commissioning and startup of a new plant.

Process Drawings **185**

Figure 13.6 Sample process flow diagram.

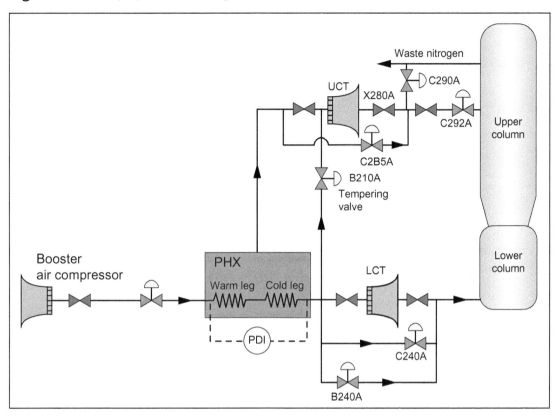

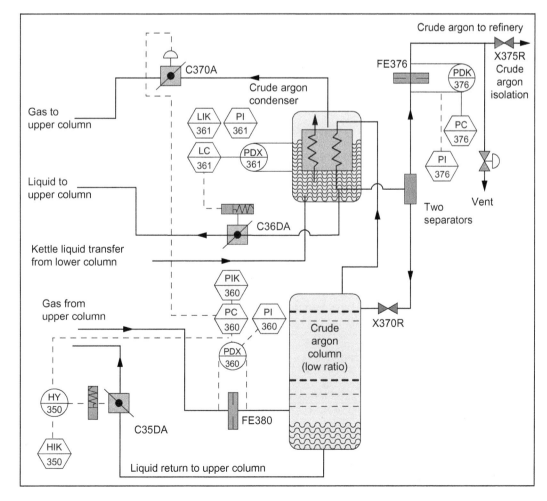

Figure 13.7 Sample piping and instrumentation diagram.

Engineering Flow Diagrams (EFDs)

Engineering flow diagrams are similar to PFDs. These drawings contain all process plant information, including equipment, flows to and from each piece of equipment, and the explanation of those flows and products. EFDs are not commonly used by process technicians.

Engineering flow diagram a high-level drawing that represents the overall process, its flow, and unit equipment and their relationships to each other; similar to a PFD.

Plot Plan Diagrams (PPDs)

A **plot plan diagram** is a high-level diagram, usually an architectural rendition of a process plant showing equipment layout, road positioning, and proposed construction on a defined scale. This drawing is created before the plant, or an addition to the plant, exists. PPDs are not regularly used by process technicians.

Plot plan diagram illustration drawn to scale, showing the layout and dimensions of equipment, units, and buildings; also called an *equipment location drawing*.

Utility Flow Diagrams (UFDs)

Utility flow diagrams provide process technicians with a P&ID-type view of the utilities used for a process. UFDs represent the way utilities connect to the process equipment, piping, and main instrumentation used to operate those utilities (see Figure 13.8).

Utility flow diagrams (UFD) drawings that show the piping and instrumentation for the utilities in a process.

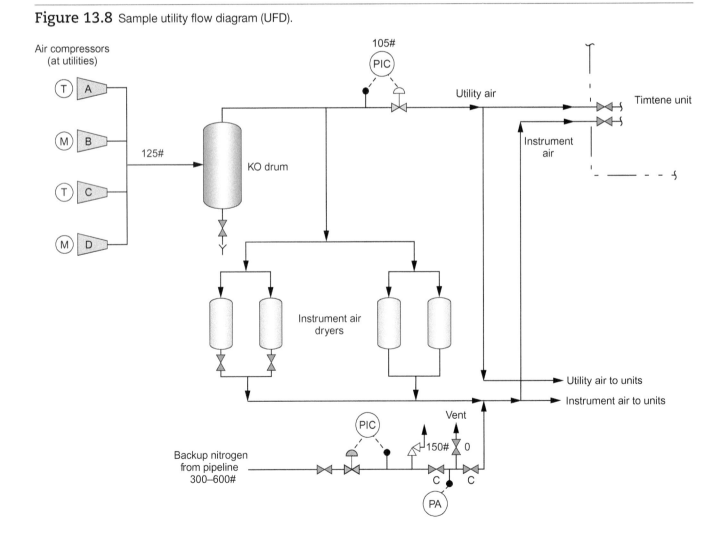

Figure 13.8 Sample utility flow diagram (UFD).

Typical utilities shown on a UFD include the following:

- Steam
- Condensate
- Cooling water
- Potable water
- Instrument air
- Plant air
- Nitrogen
- Fuel gas

Electrical Diagrams

All process facilities rely on electricity, so process technicians must understand electrical system usage and its relationship to the process equipment. **Electrical diagrams** help process technicians understand power transmission and its relation to the process. A firm understanding of these relationships is critical when performing lockout and tagout procedures (i.e., control of hazardous energy) and when monitoring electrical measurements.

Most process technicians (operators) will not work with electrical diagrams on a regular basis. Many companies have regulations that limit the level of electrical current with which the operator can interface. To interface with the electric room operations, the operator is required to have special training. Other electrical maintenance functions fall to an electrical maintenance employee.

Electrical diagrams show components and their relationships, including the following:

- Switches used to stop, start, or change the flow of electricity in a circuit
- Power sources provided by transmission lines, generators, or batteries
- Loads, the components that actually use the power
- Coils or wire used to increase the voltage of a current
- Inductors (coils of wire that generate a magnetic field and can create a brief current in the opposite direction of the original current); can be used for surge protection
- Transformers used to make changes in electrical power by means of electromagnetism
- Resistors, coils of wire used to provide resistance in a circuit
- Contacts used to join two or more electrical components
- Interlocks, jogging circuits that prevent forward and reverse motors from starting at the same time; see Figure 13.9

(Text continues on p. 190.)

Electrical diagram   a drawing that shows electrical components and their relationships.

Figure 13.9 Sample of an interlock logic system.

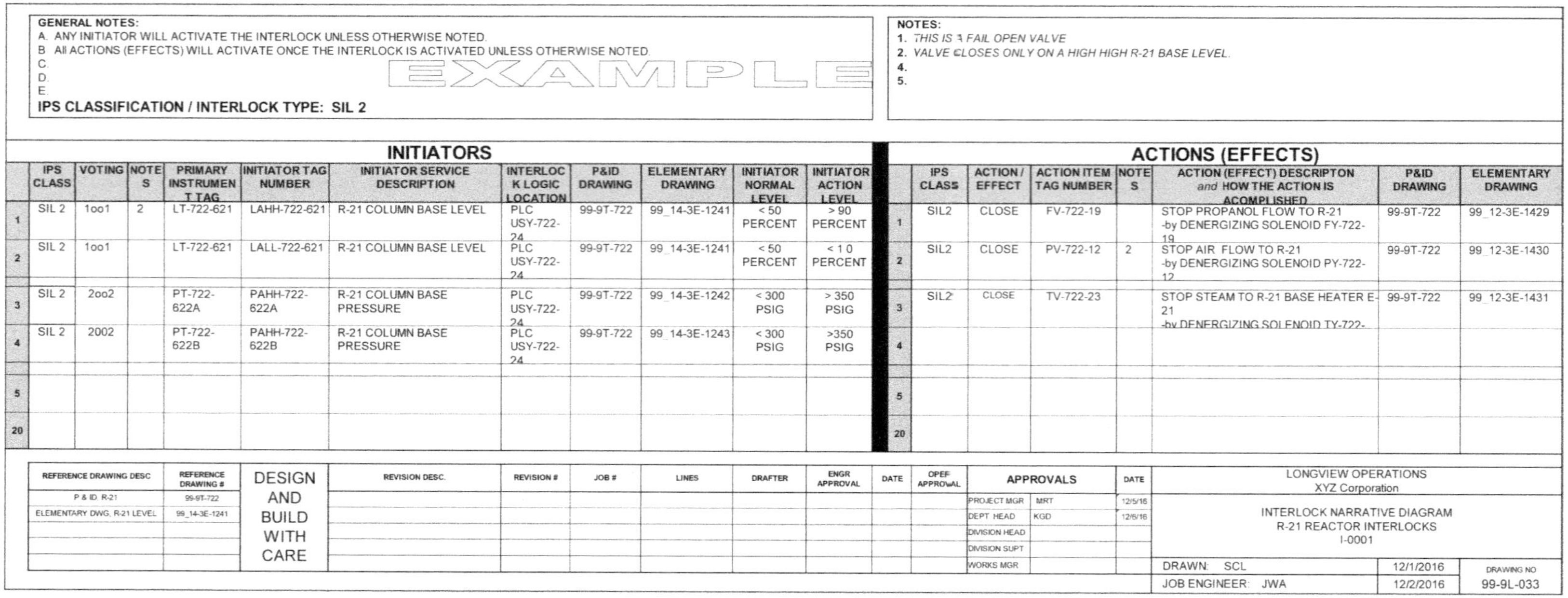
GENERAL NOTES:
A. ANY INITIATOR WILL ACTIVATE THE INTERLOCK UNLESS OTHERWISE NOTED.
B. All ACTIONS (EFFECTS) WILL ACTIVATE ONCE THE INTERLOCK IS ACTIVATED UNLESS OTHERWISE NOTED.
C.
D.
E.
IPS CLASSIFICATION / INTERLOCK TYPE: SIL 2
EXAMPLE
NOTES:
1. THIS IS A FAIL OPEN VALVE
2. VALVE CLOSES ONLY ON A HIGH HIGH R-21 BASE LEVEL.
4.
5.
INITIATORS
IPS CLASS
VOTING
NOTES
PRIMARY INSTRUMENT TAG
INITIATOR TAG NUMBER
INITIATOR SERVICE DESCRIPTION
INTERLOCK LOGIC LOCATION
P&ID DRAWING
ELEMENTARY DRAWING
INITIATOR NORMAL LEVEL
INITIATOR ACTION LEVEL
1 SIL 2 1oo1 2 LT-722-621 LAHH-722-621 R-21 COLUMN BASE LEVEL PLC USY-722-24 99-9T-722 99_14-3E-1241 < 50 PERCENT > 90 PERCENT
2 SIL 2 1oo1 LT-722-621 LALL-722-621 R-21 COLUMN BASE LEVEL PLC USY-722-24 99-9T-722 99_14-3E-1241 < 50 PERCENT < 1 0 PERCENT
3 SIL 2 2oo2 PT-722-622A PAHH-722-622A R-21 COLUMN BASE PRESSURE PLC USY-722-24 99-9T-722 99_14-3E-1242 < 300 PSIG > 350 PSIG
4 SIL 2 2002 PT-722-622B PAHH-722-622B R-21 COLUMN BASE PRESSURE PLC USY-722-24 99-9T-722 99_14-3E-1243 < 300 PSIG >350 PSIG
5
20
ACTIONS (EFFECTS)
IPS CLASS
ACTION / EFFECT
ACTION ITEM TAG NUMBER
NOTES
ACTION (EFFECT) DESCRIPTON and HOW THE ACTION IS ACOMPLISHED
P&ID DRAWING
ELEMENTARY DRAWING
1 SIL2 CLOSE FV-722-19 STOP PROPANOL FLOW TO R-21 -by DENERGIZING SOLENOID FY-722-19 99-9T-722 99_12-3E-1429
2 SIL2 CLOSE PV-722-12 2 STOP AIR FLOW TO R-21 -by DENERGIZING SOLENOID PY-722-12 99-9T-722 99_12-3E-1430
3 SIL2 CLOSE TV-722-23 STOP STEAM TO R-21 BASE HEATER E-21 -by DENERGIZING SOLENOID TY-722- 99-9T-722 99_12-3E-1431
4
5
20
REFERENCE DRAWING DESC
REFERENCE DRAWING #
P & ID R-21 99-9T-722
ELEMENTARY DWG. R-21 LEVEL 99_14-3E-1241
DESIGN AND BUILD WITH CARE
REVISION DESC.
REVISION #
JOB #
LINES
DRAFTER
ENGR APPROVAL
DATE
OPER APPROVAL
APPROVALS
DATE
PROJECT MGR MRT 12/5/16
DEPT HEAD KGD 12/6/16
DIVISION HEAD
DIVISION SUPT
WORKS MGR
LONGVIEW OPERATIONS
XYZ Corporation
INTERLOCK NARRATIVE DIAGRAM
R-21 REACTOR INTERLOCKS
I-0001
DRAWN: SCL 12/1/2016
JOB ENGINEER: JWA 12/2/2016
DRAWING NO
99-9L-033

SIS INTERLOCK (SINGLE ONLY!)

A.

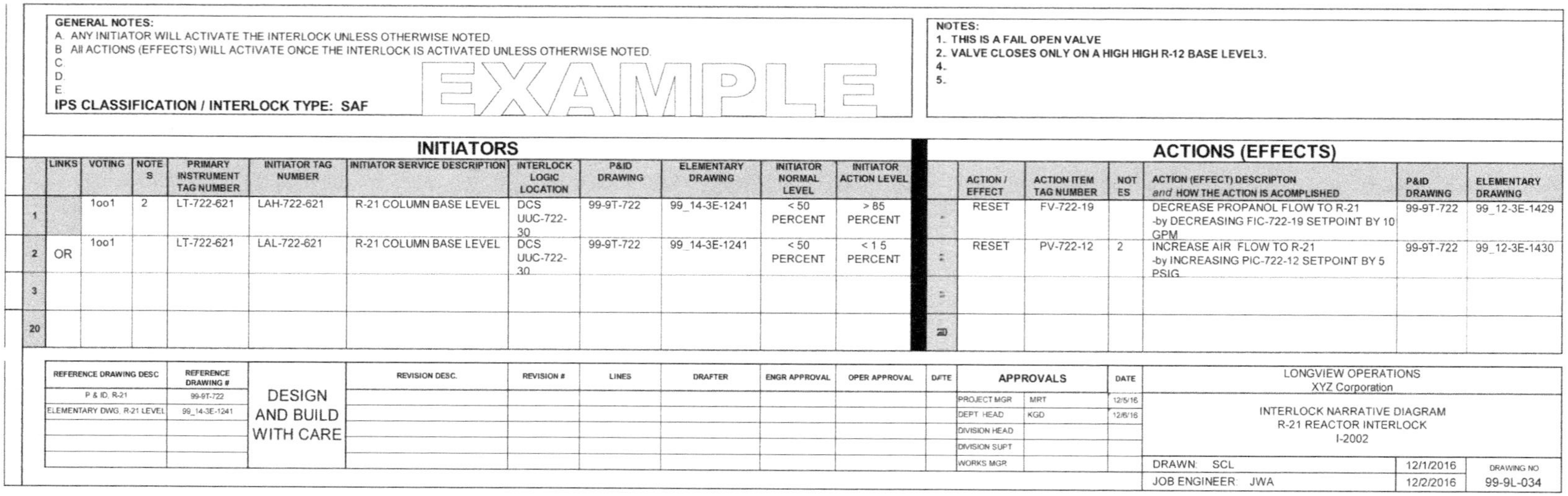
GENERAL NOTES:
A. ANY INITIATOR WILL ACTIVATE THE INTERLOCK UNLESS OTHERWISE NOTED.
B. All ACTIONS (EFFECTS) WILL ACTIVATE ONCE THE INTERLOCK IS ACTIVATED UNLESS OTHERWISE NOTED.
C.
D.
E.
IPS CLASSIFICATION / INTERLOCK TYPE: SAF
EXAMPLE
NOTES:
1. THIS IS A FAIL OPEN VALVE
2. VALVE CLOSES ONLY ON A HIGH HIGH R-12 BASE LEVEL3.
4.
5.
INITIATORS
LINKS
VOTING
NOTES
PRIMARY INSTRUMENT TAG NUMBER
INITIATOR TAG NUMBER
INITIATOR SERVICE DESCRIPTION
INTERLOCK LOGIC LOCATION
P&ID DRAWING
ELEMENTARY DRAWING
INITIATOR NORMAL LEVEL
INITIATOR ACTION LEVEL
1 1oo1 2 LT-722-621 LAH-722-621 R-21 COLUMN BASE LEVEL DCS UUC-722-30 99-9T-722 99_14-3E-1241 < 50 PERCENT > 85 PERCENT
2 OR 1oo1 LT-722-621 LAL-722-621 R-21 COLUMN BASE LEVEL DCS UUC-722-30 99-9T-722 99_14-3E-1241 < 50 PERCENT < 1 5 PERCENT
3
20
ACTIONS (EFFECTS)
ACTION / EFFECT
ACTION ITEM TAG NUMBER
NOTES
ACTION (EFFECT) DESCRIPTON and HOW THE ACTION IS ACOMPLISHED
P&ID DRAWING
ELEMENTARY DRAWING
RESET FV-722-19 DECREASE PROPANOL FLOW TO R-21 -by DECREASING FIC-722-19 SETPOINT BY 10 GPM 99-9T-722 99_12-3E-1429
RESET PV-722-12 2 INCREASE AIR FLOW TO R-21 -by INCREASING PIC-722-12 SETPOINT BY 5 PSIG 99-9T-722 99_12-3E-1430
REFERENCE DRAWING DESC
REFERENCE DRAWING #
P & ID R-21 99-9T-722
ELEMENTARY DWG. R-21 LEVEL 99_14-3E-1241
DESIGN AND BUILD WITH CARE
REVISION DESC.
REVISION #
LINES
DRAFTER
ENGR APPROVAL
OPER APPROVAL
DATE
APPROVALS
DATE
PROJECT MGR MRT 12/5/16
DEPT HEAD KGD 12/6/16
DIVISION HEAD
DIVISION SUPT
WORKS MGR
LONGVIEW OPERATIONS
XYZ Corporation
INTERLOCK NARRATIVE DIAGRAM
R-21 REACTOR INTERLOCK
I-2002
DRAWN: SCL 12/1/2016
JOB ENGINEER: JWA 12/2/2016
DRAWING NO
99-9L-034

NON SIS INTERLOCK DWG SINGLE

B.

GENERAL NOTES:
A. ANY INITIATOR WILL ACTIVATE THE INTERLOCK UNLESS OTHERWISE NOTED.
B. ALL ACTIONS (EFFECTS) WILL ACTIVATE ONCE THE INTERLOCK IS ACTIVATED UNLESS OTHERWISE NOTED.
C. EMERGENCY SHUT DOWN (ESD) INTERLOCKS ARE USED TO MANUALLY INITIATE A PROCESS SHUTDOWN DUE TO AN EXTERNAL EMERGENCY (LOSS OF STEAM, LOSS OF FEEDSTOCK)
D.
E.

IPS CLASSIFICATION / INTERLOCK TYPE: SEE BELOW (SAF = SAFETY, PRC = PROCESS CONTROL, EPP = EQUIPMENT/PROPERTY PROTECTION, ESD = EMERGENCY SHUT DOWN)

EXAMPLE

NOTES:
1. THIS IS A FAIL OPEN VALVE
2. VALVE CLOSES ONLY ON A HIGH HIGH R-12 BASE LEVEL3.
3. FLOW MUST BE LOW FOR >15 SECONDS TO TRIGGER INITIATOR
4. CONTROLLER REVERTS TO -3% OUTPUT AFTER 15 MINUTES
5.

INITIATORS

IPS CLASS / INTERLOCK TYPE	INTERLOCK NUMBER	INTERLOCK NAME	LINKS	VOTING	NOTES	PRIMARY INSTR TAG NUMBER	INITIATOR TAG NUMBER	INITIATOR SERVICE DESCRIPTION	INTERLOCK LOGIC LOCATION	P&ID DRAWING	ELEMENTARY DRAWING	INITIATOR NORMAL LEVEL	INITIATOR ACTION LEVEL
SAF	TX99 1-2002	R-21 LEVEL SHUT DOWN		1oo1	2	LT-722-621	LAH-722-621	R-21 COLUMN BASE LEVEL	DCS UUC-722-30	99-9T-722	99_14-3E-1241	<50 PERCENT	>85 PERCENT
			OR	1oo1		LT-722-621	LAL-722-621	R-21 COLUMN BASE LEVEL	DCS UUC-722-30	99-9T-722	99_14-3E-1241	<50 PERCENT	<15 PERCENT
PRC	TX99 1-2003	R-21 PRESSURE SHUT DOWN		2oo2		PT-722-622A	PAH-722-622A	R-21 COLUMN BASE PRESSURE	DCS UUC-722-30	99-9T-722	99_14-3E-1242	<300 PSIG	>325 PSIG
			AND	2oo2		PT-722-622B	PAH-722-622B	R-21 COLUMN BASE PRESSURE	DCS UUC-722-30	99-9T-722	99_14-3E-1243	<300 PSIG	>325 PSIG
			OR	2oo2		PT-722-622A	PAL-722-622A	R-21 COLUMN BASE PRESSURE	DCS UUC-722-30	99-9T-722	99_14-3E-1242	>250 PSIG	<200 PSIG
			AND	2oo2		PT-722-622B	PAL-722-622B	R-21 COLUMN BASE PRESSURE	DCS UUC-722-30	99-9T-722	99_14-3E-1243	>250 PSIG	<200 PSIG
			OR	1oo1		PT-722-623	PAHH-722-623	R-21 COLUMN TOP PRESSURE	DCS UUC-722-30	99-9T-722	99_14-3E-1244	<275 PSIG	>290 PSIG
EPP	TX99 1-2004	R-21 BASE HEATER E-21 SHUT DOWN		1oo1	3	FT-722-624	FAL-722-624	R-21 BASE HEATER E-21 FEED FLOW	DCS UUC-722-30	99-9T-722	99_14-3E-1245	<50 GPM	<25 GPM
			OR			TT-722-23	TAH-722-23	R-21 COLUMN BASE TEMPERATURE	DCS UUC-722-30	99-9T-722	99_14-3E-1246	<325 DEG F	>350 DEG F
ESD	TX99 1-2005	R-21 EMERGENCY SHUT DOWN		1oo1		HS-722-621A		R-21 COLUMN EMERGENCY SHUT DOWN	DCS UUC-722-30	99-9T-722	99_14-3E-1247	NOT ACTIVATED	ACTIVATED
						HS-722-621B		R-21 COLUMN EMERGENCY SHUT DOWN	PANEL BOARD PUSH BUTTON	99-9T-722	99_14-3E-1248	NOT ACTIVATED	ACTIVATED

ACTIONS (EFFECTS)

#	ACTION / EFFECT	ACTION ITEM TAG NUMBER	NOTES	ACTION (EFFECT) DESCRIPTION and HOW THE ACTION IS ACCOMPLISHED	P&ID DRAWING	ELEMENTARY DRAWING
1	RESET	FV-722-19		DECREASE PROPANOL FLOW TO R-21 -by DECREASING FIC-722-19 SETPOINT BY 10 GPM	99-9T-722	99_12-3E-1429
2	RESET	PV-722-12	2	INCREASE AIR FLOW TO R-21 -by INCREASING PIC-722-12 SETPOINT BY 5 PSIG	99-9T-722	99_12-3E-1430
7	CLOSE	FV-722-19		STOP PROPANOL FLOW TO R-21 -by DENERGIZING SOLENOID FY-722-19	99-9T-722	99_12-3E-1429
8	CLOSE	PV-722-12		STOP AIR FLOW TO R-21 -by DENERGIZING SOLENOID PY-722-12	99-9T-722	99_12-3E-1430
9	CLOSE	TV-722-23		STOP STEAM TO R-21 BASE HEATER E-21 -by DENERGIZING SOLENOID TY-722-23	99-9T-722	99_12-3E-1431
13	CLOSE	TV-722-23		STOP STEAM TO R-21 BASE HEATER E-21 -by DENERGIZING SOLENOID TY-722-23	99-9T-722	99_12-3E-1431
18	CLOSE	FV-722-19		STOP PROPANOL FLOW TO R-21 -by DENERGIZING SOLENOID FY-722-19	99-9T-722	99_12-3E-1429
19	CLOSE	PV-722-12		STOP AIR FLOW TO R-21 -by DENERGIZING SOLENOID PY-722-192	99-9T-722	99_12-3E-1430
20	CLOSE	TV-722-23		STOP STEAM TO R-21 BASE HEATER E-21 -by DENERGIZING SOLENOID TY-722-23	99-9T-722	99_12-3E-1431
21	CLOSE	TV-722-24		STOP PROPIONALDEHYDE FLOW FROM R-21 -by DENERGIZING SOLENOID FY-722-24	99-9T-722	99_12-3E-1432
22	OPEN	TV-722-25	1, 4	FULL COOLING TO FINAL PRODUCT EXCH EX-21A -by PLACING TIC-722-25 IN MAN MODE & 105% OUTPUT	99-9T-722	99_12-3E-1431

LONGVIEW OPERATIONS
XYZ Corporation
INTERLOCK NARRATIVE DIAGRAM
R-21 REACTOR INTERLOCKS
I-2002, I-2003, I-2004, I-2005

DRAWING NO: 99-9L-035 REV 0

NON SIS INTERLOCK MULTIPLE

C.

Two specific types of electrical diagrams are wiring diagrams and schematics.

Wiring diagram a drawing that shows electrical components in their relative position in the circuit and all connections in between.

WIRING DIAGRAMS **Wiring diagrams** show electrical components in their relative position in the circuit and all connections between components. Process technicians use wiring diagrams to determine specific information about electrical components, how they are connected, and the physical location of these components (see Figure 13.10).

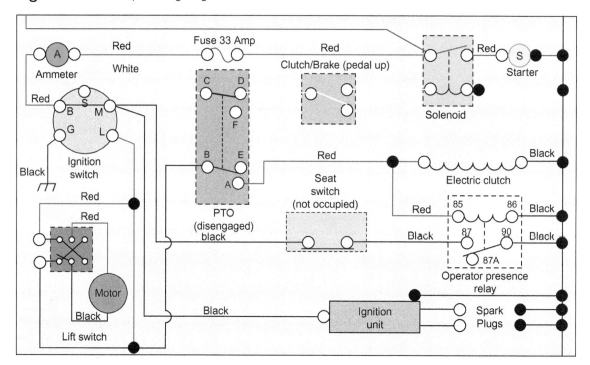

Figure 13.10 Sample wiring diagram.

Single line electrical diagrams illustrate the path of electricity from any distribution source (such as a hydroelectric power plant or substation) to a distribution point (such as a substation, a motor control center, or a distribution panel).

Schematic a drawing that shows the direction of current flow in a circuit, typically beginning at the power source.

SCHEMATICS **Schematics** show the direction of current flow in a circuit, typically beginning at the power source. Process technicians use schematics to visualize how current flows between two or more circuits. Schematics also help electricians detect potential trouble spots in a circuit.

Isometric Drawings

Isometric drawings drawings that show objects as they would be seen by the viewer (like a 3-D drawing, the object has depth and height). Isometrics are drawn on graph paper with equipment shown in relation to compass points and line relationships to the equipment.

Cutaway drawing that shows internal elements and structures.

Isometric drawings show objects, such as equipment, as they would appear to the viewer. In other words, they are like a 3-D drawing that appears to come off the page (see Figure 13.11). These drawings are created using points of the compass to show equipment direction as it would be placed in the field. Isometric drawings also might contain **cutaway** views to show the inner workings of an object.

Isometric drawings show the three sides of the object that can be seen, with the object appearing at a 30-degree angle with respect to the viewer. All vertical lines appear vertical and are parallel to one another. All horizontal lines appear at a 30-degree angle and are parallel to one another. Lines and vessels are drawn to compass direction. A compass marked with north is always on the diagram. It is assumed the other cardinal points can be figured out. On isometric drawings, up relates to elevation. A vertically placed line means the line rises upward. Horizontal planes are shown at 30 degrees. Learning these guidelines will help the student understand the drawing relationship of height and direction of piping and equipment.

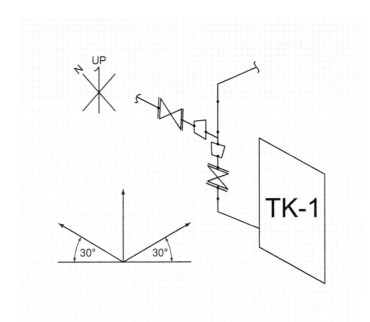

Figure 13.11 Sample isometric drawing. These are often done on graph paper. The compass indicates north.

An isometric **simulation** typically is used only during new unit construction. Process technicians are rarely exposed to isometric drawings unless a new unit is being built. However, such drawings can prove useful to new process technicians as they learn to identify equipment and understand its inner workings.

Other Drawings

Along with the drawings already mentioned, process technicians might encounter other types of drawings, such as:

- **Elevation diagrams** that represent the relationship of equipment to ground level and to other structures
- **Equipment location diagrams** that show the relationship of units and equipment to a facility's boundaries
- **Loop diagrams** that show all components and connections between a specific instrumentation loop, including its interface with a control room.

Simulation realistic three-dimensional representation.

Elevation diagram a drawing that represents the relationship of equipment to ground level and other structures.

Equipment location diagram a drawing that shows the relationship of units and equipment to a facility's boundaries.

Loop diagram a drawing that shows all components and connections between instrumentation and a control room.

Summary

Process drawings provide process technicians with visual descriptions and explanations of processes, equipment, and other important items in a facility. Without process drawings, it can be difficult for process technicians to understand and operate processes. It would be even more difficult to make repairs.

Each of the many different types of process drawings represents a different aspect of the process and different levels of detail. Each contains a title block, application block, symbols, and legend.

The purpose of diagrams and drawings is to simplify, to explain, and to standardize information.

The two most common types of drawings are process flow diagrams (PFDs) and piping and instrumentation diagrams (P&IDs). Drawings show directional process flow, equipment used in processes, interrelationships of parts of processes, and more. They can help the process technician identify points to monitor in a process and avoid potential problems in process units. Looking at combinations of these drawings gives the process technician a more complete picture of the processes at a facility.

Checking Your Knowledge

1. Define the following terms:
   a. block flow diagram (BFD)
   b. electrical diagram
   c. elevation diagram
   d. equipment location diagram
   e. isometric diagram
   f. plan drawing
   g. engineering flow drawing
   h. loop diagram
   i. piping and instrumentation diagram (P&ID)
   j. symbol
   k. utility flow diagram (UFD)
   l. application block

2. What are the two most common types of drawings in the process industries?

3. Which of the following drawing types is the simplest?
   a. PFD
   b. BFD
   c. P&ID
   d. UFD

4. Name at least three components common to all process drawings.

5. What is a PFD?
   a. piping flow diagram
   b. process fixtures diagram
   c. pipe fixture drawing
   d. process flow diagram

6. Where are continuation arrows found on a drawing, and what do they represent?

7. What is a vital part of a PFD?
   a. blocks indicating a part of the process
   b. pressure and temperature variable listings
   c. elevation markers
   d. utility costs

8. UFDs are most similar to what other type of drawing used in a process?
   a. BFD
   b. schematics
   c. P&ID
   d. PFD

9. A P&ID typically includes which of the following items? Select all that apply.
   a. instrumentation
   b. 30-degree angle perspective
   c. piping specifications
   d. equipment numbers
   e. piping and flow arrows

NOTE: Answers to Checking Your Knowledge questions are in Appendix I.

Student Activities

1. Using a P&ID provided, identify the component blocks. Label each part, and discuss.
2. Match the symbol to the appropriate equipment name.

Symbol	Equipment Name
(1)	a. Motor and compressor
(2)	b. Furnace
(3)	c. Motor
(4)	d. Pump
(5)	e. Turbine
(6)	f. Gate
(7)	g. Heat exchanger
(8)	h. Pneumatic control valve

PART 4 Equipment Used in Process Technology

Chapter 14
Piping and Valves

Objectives

Upon completion of this chapter, you will be able to:

14.1 Describe the purpose and function of piping and valves in the process industries. (NAPTA Piping 1) p. 195

14.2 Identify the different materials used to manufacture piping and valve components. (NAPTA Piping 2) p. 196

14.3 Identify the different types of piping and valve connecting methods. (NAPTA Piping 3) p. 196

14.4 Identify the different types of pipe fittings used in the industry and their application. (NAPTA Piping 4) p. 198

14.5 Identify the different types of valves used in the industry and their application. (NAPTA Piping 5) p. 199

14.6 Discuss the hazards associated with the improper operation of a valve. (NAPTA Piping 6) p. 206

14.7 Describe the monitoring and maintenance activities associated with piping and valves. (NAPTA Piping 7, 8) p. 206

Key Terms

Alloy—a compound mixture composed of two or more metals or a mixture of a metal and another element that are mixed together when molten to form a solution; not a chemical compound, **p. 196.**

Ball valve—a flow-regulating device that uses a flow-control element shaped like a hollowed-out ball, attached to an external handle, to increase or decrease flow; a ball valve requires only a quarter turn to go from fully open to fully closed, **p. 200.**

Braze—to join two pieces of metal together by melting an alloy (filler) that bonds the metals together; a brazed joint can be stronger than the metals but is not stronger than a welded joint, **p. 198.**

Butt weld—a type of weld used to connect two pipes of the same diameter that are butted against each other, **p. 198.**

Butterfly valve—a flow-regulating device that uses a disc-shaped flow-control element to increase or decrease flow; requires a quarter turn to go from fully open to fully closed, **p. 201.**

Check valve—a type of valve that allows flow in only one direction and is used to prevent reversal of flow in a pipe, **p. 201**.

Diaphragm valve—a flow-regulating device that uses a flexible, chemical resistant, rubber-type diaphragm attached to a stem that closes onto a weir located in the valve, **p. 202**.

Fitting—a piping system component used to connect two or more pieces of pipe together, **p. 198**.

Flange—a type of pipe connection, glued, welded, or threaded, consisting of two matching circular plates that are joined together with bolts, **p. 197**.

Gasket—a flexible material placed between the two surfaces of a flange to seal against leaks, **p. 197**.

Gate valve—a positive shutoff valve using a gate or guillotine that, when moved between two seats, causes tight shutoff, **p. 203**.

Globe valve—a type of valve that uses a plug and seat to regulate the flow of fluid through the valve body, which is shaped like a sphere or globe, **p. 203**.

HDPE—high-density polyethylene; a plastic material used to create water pipes and drains, **p. 198**.

Plug valve—a flow-regulating device that uses a flow-control element shaped like a hollowed-out plug, attached to an external handle, to increase or decrease flow; requires a quarter turn to go from fully open to fully closed, **p. 201**.

PVC—polyvinyl chloride; a plastic-type material that can be used to create cold water pipes and drains and other low-pressure applications, **p. 198**.

Relief valve—a safety device designed to open slowly as the pressure of a fluid in a closed vessel exceeds a preset level; can be used in services where liquid expands to create a gas, **p. 204**.

Safety valve—a safety device designed to open quickly as the pressure of a fluid in a closed vessel exceeds a preset level; typically used in gas service, **p. 205**.

Socket weld—a type of weld used to connect pipes and fittings when one pipe is small enough to fit snugly inside the other, **p. 198**.

Solder—a metallic compound that is melted and applied in order to join together and seal the joints and fittings in tubing systems and electrical components; soldered joints are not as strong as brazed joints, **p. 198**.

Threaded (screwed) pipe—piping that is connected using male and female threads, **p. 197**.

Throttling—a condition in which a valve is partially opened or partially closed in order to restrict or regulate the amount of flow, **p. 200**.

Valve—a piping system component used to control the flow of fluids through a pipe, **p. 199**.

Introduction

Piping and valves are the most prevalent pieces of equipment in the process industries. Some estimates are that piping makes up 30 to 40 percent of the initial investment when creating a new plant.

14.1 Purpose and Function of Piping and Valves

In every plant, you will see piping which connects different pieces of equipment and processing areas. These pipes carry chemicals and other materials into and out of various processes and equipment.

When building a plant, it is important to select proper construction materials and connectors, because some materials and connectors are not adequate for certain processes, pressures, or temperatures. Improper operation or improper material selection can lead to leaks, wasted product, or hazardous conditions.

14.2 Construction Materials in Piping and Valves

Industrial pipes and valves can be made of many different materials such as carbon steel, stainless steel, other alloy steel, iron, exotic metals, and plastics. The most common type of piping, however, is carbon steel. It is appropriate for a wide range of temperatures and is relatively economical.

When piping and valve systems are designed for a particular process, designers must be familiar with the process and the substances that will pass through the pipes. Specifically, they need to know the temperature of the substance, its viscosity, how much pressure it exerts, and how flammable, corrosive, or reactive it is.

Some metals become brittle at extremely low temperatures. Others are weakened by high temperatures, high pressures, the corrosive effects of process substances such as strong acids or bases, or the erosive effects of high velocity fluids.

Although some construction materials are pure metals, others are made of alloys. An **alloy** is a compound composed of two or more metals that are mixed together when molten to form a solution. For example copper and tin are melted and mixed together to make bronze alloy. Alloys improve the properties of single component metals and provide special characteristics needed in specific applications.

Table 14.1 contains a list of common construction materials and their applications.

Alloy a compound composed of two or more metals that are mixed together when molten to form a solution; not a chemical compound.

Table 14.1 Common Piping and Valve Construction Materials and Their Applications

Construction Material	Temperature Ranges	Description
Carbon steel	−20 degrees F to 800 degrees F (−29 to 427 degrees C)	The most commonly used construction material because of its flexibility, weldability, strength, and relatively low cost
Stainless steel	−150 degrees F to 1400+ degrees F (−101 to 760 degrees C)	Less brittle than carbon steel at extremely low temperatures; appropriate for use in high temperature applications; more corrosion-resistant than carbon steel, especially in acid service
Brass, bronze, and copper	−50 degrees F to 450 degrees F (−46 to 232 degrees C)	Used for corrosion resistance at low or moderate temperatures; excellent heat transfer characteristics
Alloys	Varies by alloy	Can be used in high-temperature applications (above 800 degrees F), such as in furnace tubes and highly corrosive service; high cost
Plastics	Varies by plastic	Used for low pressure applications and corrosive services; easy to install, low cost, and lightweight

14.3 Connecting Methods for Piping and Valves

Pipes and valves can be connected together in a variety of ways (Figure 14.1). They can be threaded (screwed), flanged, or bonded (e.g., welded, glued, soldered, or brazed).

The factor that determines which connection type is the most appropriate is the purpose of the pipe. For example, if the pipe is used for low-pressure water service, a threaded joint might be appropriate. These joints are cheaper and easier to install (and uninstall) than welded or flanged joints.

If the pipe is used for a high-pressure, flammable, or corrosive service, a welded joint would be a better option because screwed joints, and flanged joints with gaskets, are more likely to leak. Welded joints are usually the connection method of choice for critical service piping.

Figure 14.1 Examples of screwed, flanged, welded, and bonded connections.

Screwed

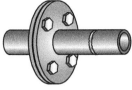
Flanged

Welded
(butt weld shown)

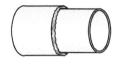
Bonded

Threaded (Screwed)

Screw-type connections involve the joining together of two pipes through a series of tapered threads such as the ones shown in Figure 14.2. In a screw-type connection, the **threaded (screwed) pipe** is cut with external "male" threads and the connector is cut with internal "female" threads so the two join together. When these threads are cut, they generally are cut with precision to ensure the two pieces fit together tightly to avoid leaks. However, this tightness can make it difficult to join the two pieces together. That is why threading compound or Teflon tape is often employed. These materials lubricate the joints, facilitate the connection, and provide a flexible connection to seal against leaks.

Threaded pipes have a portion of the pipe wall removed to create the threaded section. This affects the maximum allowable working pressure (MAWP) of the pipe. Because of this, certain applications, like high-pressure, flammable, and toxic service, do not use threaded pipe larger than 2 inches (5 cm) in diameter. Threaded pipe with diameter less than 2 inches is commonly used in most industries. Threaded connections are more prone to leak.

Threaded (screwed) pipe piping that is connected using male and female threads.

Figure 14.2 Threaded (screwed) connection.

Screwed

Flanges

Flanged connections, such as the one shown in Figure 14.3, are typically used in instances where the piping might need to be disconnected from another pipe or a piece of equipment.

A **flange** is a type of pipe connection in which two matching plates are joined together with bolts. Between the two plates is a **gasket**. As the bolts are tightened, the gasket is compressed between the two plates. This compression increases the tightness of the seal and prevents leakage.

Flange a type of pipe connection, glued, welded, or threaded, consisting of two matching circular plates that are joined together with bolts.

Gasket a flexible material placed between the two surfaces of a flange to seal against leaks.

Figure 14.3 Flanged connection.
CREDIT: photostock77/Shutterstock.

Did You Know?

The thickness of a flange and the number of bolts it contains can be a good general indicator for process technicians.

The thicker the flange and the more bolts used to connect it, the higher the pressure rating of the piping.

CREDIT: LightCooker/Fotolia.

Welds

Welding materials are made of similar metallic compounds. In order to create welded joints, welding material must be melted and applied to the pipes being connected.

If the pipes are the same diameter, a **butt weld** is used. If one pipe is small enough to fit snugly inside the other, a **socket weld** is used. Figure 14.4 shows examples of a butt and a socket weld.

Butt weld a type of weld used to connect two pipes of the same diameter that are butted against each other.

Socket weld a type of weld used to connect pipes and fittings when one pipe is small enough to fit snugly inside the other.

Figure 14.4 Butt and socket welds.
CREDIT: *(left)* Roman 23203/Fotolia; *(right)* LightCooker/Fotolia.

PVC polyvinyl chloride; a plastic-type material that can be used to make cold water pipes and drains and other low-pressure applications.

HDPE high-density polyethylene; a plastic material used to create water pipes and drains.

Solder a metallic compound that is melted and applied in order to join together and seal the joints and fittings in tubing systems and electrical components. A soldered joint is less strong than a brazed joint.

Braze to join two pieces of metal together by melting an alloy (filler) that bonds the metals together; a brazed joint can be stronger than the metals but is not stronger than a welded joint.

Bonds

Bonded pipe joints, such as the one show in Figure 14.5, can be glued, brazed, or soldered.

Gluing is fusing joints together with glue. Glued joints are typically found on plastic lines and pipes (e.g., pipes made of PVC or HDPE). **PVC** (polyvinyl chloride) is a plastic-type material that can be used for cold-water system pipes and drains and other low-pressure applications. **HDPE**, high-density polyethylene, is a plastic material also used in water pipes and drains. Process piping can also be joined using hydraulic pressure rams. This bond has the same integrity as a weld.

Soldering and brazing are methods of fusing joints together with molten metal. **Solder** is a metallic compound that is melted and applied in order to join and seal the joints and fittings in tubing systems and electrical components. To **braze** is to solder together using a hard solder with a high melting point. Brazing creates a stronger joint than soldered applications.

The method used for joining pipes and fittings is determined by the materials being used and their applications.

Figure 14.5 Establishing a bonded joint.
CREDIT: Leonid Eremeychuk/Shutterstock.

14.4 Fitting Types

Fitting a piping system component used to connect two or more pieces of pipe together.

A **fitting** is a piping system component used for many applications such as:

- Connecting two or more pieces of pipe together
- Reducing or increasing the size of piping
- Closing or blocking the end of piping
- Changing the direction of flow

The process industries use many different types of fittings.

Figure 14.6 shows examples of some of the most common fittings. Table 14.2 lists their applications.

Figure 14.6 Different types of fittings.

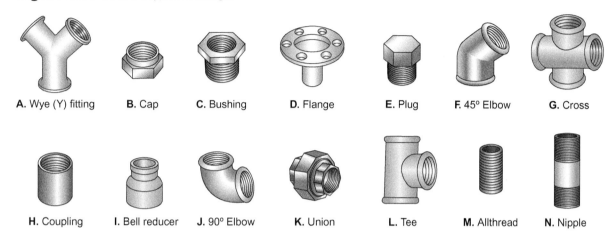

A. Wye (Y) fitting   B. Cap   C. Bushing   D. Flange   E. Plug   F. 45° Elbow   G. Cross

H. Coupling   I. Bell reducer   J. 90° Elbow   K. Union   L. Tee   M. Allthread   N. Nipple

Table 14.2 Different Types of Fittings and Their Applications

Fitting Name	Description and Application
A. Wye (Y) fitting	A pipe fitting that can be used as a sampling point, a blowdown point, or a strainer to trap particles before entering process equipment
B. Cap	A pipe fitting that fits over the open end of a pipe to seal it
C. Bushing	A pipe fitting that is threaded on both the internal and external surfaces; it is used to join two pipes of differing sizes
D. Flange	A pipe fitting that consists of two plates and a gasket joined together with bolts
E. Plug	A pipe fitting that fits inside the open end of a pipe to seal it
F. Elbow	A pipe fitting with a 45- or 90-degree angle that is used to change the direction of flow
G. Cross	A pipe fitting that allows four pipes to be connected together at 90-degree angles; also commonly called a 4-way
H. Coupling	A short piece of pipe or "collar" used to join two lengths of pipe
I. Bell reducer	A pipe fitting that is used to connect two pipes of different diameters. Its purpose is to increase or decrease the size of the piping run.
J. Union	A pipe fitting that joins two sections of threaded pipe but allows them to be disconnected without cutting or disturbing the position of the pipe
K. Tee	A T-shaped pipe fitting that is used to allow flow to two different pipes that are 90 degrees apart
L. Allthread	A short length of pipe (usually less than 6 inches [15 cm]) with threads throughout
M. Nipple	A short length of pipe (usually less than 6 inches [15 cm]) with or without threads on both ends (threads not needed with welded connections)

14.5 Valve Types

A **valve** is a piping system component used to control the flow of fluids through a pipe. Valves work to control, throttle, or stop the flow.

Valves are generally divided into three categories:

- Manually operated or actuated valves (see Valve Actuators below)
- Check valves
- Pressure relief valves

Manually operated valves have a handle or a hand wheel that is turned to open or close the valve.

Valve a piping system component used to control the flow of fluids through a pipe.

Many different types of manual valves are used in the process industries. Some of the most common valves include:

- Ball
- Plug
- Butterfly
- Diaphragm
- Gate
- Globe

Check valves are used in service as single-direction flow control. They operate on the difference in pressure upstream and downstream of the valve to prevent backflow into the system.

Relief valves (RV) and pressure safety valves (PSV) are mechanisms that automatically operate, or open, to release pressure when it exceeds operating values. Relief valves are slow acting and used in liquid service. Pressure safety valves are fast acting and used in gas (or vapor) service.

When determining which type of valve should be used, engineers must consider the substances that will pass through it and how and where the valve will be used in the process.

For example, if the fluid passing through the valve is very thick (viscous) or corrosive, then a diaphragm valve would be a good choice, because other valve types do not perform well with these sorts of substances.

If the valve will be used for throttling then a globe valve would be a good choice. **Throttling** is a condition in which a valve is partially opened or closed in order to restrict or regulate the amount of flow. Many other types of valves (e.g., gate valves) can be damaged by throttling, but because of their design, globe valves are not.

Control valves are another common type of valve used in the process industries. Control-valve applications are used within an instrumentation control loop or as part of a process unit's isolation procedure. These applications will be discussed in more depth in the Instrumentation course.

Process technicians need to be familiar with each of the different valve types and the maintenance and operating characteristics of each.

Ball Valve

A **ball valve** is a flow-regulating device that uses a flow-control element shaped like a hollowed-out ball, attached to an external handle, to increase or decrease flow. Ball valves are quarter-turn valves, meaning that turning the valve's stem a quarter of a turn brings it to a fully open or fully closed position. In comparison, valves such as gate valves require multiple turns of the hand wheel to open or close fully. Figure 14.7 shows an open and a closed ball valve.

Ball valves are typically used for on/off service. When a ball valve is open, the hollowed-out portion of the ball (sometimes referred to as the port) lines up perfectly with the inner diameter of the pipe. When a ball valve is closed, the port aligns with the wall of the pipe.

Figure 14.7 A. Ball valve open. B. Ball valve closed.

CREDIT: Dmitry Syechin/Fotolia.

A.   B.

Throttling a condition in which a valve is partially opened or partially closed in order to restrict or regulate the amount of flow.

Ball valve a flow-regulating device that uses a flow-control element shaped like a hollowed-out ball, attached to an external handle, to increase or decrease flow; a ball valve requires only a quarter turn to go from fully open to fully closed.

Plug Valve

A **plug valve** is a flow-regulating device that uses a flow-control element shaped like a hollowed-out plug, attached to an external handle, to increase or decrease flow.

Plug valves are almost identical to ball valves. Both are quarter-turn valves that use a hollowed-out object to control flow. Figure 14.8 shows a cutaway of a partially opened and a closed plug valve. Note that position of the flow indicator gives information about how far open or shut the valve is.

Plug valves are designed for on/off service and are well suited for certain types of applications such as slurry, boiler feed water, and fuel gas.

Ball valves and plug valves are used in many of the same service applications.

Plug valve a flow-regulating device that uses a flow-control element shaped like a hollowed-out plug, attached to an external handle, to increase or decrease flow; requires a quarter turn to go from fully open to fully closed.

Figure 14.8 Plug valve. **A.** Partially open (see arrow). **B.** Closed.

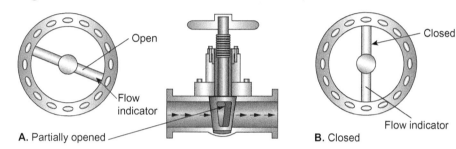

Butterfly Valve

A **butterfly valve** is a flow-regulating device that uses a disc-shaped flow-control element to increase or decrease flow. Like ball valves, butterfly valves can be fully opened or closed by turning the valve handle one-quarter of a turn.

The design of these valves makes them most suitable for low-temperature, low-pressure applications such as cooling water systems. Figure 14.9 shows a cutaway of a butterfly valve.

Unlike many other valves, butterfly valves can be used for throttling. However, the throttling capabilities of a butterfly valve are not uniform or exact (e.g., opening the valve halfway might provide a flow rate that is near maximum capacity).

Butterfly valve a flow-regulating device that uses a disc-shaped flow-control element to increase or decrease flow; requires a quarter turn to go from fully open to fully closed.

Figure 14.9 Butterfly valve.
CREDIT: freeman98589/Fotolia.

Check Valve

A **check valve** allows flow only in one direction. These valves eliminate backflow into the system, thereby preventing equipment damage and contamination of the process.

Check valves can be composed of many different materials and can be used in a wide variety of applications. The most common types of check valves include swing check, lift check, and ball check valves. Figure 14.10 shows an example of a swing check valve.

Check valve a type of valve that allows flow in only one direction and is used to prevent reversal of flow in a pipe.

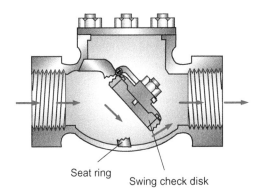

Figure 14.10 Swing check valve.

Did You Know?

The human heart has valves that function like check valves.
Without these valves, blood would not circulate properly.

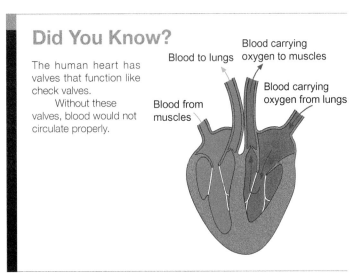

In a swing check valve, the valve disc swings open as the fluid moves through the valve and is forced closed if the fluid changes directions (backward flow).

In a lift check valve (Figure 14.11), a disc-shaped flow-control element controls the flow of the fluid. The disc is lifted off its seat to allow flow in one direction. It rests back on its seat to prevent flow in a backward direction.

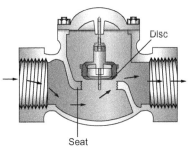

Figure 14.11 Lift check valve.

In a ball check valve (Figure 14.12), a ball- or sphere-shaped flow-control element is used. Forward flow pushes the ball off the valve seat; backflow pressure causes the ball to drop back down onto the seat, sealing the valve.

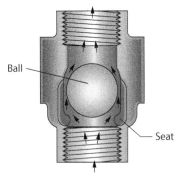

Figure 14.12 Ball check valve.

Diaphragm Valve

Diaphragm valve a flow-regulating device that uses a flexible, chemical resistant, rubber-type diaphragm attached to a stem that closes onto a weir located in the valve.

A diaphragm valve is a flow-regulating device that uses a flexible, chemical resistant, rubber-type diaphragm to control flow instead of a typical flow-control element. In this type valve, the diaphragm seals the parts above it (e.g., plunger) from the process fluid. Figure 14.13 shows an example of a diaphragm valve.

The diaphragm valve's unique design enables it to work well with process substances that are exceptionally sticky, viscous, or corrosive. However, it is not adequate for applications with high pressure or excessive temperature.

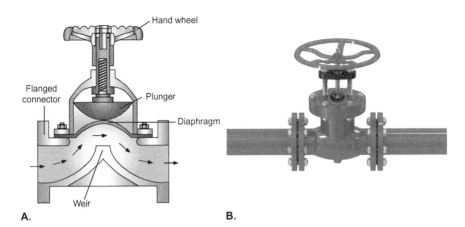

Figure 14.13 Diaphragm valve. **A.** Cutaway. **B.** Sample diaphragm valve.

CREDIT: B. alexlmx/Fotolia.

Gate Valve

A **gate valve** is a positive shutoff valve that uses a gate or guillotine to stop flow. When the gate is moved between two seats, it causes tight shutoff.

Gate valves are the most common type of valve in process industries. They are designed for on/off service and are not intended for throttling. Figure 14.14 shows a cutaway of a gate valve.

Gate valve a positive shutoff valve using a gate or guillotine that, when moved between two seats, causes tight shutoff.

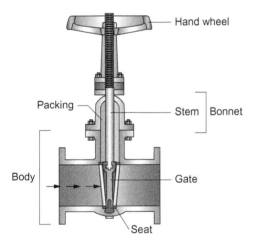

Figure 14.14 Gate valve.

In a gate valve, a wedge-shaped disc, or "gate," is lowered into the body of the valve with a hand wheel. The body of the valve contains two seat rings that seal around the gate when it is fully closed, completely blocking flow.

Because of their design, gate valves should not be used for throttling. Throttling can cause metal erosion and seat damage, which would prevent the valve from sealing properly.

Did You Know?

Throttling a gate valve can cause the flow-control element to vibrate back and forth and cause the valve to wear excessively. This back-and-forth movement causes a sound, referred to as "valve chatter," that process technicians can monitor.

CREDIT: Aleksandrs Bondars/Shutterstock.

Globe Valve

A **globe valve** uses a plug to regulate the flow of fluid through the valve body, which is shaped like a sphere or globe. This type of valve is designed to regulate flow in one direction and can be used in throttling service. Globe valves are very common in process industries. Figure 14.15 shows a cutaway of a globe valve.

Globe valve a type of valve that uses a plug and seat to regulate the flow of fluid through the valve body, which is shaped like a sphere or globe.

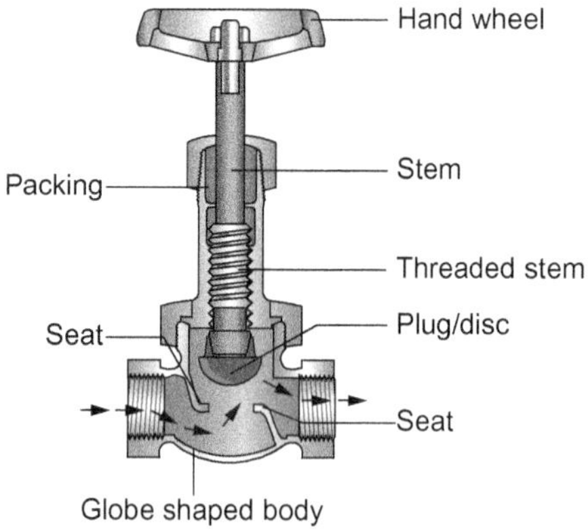

Figure 14.15 Globe valve.

In a globe valve, fluid flow is increased or decreased by raising or lowering the disc or flow-control element. These flow-control elements come in a variety of shapes, including ball, cylindrical, and needle discs. The seat in these valves is designed to accommodate the shape of the disc. Figure 14.16 shows an example of these different disc shapes.

Ball-shaped Cylinder-shaped Needle-shaped

Figure 14.16 Globe valve disc designs.

Relief and Safety Valves

Safety and relief valves are used to protect equipment and personnel from overpressurization in the system. Both these valves are designed to respond to pressure in a line, vessel, or other equipment that exceeds a preset threshold. When this happens, they open and discharge to a collection system (or to the atmosphere if nonflammable). Neither type of pressure relief valve would send liquid directly to a flare or vent header. The liquid would be released to a vessel where it is vaporized before being burned at a flare tip.

Relief and safety valves differ in the type of service they are intended for (liquid vs. gas) and in the speed at which they open.

RELIEF VALVES A **relief valve** is a safety device designed to open slowly if the pressure of a liquid in a closed vessel exceeds a preset level. These valves open more slowly and with less volume than safety valves because liquids are virtually noncompressible. With non-compressible substances, a small release is all that is required to correct overpressurization. Figure 14.17 shows an example of a relief valve.

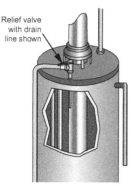

Did You Know?

The hot water heater in your home contains a relief valve.
 This valve prevents tank overpressurization and explosion.

Relief valve a safety device designed to open slowly as the pressure of a fluid in a closed vessel exceeds a preset level; can be used in services where liquid expands to create a gas.

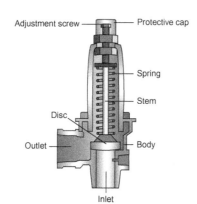

Figure 14.17 Relief valve.

Because of the slow speed at which they open and their small outlet port, relief valves are not appropriate for gas service.

In a relief valve, a flow-control disc is held in place by a spring. When the pressure in the system exceeds the threshold of the spring, the valve is forced open (proportional to the increase in pressure) and liquid is allowed to escape into a containment receptacle, flare system, or other safety system. As the pressure drops below the threshold, the spring gradually forces the flow-control element back into the seat, thereby resetting the valve.

SAFETY VALVES A **safety valve** is designed to open quickly if the pressure of a gas in a closed vessel exceeds a preset level. These valves open more quickly, release more volume, and generally have larger outlets than relief valves because gases are highly compressible. Compressible substances require a much larger release to correct overpressurization. Figure 14.18 shows an example of a safety valve.

Did You Know?

Process technicians often use the slang term *pop valve* or *pop off valve* to refer to a safety valve.

This is because safety valves open quickly or "pop off" when the pressure threshold has been exceeded (as opposed to relief valves, which open slowly).

CREDIT: dcwcreations/Shutterstock.

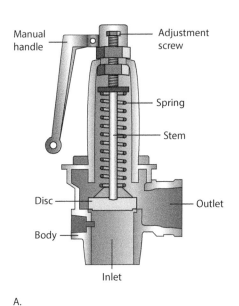

A.

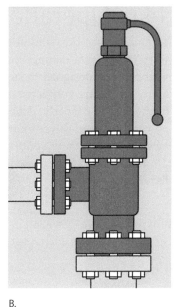

B.

Figure 14.18 Safety valve with manual pressure-relief handle.

CREDIT: B. AlexanderZam/Shutterstock.

Safety valve a safety device designed to open quickly as the pressure of a fluid in a closed vessel exceeds a preset level; typically used in gas service.

Safety valves are designed to operate quickly and prevent overpressurization that can cause equipment damage or injury to personnel. The ASME code requires safety relief valves with levers be used in air, steam, and hot water [over 140° F (60° C)] services. Levers are used for testing purposes, to remove anything trapped in the seat, and to test functionality of the valve. This testing is completed by trained and authorized personnel. Company or site policy dictates the testing method of these valves.

When a safety valve opens, the excess pressure is vented either to a flare system or through a large exhaust port into the atmosphere (depending on the substance being vented).

Some safety valves will reset themselves after being activated. Others must be taken to a shop to be reset manually.

The maintenance of safety and relief valves must be performed by certified personnel.

Did You Know?

The first safety valve was invented in 1681 by Frenchman Denis Papin for use on his invention "Papin's Bone Digester," a steam cooking machine used to soften bones in order to extract the jelly.

CREDIT: Heritage Images/Contributor/Getty Images.

Valve Actuators

Valves can be operated, or actuated, by several methods. Valves can be operated by hand, by instrument signal (pneumatic or electronic), by electrical solenoid, by motor, or by hydraulic signal. As mentioned, valves designed for pressure relief are actuated by overcoming tension on a spring when process pressure exceeds design values.

Valves operated by pneumatic signal are usually part of control-loop instrumentation within a process unit. Valves actuated by air, electricity, solenoid, or hydraulics also might be part of an automated unit that operates through sequential steps to complete a process function (e.g. air drying unit). More detailed information will be discussed in the instrumentation chapter of this textbook and in the instrumentation course.

Some manually operated valves with hand wheels, such as gate and diaphragm, also are classified as either rising stem or non-rising stem. In a rising stem valve, the stem rises through the hand wheel when opened. In a non-rising stem valve, the hand wheel and stem rise together. The process technician must know which application is being used to ensure the valve is either in the open, or the closed position, as appropriate for its current operation. Symbols for valves are shown at the end of this chapter.

14.6 Operational Hazards

Technicians should avoid using excessive force when opening or closing valves, as this can warp the valve or damage the seat, preventing a good seal. However, properly sized valve wrenches are designed to supply additional force when needed to open or close a valve. Many valves are very large in diameter and require more than one person (teamwork) to open or close properly. Other tools are used in these type operations such as pneumatic power wrenches. Valves located in areas not readily accessible to the process technician can have extensions that fit onto the valve wheel for use in opening and closing the valve.

Many hazards are associated with piping and valves. Table 14.3 lists some of these hazards and their impacts.

Table 14.3 Hazards Associated with Improper Valve Operation

Improper Operation	Possible Impacts			
	Individual	Equipment	Production	Environment
Throttling a valve that is not designed for throttling		Valve damage to the disc and seat to the point that it will not seat and stop flow, even when closed	Off-spec product because of improper flows	
Use of excessive force when opening or closing a valve		Damage to the valve seat, the packing, or the valve stem; causes leakage and makes the valve difficult to open or close	Off-spec product because of improper flows	
Failure to clean and lubricate valve stems	Injuries as a result of a valve wrench (because the valve is difficult to open) slipping off the valve handle	Valve stem seizure or thread damage makes the valve difficult to open and close		
Improperly closing a valve on a high-pressure line	Injury because of equipment overpressurization	Equipment, damage (e.g., overpressurizing a pump)		Leak to the environment
Failure to wear proper protective equipment when operating valves in high-temperature, high-pressure, hazardous, or corrosive service	Burns or other serious injuries			

14.7 Monitoring and Maintenance Activities

When monitoring and maintaining piping and valves, process technicians must always remember to look, listen, and check for the items shown in Table 14.4. Note that instrumentation control valves and their operation will be addressed in the instrumentation chapter of this textbook and further information given in the instrumentation course.

Table 14.4 Monitoring and Maintenance Activities for Valves

Look	Listen	Check
■ Look at valves to make sure there are no leaks. ■ Examine valve for excessive wear. ■ Check to make sure valve stems are properly lubricated.	■ Listen for abnormal noises (e.g., valve "chatter").	■ Feel to make sure the valve is not being overly tightened.
Verify valve position		
■ Bypass valve closed or open depending on current operation ■ Rising stem/nonrising stem, visually ensure valve is in correct position ■ Quarter-turn valve indicators match current operation of open or closed		The correct type valve is used in the correct way. (i.e. gate valve not being throttled for use to control flow)
Adjust packing		
■ Valve packing is adjusted to vendor specifications for the specific type valve design. Packing leakage can be allowed for certain type processes; the EPA has strict regulations in regard to this specification. ■ Valve packing adjustment is typically a function and responsibility of maintenance.	If fluid is escaping around the valve stem, the packing could be too loose.	If the valve is hard to operate, the packing could be too tight.
Labeling		
■ Check for directional arrows on the valve to ensure it is directionally correct in the line. ■ Check the valve label for the correct information when isolating for repair or replacement, or when placing into service.		

Failure to perform proper maintenance and monitoring could affect the process and result in equipment damage.

Piping and Valve Symbols

There are many symbols associated with piping and valves. Although standards do exist, symbols can vary slightly from plant to plant. Figure 14.19 shows some of the more commonly used symbols.

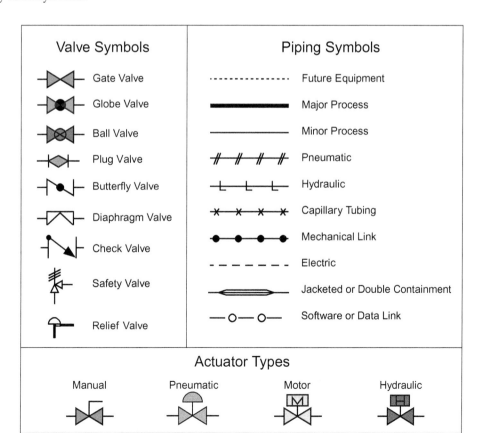

Figure 14.19 Common piping and valve symbols.

Summary

Piping and valves are the most prevalent pieces of equipment in the process industries. Their main purpose is to carry chemicals and other materials into and out of other process equipment.

Piping and valves can be made of many different materials, ranging from carbon steel to plastic. Material selection is based on process characteristics (such as temperature, pressure, corrosiveness, and erosive properties).

Pipes can be connected using a variety of methods. They can be screwed, flanged, or bonded. Many types of fittings are used to connect piping, depending on the service usage and requirement.

It is important to operate valves properly. Throttling valves that should not be throttled, using excessive force to open or close a valve, or failing to clean and lubricate valve stems can cause excessive wear and damage the valve. Over time, this damage can cause the valve to leak, seize up, or fail to open or close completely.

Improperly closing a valve on a high-pressure line (e.g., blocking or "deadheading" a pump) or operating a steam-filled valve without proper protective gear can cause equipment damage and personal injury.

When making rounds, process technicians should always inspect valves for leaks and perform proper maintenance and lubrication procedures to prevent damage or excessive wear. Technicians also should listen for abnormal noises or "valve chatter" that could be an indication of improper throttling or the possibility of internal damage to valve components. These activities are critical to protecting lives, the environment, and equipment.

Process technicians should be familiar with the various symbols associated with piping and valves and be able to identify them on plant diagrams.

Checking Your Knowledge

1. Define the terms *alloy, valve,* and *throttling*.
2. Explain why design engineers must understand all aspects of a process when selecting construction materials.
3. If you wanted to connect a piece of pipe to a piece of equipment and you knew you were going to have to disconnect it several times over the next few months, which of the following connection methods would you select? Why?
   a. screwed pipe connection
   b. flanged pipe connection
   c. bonded pipe connection
   d. union connection
4. If you were going to connect two high-pressure, critical service pipes, would it be better to use a screw-type connection or a welded connection? Why?
5. If you wanted to connect two pipes of the same diameter, would you use a socket weld or a butt weld?

6. Give the proper name for each of the following fittings.

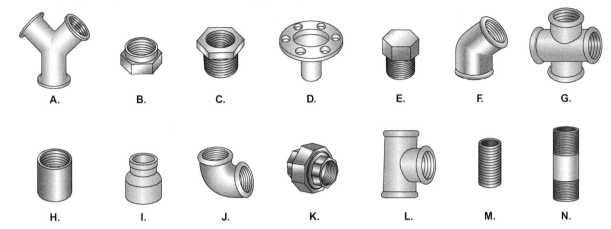

7. *(True or False)* Relief valves are designed to open quickly.

8. *(True or False)* Safety valves are designed for gas service.

9. What can happen to a valve over time if a process technician throttles it and it was not designed for throttling?

10. Why should a process technician refrain from using excessive force when opening or closing a valve?

11. List three (3) valves that would be classified to open/close with a 90-degree turn of the hand wheel.

12. Match the valve type with its description.

Valve Type	Description
I. Ball	a. A safety device designed to open quickly if the pressure of a gas exceeds a preset threshold
II. Butterfly	b. Uses a disc-shaped flow-control element to increase or decrease flow
III. Check	c. Uses a hollowed-out plug to increase or decrease flow
IV. Diaphragm	d. Uses a rubber-type diaphragm to control flow
V. Gate	e. Uses a metal gate to block the flow of fluids
VI. Globe	f. Uses a hollowed-out ball to increase or decrease flow
VII. Plug	g. A safety device designed to open slowly if the pressure of a liquid exceeds a preset level
VIII. Relief	h. Uses a spherical or globe-shaped plug to increase or decrease fluid flow
IX. Safety	i. Used to prevent accidental backflow

NOTE: Answers to Checking Your Knowledge questions are in Appendix I.

Student Activities

1. Look around your house and identify at least five valves (e.g., the valve under your kitchen sink). Tell where each valve is located and try to identify what type of valve you think it might be.

2. Given a drawing of a valve, identify the components (e.g., seat, stem, flow-control element).

3. Given a piping and instrumentation diagram (see Chapter 13), tell how many of the following valves are present:
  a. gate
  b. globe
  c. ball
  d. butterfly
  e. diaphragm
  f. check
  g. safety
  h. relief

4. Work in teams using a given piping and instrumentation diagram. Identify all valves in the following categories:
  a. manual
  b. pneumatic
  c. motor driven
  d. hydraulic

Chapter 15
Vessels

Objectives

Upon completion of this chapter, you will be able to:

15.1 Describe the purpose or function of vessels (tanks, drums, cylinders, dryers, filters, reactors, and bins/hoppers) in the process industries. (NAPTA Vessels 1) p. 213

15.2 Explain the relationship of pressure to the vessel shape and wall thickness. (NAPTA Vessels 2) p. 214

15.3 Define and provide examples of the following components as they relate to vessels:

desiccant

floating roof

articulated drain

blanketing

spherical tank

foam chamber

sump

mixer/agitator

gauge hatch

manway

heat tracing system (steam or electrical)

vapor recovery system

vortex breaker

baffle

weir

boot

mist eliminator

vane separator (NAPTA Vessels 4) p. 215

15.4 Describe the purpose of dikes, firewalls, and containment walls around vessels. (NAPTA Vessels 3) p. 220

15.5 Identify and describe the various types of reactors and their purpose. (NAPTA Vessels 7) p. 220

15.6 Identify possible hazards associated with vessels, including the following:

improper valve lineup

loss of nitrogen flow

cross contamination

failure of vent system

leaks/spills

chemical reactions (such as corrosion, pH, etc.) (NAPTA Vessels 8) p. 222

15.7 Describe the monitoring and maintenance activities associated with vessel operations. (NAPTA Vessels 5, 6) p. 223

Key Terms

American Society of Mechanical Engineers (ASME)—organization that provides laws of regulation for boilers and pressure vessels, **p. 214.**

Articulated drain—a hinged drain, attached to the roof of an external floating roof tank, that moves up and down as the roof and the fluid levels rise and fall, **p. 215.**

Atmospheric tank—an enclosed vessel that operates at atmospheric pressure; usually cylindrical in shape, equipped with either a fixed or floating roof, and containing nontoxic vapor liquids, **p. 214.**

Baffle—a metal plate, placed inside a tank or other vessel, that is used to alter the flow of chemicals or facilitate mixing, **p. 218.**

Batch reaction—a carefully measured and controlled process in which raw materials (reactants) are added together to create a reaction that makes a single quantity (batch) of the final product, **p. 220.**

Bin/hopper—a vessel that typically holds dry solids, **p. 214.**

Blanketing—the process of putting an inert gas, usually nitrogen, into the vapor space above the liquid in a tank to prevent air leakage into the tank, **p. 215.**

Boot—a section in the lowest area of a process drum where water or other liquid is collected and removed, **p. 219.**

Containment wall—an earthen berm or constructed wall used to protect the environment and people against tank failures, fires, runoff, and spills; also called a *bund wall, bunding, dike,* or *firewall,* **p. 220.**

Continuous reaction—a chemical process in which raw materials (reactants) are continuously being fed in and products are continuously being formed and removed from the reactor vessel, **p. 220.**

Cylinder—a vessel that can hold extremely volatile or high-pressure materials, **p. 215.**

Desiccant—a specialized substance contained in a dryer that removes hydrates (moisture) from the process stream, **p. 213**

Drum—a specialized type of storage tank or intermediary process vessel, **p. 214.**

Dryer—a vessel containing desiccant and screens across which process streams flow to have moisture (hydrates) removed, **p. 214**

Filter—a device used to remove liquid, gas, or solid particulates from the process stream, **p. 214.**

Fixed bed reactor—a reactor vessel in which the catalyst bed is stationary and the reactants are passed over it; in this type of reactor, the catalyst occupies a fixed position and is not designed to leave the reactor, **p. 221.**

Floating roof—a type of vessel covering (steel or plastic), used on storage tanks, that floats upon the surface of the stored liquid and is used to decrease vapor space and reduce potential for evaporation, **p. 215.**

Fluidized bed reactor—a reactor that uses high-velocity fluid to suspend or fluidize solid catalyst particles, **p. 222.**

Foam chamber—a reservoir and piping installed on liquid storage vessels and containing fire-extinguishing chemical foam, **p. 216.**

Gauge hatch—an opening on the roof of a tank that is used to check tank levels and obtain samples of the tank contents, **p. 217.**

Heat tracing—a coil of heated wire or tubing that adheres to or is wrapped around a pipe in order to increase the temperature of the process fluid, reduce fluid viscosity, and facilitate flow, **p. 218.**

Manway—an opening in a vessel that permits entry for inspection and repair, **p. 217.**

Mist eliminator—a device in the top of a tank, composed of mesh, vanes, or fibers, that collects droplets of mist (moisture) from gas to prevent it from leaving the tank and moving forward with the process flow, **p. 220.**

Mixer—a device used to mechanically combine chemicals or other substances; also known as an agitator, **p. 217.**

Pressurized tank—an enclosed vessel in which a greater-than-atmospheric pressure is maintained, **p. 214.**

Reaction furnace—a reactor that combines a firebox with tubing to provide heat for a reaction that occurs inside the tubes, **p. 222.**

Reactor—a vessel in which chemical reactions are initiated and sustained, **p. 220.**

Spherical tank—a type of pressurized storage tank that is used to store volatile or highly pressurized material; also referred to as "round" tanks, **p. 215.**

Stirred tank reactor—a reactor vessel that contains a mixer or agitator to improve mixing of reactants, **p. 221.**

Sump—an area of temporary storage located at the bottom of a tank from which undesirable material is removed, **p. 216.**

Tank—a large container or vessel for holding liquids and/or gases, **p. 213.**

Tubular reactor—a continuously flowed vessel in which reactants are converted in relation to their position within the reactor tubes, not influenced by residence time in the reactor, **p. 222.**

Vane separator—a device, composed of metal vanes, used to separate liquids from gases or solids from liquids, **p. 220.**

Vapor recovery system—the process of recapturing vapors by methods such as chilling or scrubbing; vapors are then purified, and the vapors or products are sent back to the process, sent to storage, or recovered, **p. 218.**

Vessel—a container in which materials are processed, treated, or stored, **p. 212.**

Vortex—the cone formed by a swirling liquid or gas, **p. 218.**

Vortex breaker—a metal plate, or similar device, placed inside a cylindrical, cone-shaped, or other type operating unit, which prevents a vortex from being created as liquid is drawn out of the vessel, **p. 218.**

Weir—a flat or notched dam or barrier to liquid flow that is normally used either for the measurement of fluid flows or to maintain a given depth of fluid as on a tray of a distillation column, in a separator, or other vessel, **p. 219.**

Introduction

Vessel a container in which materials are processed, treated, or stored.

Vessels (Figure 15.1) are a vital part of the operational units in the process industries. A **vessel** is a container in which materials are processed, treated, or stored. Without this type of

Figure 15.1 Types of vessels.

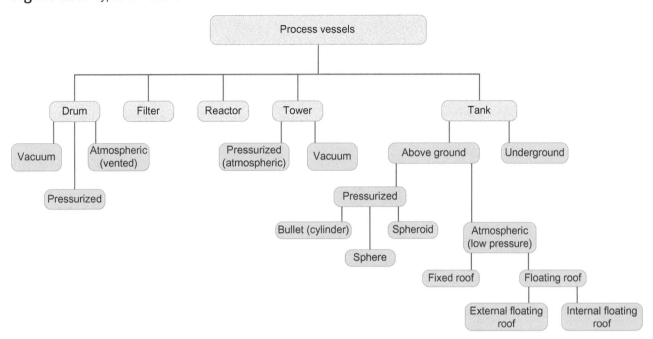

equipment, the process industries would be unable to create and store large amounts of product. Vessels include tanks, towers, reactors, drums, dryers, cylinders, hoppers, bins, and other similar containers that are used to process or store materials. Vessels vary greatly in design (e.g., size and shape) based on the requirements of the process. Factors that affect vessel design include pressure requirements, type of product contained in the vessel (liquid, gas, or solid), temperature requirements, corrosion factors, and volume.

By using tanks to store products in large quantities, companies can maximize both cost effectiveness and efficiency. Drums are used both for process mainstream applications and for auxiliary process applications. Reactors are used to provide the chemical reactions that some processes require to create the end product. Dryers use **desiccants** (specialized substances contained in a dryer that remove hydrates from the process stream). Hydrates (moisture) in a process stream could create a bottleneck in downstream operations. Towers, often the tallest vessels in a plant, are where separation or stripping processes take place. Vessels can operate 24 hours per day, and in all seasons. Many process fluids must maintain a certain temperature to ensure flow ability and stability. For this reason, the vessels that contain these fluids must be protected, particularly in areas of the country that have extremely low temperatures, such as Alaska, and in the winter months in the lower 48 states. Heat tracing, either steam or electrical in design, is placed on strategic areas of the vessel to ensure the contents remain at a constant temperature.

Desiccant a specialized substance contained in a dryer that removes hydrates (moisture) from the process stream

15.1 Purpose of Vessels

Vessels are used to carry out process operations such as distillation, drying, filtration, stripping, and reaction. These operations usually involve many different types of vessels, ranging from large towers to small additive and waste collection drums. Vessels are also used to provide intermediate storage between processing steps. They can provide residence time for reactions to complete or for contents to settle.

A storage **tank** is a common type of vessel, because every process requires containers to hold feed and other materials. Tanks are used for raw materials and additives (process inputs), intermediate (not-yet-finished) products, final products (process outputs), and wastes (recoverable or nonrecoverable off-spec products and by-products).

Tank a large container or vessel for holding liquids and/or gases.

Drum a specialized type of storage tank or intermediary process vessel.

Dryer a vessel containing desiccant and screens across which process streams flow to have moisture (hydrates) removed

Filter a device used to remove liquid, gas, or solid particulates from the process stream.

Bin/hopper a vessel that typically holds dry solids.

American Society of Mechanical Engineers (ASME) organization that provides laws of regulation for boilers and pressure vessels.

Atmospheric tank an enclosed vessel that operates at atmospheric pressure; usually cylindrical in shape, equipped with either a fixed or floating roof, and containing nontoxic vapor liquids.

A **drum** is used for the collection and separation of material in processes for storage and for the separation of materials before furthering processing. Drums operate either under greater-than-atmospheric pressure or under vacuum pressure.

A **dryer** removes moisture from process streams. Dryers ensure that hydrates do not reach the lower-temperature sections of a process plant, such as distillation, or that hydrates are kept from processes that react with moisture.

A **filter** is used to remove liquid, gas, or solid particulates from the process stream. Filters help prevent damage to downstream equipment.

Bins or **hoppers** typically hold dry solids.

15.2 Types of Tanks

Tanks can reside above or below ground depending on service and location area. In the design and manufacture of tanks, key factors affect wall thickness, materials of construction, and the shape of the tank. These factors are pressure, temperature, and chemical properties. Tanks might be pressurized or operate at atmospheric pressure, depending on contents. Process tank specifications and operation are governed by specification codes from the **ASME** and regulations set by the EPA.

Gasoline (petroleum) storage tanks located at service stations are a good example of underground tanks having problems with corrosion. In the early 1980s, these tanks began to leak, contaminating soil and causing fumes to rise in nearby buildings. As a result, the EPA passed laws in 1988 regulating those tanks. Many are still being removed or replaced.

An **atmospheric tank** is an enclosed vessel in which atmospheric pressure is maintained (i.e., these tanks are neither pressurized nor placed under a vacuum; they maintain the same pressure as the air around them). Atmospheric tanks are usually cylindrical (round) in shape and are equipped with fixed or floating roofs, or both. Figure 15.2 shows an example of an atmospheric tank.

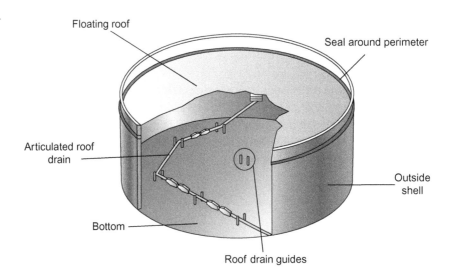

Figure 15.2 Example of a floating roof atmospheric tank.

Atmospheric tanks are usually made of steel plates that are welded together in large sections. Because they do not seal as tightly as pressurized tanks, atmospheric tanks are appropriate only for substances that do not contain toxic vapors or high vapor pressure liquids.

Pressurized tank an enclosed vessel in which a greater-than-atmospheric pressure is maintained.

A **pressurized tank** is an enclosed vessel in which a pressure greater than atmospheric pressure is maintained. Figure 15.3 shows examples of common pressurized tanks.

A. Spherical

B. Cylindrical (Bullet)

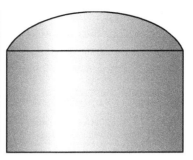
C. Hemispheroid

Figure 15.3 Spherical, cylindrical (bullet), and hemispheroid tanks.

CREDIT: **A.** Arcady/Fotolia (*sphere*). **B.** supakitmodn/Fotolia (bullet).

The most common types of pressurized tanks are:

- **Spherical tank**—a type of pressurized storage tank that is used to store volatile or highly pressurized material. Spherical tanks are sometimes referred to as "round" tanks. Their shape allows even distribution of pressure throughout the vessel. Figure 15.4 shows an example of a spherical tank.
- Cylindrical (bullet) tank—used for moderately pressurized contents; rounded ends distribute pressure more evenly than can be done in a nonrounded tank. A **cylinder** is a vessel that can hold extremely volatile or high-pressure materials (such as propane).
- Hemispheroid tank—used for low-pressure substances.

Did You Know?

The spherical tank is commonly called a "Hortonsphere." It was invented by Horace Ebenezer Horton, and the first field tanks were built in Port Arthur, Texas in 1923.

Spherical tank a type of pressurized storage tank that is used to store volatile or highly pressurized material; also referred to as "round" tanks.

Cylinder a vessel that can hold extremely volatile or high-pressure materials.

Figure 15.4 Spherical tank.
CREDIT: muratart/Fotolia.

15.3 Common Components of Vessels

Process technicians will work in various positions on the job site and should be familiar with the many different types of process vessels and their usage.

Tanks are used in many services, including auxiliary processes associated with the main process stream. The nature of the service dictates the type of tank used, its material of fabrication, its size, and regulatory ASME rating. Some of the major components of tanks are:

- **Floating roof**—a steel or plastic roof used on storage tanks; it floats upon the surface of the stored liquid and is used to decrease vapor space and reduce potential for evaporation. Floating roofs can be either internal or external. They use a flexible seal to prevent leakage. Because of their design, floating roofs are not appropriate for pressurized fluids.
- **Articulated drain**—hinged drain, attached to the roof of an external floating roof tank, which removes water from the roof. An articulated drain moves up and down as the roof and the fluid levels rise and fall.
- **Blanketing**—the process of putting an inert gas, usually nitrogen, into the vapor space above the liquid in a tank to prevent air leakage into the tank.

Floating roof a type of vessel covering (steel or plastic), used on storage tanks, that floats upon the surface of the stored liquid and is used to decrease vapor space and reduce potential for evaporation.

Articulated drain a hinged drain, attached to the roof of an external floating roof tank, that moves up and down as the roof and the fluid levels rise and fall.

Blanketing the process of putting an inert gas, usually nitrogen, into the vapor space above the liquid in a tank to prevent air leakage into the tank.

Blanketing a tank reduces the amount of oxygen present and decreases the risk of fire and explosion. Blankets also reduce the risk of tank *implosion* (collapse) by preventing a vacuum from being created as the tank is being emptied. Figure 15.5 shows an example of a tank blanketed with an inert gas.

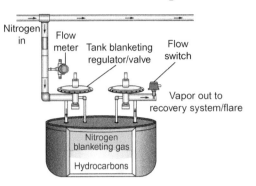

Figure 15.5 Tank blanketed with inert gas.

Foam chamber a reservoir and piping installed on liquid storage vessels and containing fire-extinguishing chemical foam.

- **Foam chamber**—a reservoir and piping installed on liquid storage vessels, which contain chemical foam used to extinguish fires within a tank. Figure 15.6 shows an example of a foam chamber.

Figure 15.6 Foam chamber putting out a fire.

Sump an area of temporary storage located at the bottom of a tank from which undesirable material is removed.

Did You Know?

In some regions of the United States and on oil rigs, sumps are sometimes referred to as "possum bellies."

- **Sump**—an area located at the bottom of a vessel that receives and temporarily stores drainage from the lowest point in that vessel. Figure 15.7 shows an example of a sump.

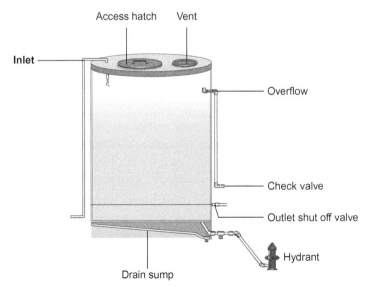

Figure 15.7 Sump.

- **Mixer**—a device used to stir chemicals or other substances; mixers can be mounted vertically or horizontally in a vessel. Mixers can be operated at the tank site using motors located outside the vessel or be started remotely, depending on the type of service. Figure 15.8 shows an example of a mixer.

Mixer a device used to mechanically combine chemicals or other substances; also known as an agitator.

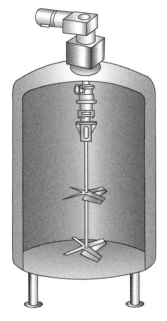

Figure 15.8 Mixer.

- **Gauge hatch**—an opening on the roof of a tank that is used to check tank levels and obtain samples of the tank contents. Figure 15.9 shows an example of a gauge hatch.

Gauge hatch an opening on the roof of a tank that is used to check tank levels and obtain samples of the tank contents.

Figure 15.9 Gauge hatch.

- **Manway**—an opening in a vessel that permits human entry for inspection and maintenance or repair. Most manways are flanged openings located on the roof or side of a tank. Figure 15.10 shows an example of a manway with bolts and nuts removed in preparation for entry.

Manway an opening in a vessel that permits entry for inspection and repair.

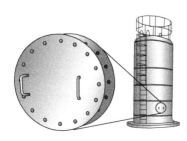

A.

B.

Figure 15.10 Manway. **A.** Illustration of manway entry port. **B.** Photo of manway in lower section of the vessel (see arrow).

CREDIT: B. wittybear/Fotolia.

Heat tracing a coil of heated wire or tubing that adheres to or is wrapped around a pipe in order to increase the temperature of the process fluid, reduce fluid viscosity, and facilitate flow.

Vapor recovery system the process of recapturing vapors by methods such as chilling or scrubbing. They are then purified, and the vapors or products are sent back to the process, sent to storage, or recovered.

- **Heat tracing**—a coil of heated wire or tubing that adheres to or is wrapped around a pipe in order to maintain or increase the temperature of the process fluid, reduce fluid viscosity, and facilitate flow. Both steam and electrical types of heat tracing are used. Process piping using a heat tracing system is covered with thermal insulation to help maintain its temperature and also to protect personnel from burn injury.

- **Vapor recovery system**—the process of capturing evaporated fluid or vapors in the tank for recycling into the system or for reuse in another area. This method employs processes such as chilling or scrubbing. The recaptured materials are then purified, and the vapors or products are sent back to the process, sent to storage, or recovered. Figure 15.11 shows an example of a vapor recovery system.

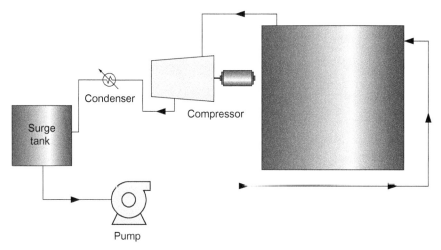

Figure 15.11 Vapor recovery system.

Vortex breaker a metal plate, or similar device, placed inside a cylindrical, cone-shaped, or other type operating unit that prevents a vortex from being created as liquid is drawn out of the vessel.

Vortex the cone formed by a swirling liquid or gas.

- **Vortex breaker**—a metal plate, or similar device, placed into the drain line of a cylindrical or cone-shaped vessel. It prevents a vortex from being created as liquid is drawn out of the tank. A **vortex** is the cone formed by a swirling liquid or gas. Vortex breakers are used to create a more linear flow from the tank and to prevent pump cavitation. Vortex breakers are also used in other vessels that need to maintain a constant flow through the equipment. Figure 15.12 shows an example of a vortex breaker.

Figure 15.12 Vortex breaker.

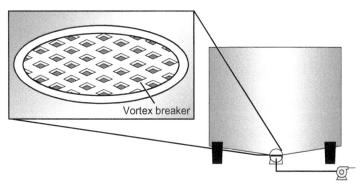

Baffle a metal plate, placed inside a tank or other vessel, that is used to alter the flow of chemicals or facilitate mixing.

- **Baffle**—a metal plate, placed at strategic positions inside a vessel, that alters the flow of the material in the vessel and can help to facilitate mixing. Figure 15.13 shows an example of a baffle.

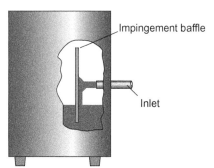

Figure 15.13 Baffle.

- **Weir**—a flat or notched dam or barrier to liquid flow that is normally used either to measure fluid flows or to maintain a given depth of fluid (setting the level). Figure 15.14 shows an example of a vessel (decanter) with a weir. In this example, anything that is above the weir flows over. Oil is lighter than water; it overflows the weir and is removed as the heavier component out the bottom. Because gasoline is also lighter than water, gasoline overflows the weir and is removed as gas out the top. This weir is used to separate three substances (oil, gasoline, and water) in a mixture.

Weir a flat or notched dam or barrier to liquid flow that is normally used either for the measurement of fluid flows or to maintain a given depth of fluid as on a tray of a distillation column, in a separator, or other vessel.

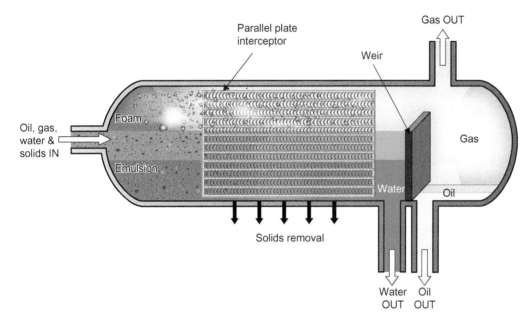

Figure 15.14 Weir in a decanter.

- **Boot**—a section at the bottom of a process vessel where water or other unwanted fluid is collected and removed. This area is located at the lowest section of the vessel. Figure 15.15 shows an example of a boot.

Boot a section in the lowest area of a process drum where water or other liquid is collected and removed.

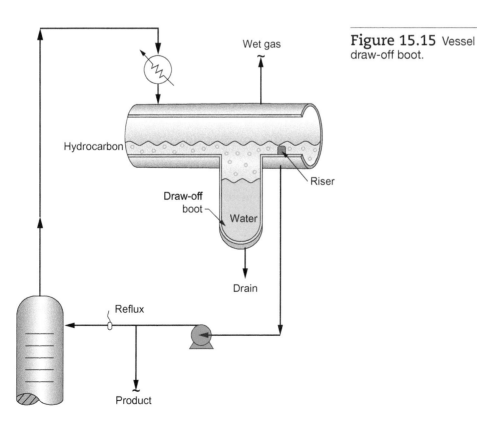

Figure 15.15 Vessel draw-off boot.

Mist eliminator a device in the top of a tank, composed of mesh, vanes, or fibers, that collects droplets of mist (moisture) from gas to prevent it from leaving the tank and moving forward with the process flow.

Vane separator a device, composed of metal vanes, used to separate liquids from gases or solids from liquids.

Containment wall an earthen berm or constructed wall used to protect the environment and people against tank failures, fires, runoff, and spills; also called a *bund wall*, *bunding*, *dike*, or *firewall*.

- **Mist eliminator**—a device placed in the top of a tank, composed of mesh, vanes, or fibers, that collects droplets of mist (moisture) from gas to prevent it from leaving the tank and moving forward with the process flow.
- **Vane separator**—a device, composed of metal vanes, used to separate liquids from gases or solids from liquids.

15.4 Containment Walls, Dikes, and Firewalls

OSHA requires all above-ground tanks to have a containment system built around them for personal and environmental safety in case of leaks, spills, or other accidents. A containment system, usually made from earth or concrete, can take the form of a dike, firewall, or containment wall. A **containment wall**, or dike, is a wall used to protect the environment and people against tank failures, fires, runoff, and spills. It can be earthen, masonry, steel, or concrete. Containment walls must be high enough to hold the entire contents of the largest tank within the enclosed area. Figure 15.16 shows an example of a containment wall. A firewall (also called a *bund*) is an earthen bank or concrete wall built around oil storage tanks to contain the oil in case of a spill or rupture.

Figure 15.16 Containment wall (also called firewall, bunding, bund wall, or dike).
CREDIT: Kat72/Shutterstock.

A containment system serves several functions:

- Contains chemical spills in a small area in order to minimize safety risks and trap contaminants and/or hazardous materials before they can spread to other areas
- Protects the soil, water, and the environment from contaminants
- Protects humans from potential hazards (e.g., chemical release)
- Contains wastewater and contaminated rainwater until it can either be drained into a proper process sewer system line or be vacuumed from the area
- Protects against the spread of fires

With certain types of tanks, in the event of heavy rainfall, technicians must open a valve and drain the liquid from within the containment wall to prevent the tank from floating. This rainfall buildup must be tested to be free of contaminants prior to draining.

Reactor a vessel in which chemical reactions are initiated and sustained.

Batch reaction a carefully measured and controlled process in which raw materials (reactants) are added together to create a reaction that makes a single quantity (batch) of the final product.

Continuous reaction a chemical process in which raw materials (reactants) are continuously being fed in and products are continuously being formed and removed from the reactor vessel.

15.5 Reactors: Purpose and Types

A **reactor** is a vessel in which chemical reactions are initiated and sustained. Within a reactor, raw materials are combined at various flow rates, pressures, and temperatures, and they react to form a product. These reactions can be either batch or continuous. **Batch reactions** are complete loads done start to finish and then removed. **Continuous reactions** are an ongoing process with raw materials entering and product being removed from the reactor.

Reactors come in many shapes and sizes. The most common types of reactors are stirred tank, fixed bed, fluidized bed, tubular, and reaction furnace. The type of catalyst and the properties of the reactants determine which type of reactor will be used.

- A **stirred tank reactor** (Figure 15.17) contains a mixer or agitator mounted to the tank. The shell, or jacket, of this type reactor can be heated or cooled, depending on the process and design.

Stirred tank reactor a reactor vessel that contains a mixer or agitator to improve mixing of reactants.

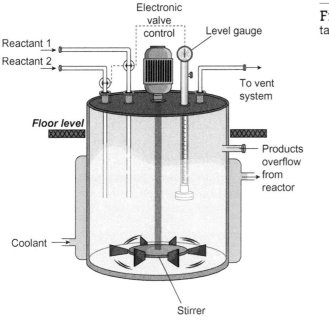

Figure 15.17 Stirred tank reactor.

- A **fixed bed reactor** (Figure 15.18) is a reactor vessel in which the catalyst bed is stationary and the reactants are passed over it. In this type of reactor, the catalyst occupies a fixed position; it is not designed to leave the reactor.

Fixed bed reactor a reactor vessel in which the catalyst bed is stationary and the reactants are passed over it.

Figure 15.18 Fixed bed reactor.

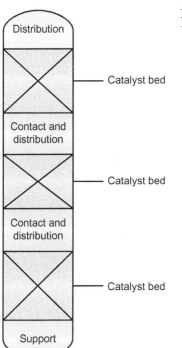

Fluidized bed reactor a reactor that uses high-velocity fluid to suspend or fluidize solid catalyst particles.

- A **fluidized bed reactor** (Figure 15.19) uses high-velocity fluid to suspend and separate solids. The reactor feed is mixed with the suspended catalyst where the reaction takes place.

Figure 15.19 Fluidized bed reactor.

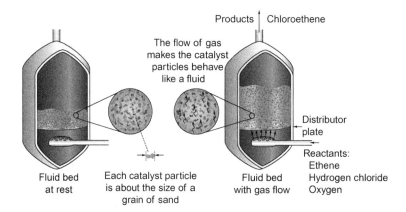

Tubular reactor a continuously flowed vessel in which reactants are converted in relation to their position within the reactor tubes, not influenced by residence time in the reactor.

Reaction furnace a reactor that combines a firebox with tubing to provide heat for a reaction that occurs inside the tubes.

- A **tubular reactor** (Figure 15.20) is a cylindrical heat exchanger used to contain a reaction. Based on process requirements, the design of a tubular reactor can range from a simple jacketed tube to a multipass shell and tube exchanger.
- A **reaction furnace** combines a firebox with tubing to provide heat for a reaction that occurs inside the tubes.

Figure 15.20 Tubular reactor.

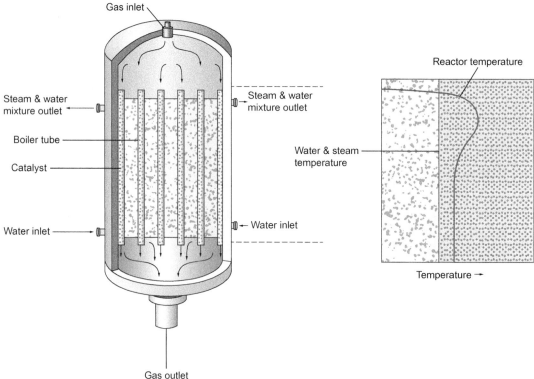

15.6 Operational Hazards

Many hazards are associated with vessels. Table 15.1 lists some of those hazards. Process technicians must stay alert to these hazards in order to protect people, the environment, and equipment.

Retraining might be required if safety measures are not upheld.

Table 15.1 Hazards Associated with Improper Operation

Improper Operation	Possible Impacts			
	Individual	Equipment	Production	Environment
Overfilling	Exposure to hazardous chemicals; possible injury	Damage to tank and equipment, especially floating roof tanks	Added cost for cleanup; lost product or raw material	Spill, possible fire, vapor release
Putting wrong or off-spec material in storage tank	Possible action related to accountability if root cause indicates operator error	Contamination of product, could create an undesired reaction between materials	Added cost to remove material, clean tank, and re-run material	Spills when removing material and cleaning tank, or unwanted chemical reactions
Misalignment of blanket system	Possible action related to accountability for actions	Loss of blanket; collapse of tank because of vacuum	Loss of production because of reduced storage	Vapor release
Misalignment of pump systems	Accountability for actions	Damaged pump	Contamination of other tanks	Spill
Pulling a vacuum on a tank while emptying	Accountability for actions	Collapse of tank because of vacuum	Loss of production because of reduced storage	Vapor release
Overpressure	Accountability for actions; exposure to hazardous chemicals; possible personal injury	Rupture of vessel	Loss of production because of reduced storage	Vapor release, fire, or explosion
Improper valve lineup	Retraining if root cause not equipment but operator error	Possible over-pressure or deadhead	Cross contamination or loss of feed stream	
Loss of nitrogen flow	Retraining if root cause not equipment but operator error	Possible vacuum and air entry into tank, possible corrosion	Possible decomposition of material	Possible vapor release
Cross contamination	Retraining in the case of misaligned valves	Possible corrosion or erosion of vessel	Possible out of specification material	
Failure of vent system	Possible personal injury if tank ruptures	Possible rupture of vessel from overpressure	Possible decomposition of material	Possible release to atmosphere or spill to the ground surface
Leaks & spills	Possible personal injury from contact with material	Vessel could have a weak spot causing a leak	Loss of production because of loss of material	Ground or air pollution because of release of material
Chemical reactions	Possible personal injury if exposed to the reaction	Possible corrosion or erosion of the vessel because of unwanted chemical reaction	Loss of production because of contaminated product	Possible environmental contamination if reaction causes a release from the vessel

15.7 Monitoring and Maintenance Activities

When monitoring and maintaining vessels, process technicians must always remember to look, listen, check, and smell for the items listed in Table 15.2.

Failure to perform proper maintenance and monitoring could affect the process and result in harm to people, the environment, and equipment.

Table 15.2 Monitoring and Maintenance Activities for Vessels

Look	Listen	Feel	Smell
▪ Monitor level ▪ Check firewalls, sumps, and drains ▪ Check auxiliary equipment associated with the vessel ▪ Check to ensure the drain remains closed ▪ Visually inspect for leaks (especially if associated with abnormal odor) ▪ Check sewer valves ▪ Use level gauges and sight glasses to monitor level ▪ Monitor pressure ▪ Inspect for corrosion and discoloration ▪ During heavy rainfall, open valves as needed to prevent vessels from floating	▪ Listen for abnormal noise	▪ Inspect for abnormal heat on vessels and piping ▪ Check for excessive vibration on pumps/mixers	▪ Be aware of abnormal odors that could indicate leakage ▪ Use sniffers to detect gas leaks and vapors

Symbols for Vessels and Reactors

In order to accurately locate vessels (e.g., tanks, drums) and reactors on a piping and instrumentation diagram (P&ID), process technicians need to be familiar with the symbols that represent different types of tanks, drums, vessels, and reactors. Figure 15.21 shows examples of vessel symbols.

Figure 15.21 Vessel symbols.

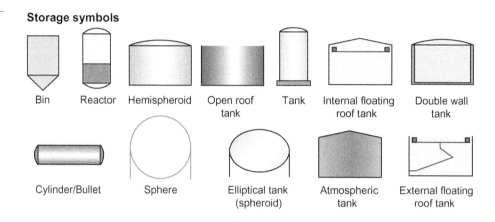

Summary

Vessels are integral to the completion of products and services associated with the process unit. Materials are processed, treated, or stored within many different types of vessels. Vessels enable the process industries to create and store large amounts of products for sale and distribution.

Vessels, such as tanks, towers, reactors, drums, dryers, filters, cylinders, hoppers, bins, and other similar containers are used to process or store materials. Vessels vary greatly in design (e.g., size and shape) based on the requirements of the process. Factors that affect vessel design may include pressure requirements, type of product contained in the vessel (liquid, gas, or solid), temperature requirements, corrosion factors, and volume.

Reactors are specialized vessels that are used to contain a controlled chemical reaction, changing raw materials into finished products. Reaction variables include temperature, pressure, time, concentration, surface area, and other factors. Like vessels, reactor designs vary widely based on the chemical reaction that must occur in the process.

The process technician must recognize and understand vessel components, including floating roof, articulated drain, blanketing, foam chamber, sump, mixer, gauge hatch, manway, vapor recovery system, vortex breaker, baffle, weir, boot, mist eliminator, and vane separator.

Some vessels have a containment system built around them for personal and environmental safety in case of leaks, spills, or other accidents. A containment system, usually made from earth or concrete, can take the form of a dike, firewall, or containment wall.

Process technicians must always remember to look, listen, feel, and smell when monitoring and maintaining vessels. They also must be aware of the hazards of improper operations of vessels in order to protect people, the environment, and equipment.

Process technicians need to be familiar with the symbols that represent different types of tanks, drums, vessels, and reactors, and be able to accurately locate vessels on a piping and instrumentation diagram (P&ID).

Checking Your Knowledge

1. Define the following terms:
   a. articulated drain
   b. baffle
   c. blanket
   d. boot
   e. containment wall
   f. floating roof
   g. gauge hatch
   h. manway
   i. mist eliminator
   j. mixer
   k. sphere
   l. sump
   m. vane separators
   n. vapor recovery
   o. vortex breaker
   p. weir

2. Which of the following tanks would the most appropriate choice for storing volatile substances under pressure? Provide the label letter and the tank name.

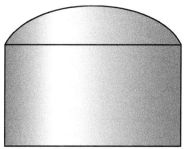

A.   B.   C.

3. *(True or False)* Atmospheric tanks are good for storing substances with toxic vapors.

4. List three purposes for the use of dikes and containment walls.

5. List at least three hazards or impacts associated with improper tank operation.

6. List at least three things a process technician should listen, feel, and smell for during normal monitoring and maintenance.

NOTE: Answers to Checking Your Knowledge questions are in Appendix I.

Student Activities

1. Given a picture of a high pressure storage tank, identify the following components:
   a. blanket
   b. manway
   c. spherical tank
   d. vapor recovery system
   e. vortex breaker

2. Describe the following types of reactors, including design, purpose, and how they work:
   a. stirred tank
   b. fixed bed
   c. fluidized bed
   d. tubular
   e. reaction furnace

3. Given a piping and instrumentation diagram (P&ID), identify vessels found on the unit.

Chapter 16
Pumps

Objectives

Upon completion of this chapter, you will be able to:

16.1 Describe the purpose or function of pumps in the process industries. (NAPTA Pumps 1) p. 227

16.2 Explain the difference between the two common types of pumps used in the process industries: positive displacement and centrifugal. (NAPTA Pumps 1, 2) p. 227

16.3 Explain the difference between the rotary and reciprocating types of positive displacement pumps and their operation. (NAPTA Pumps 5, 6) p. 228

16.4 Explain the difference between the centrifugal and axial types of dynamic pumps and their operation. (NAPTA Pumps 3, 4, 7) p. 229

16.5 Discuss the hazards associated with the improper operation of both the positive displacement and centrifugal pump. (NAPTA Pumps 8) p. 232

16.6 Describe the monitoring and maintenance activities associated with pumps. (NAPTA Pumps 9, 10) p. 234

Key Terms

Axial pump—a type of pump classified in the dynamic category that uses a propeller or row of blades to propel liquids axially along the shaft, **p. 231.**

Cavitation—a condition inside a pump in which vapor bubbles develop in the liquid being pumped and then collapse, creating vibration and loss of flow, **p. 233.**

Centrifugal force—an apparent force exerted on the fluid by the impeller rotating in a circular pattern to move it outward from the center, **p. 230.**

Centrifugal pump—a type of pump classified in the dynamic category that uses an impeller on a rotating shaft to generate centrifugal force that is converted to pressure in the volute at the discharge to move liquids outward, **p. 230.**

Dynamic pump—a category of pump that uses velocity to increase speed and converts kinetic energy to pressure at the discharge, **p. 229.**

pumps (especially very high speed and multistage pumps) can have excess vibration or undesired overheating if deadheaded. In those cases, deadheading should not be allowed to occur. Figure 16.7 shows an example of a centrifugal pump.

Figure 16.7 Centrifugal pump. **A.** Parts of a centrifugal pump. **B.** Cutaway of centrifugal pump.
CREDIT: B. smspsy / Shutterstock.

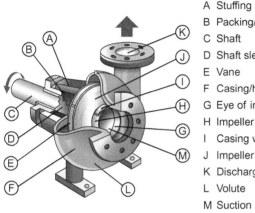

A   Stuffing box
B   Packing/seal
C   Shaft
D   Shaft sleeve
E   Vane
F   Casing/housing
G   Eye of impeller
H   Impeller
I   Casing wear ring
J   Impeller
K   Discharge/outlet
L   Volute
M   Suction intake/inlet

A.   B.

The main components of a centrifugal pump are:
- Housing (casing)
- Shaft
- Inlet (suction intake or eye)
- Outlet (discharge)
- Impeller
- Bearings and seals
- Volute

Centrifugal pumps differ from positive displacement pumps in that the amount of liquid they deliver is dependent on the discharge pressure, not the size of the chamber.

Centrifugal pumps have direct relationships among speed, velocity, pressure, and flow. As the velocity decreases, the pressure increases.

Axial Pumps

An **axial pump** uses a *propeller* or row of blades to propel liquids axially along the shaft This is in contrast to centrifugal pumps that use an impeller to force liquids to the outer wall of the chamber. Figure 16.8 shows an example of an axial pump.

Axial pump a type of pump classified in the dynamic category that uses a propeller or row of blades to propel liquids axially along the shaft.

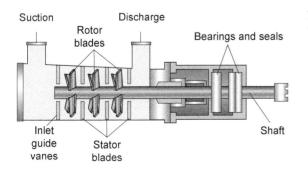

Figure 16.8 Axial pump.

The main components of an axial pump are:

- Shaft
- Inlet (intake)
- Outlet (discharge)
- Propeller (rotor blades)
- Bearings and seals

16.5 Operational Hazards

When working with pumps, process technicians should always be aware of potential hazards such as overpressurization, overheating, cavitation, vibration, and leakage. Observations are made regularly during a work shift and data are taken on major pieces of equipment and their associated systems to ensure the operating parameters are met (Table 16.1).

Table 16.1 Operational Hazards Associated with Pumps

Improper Operation	Possible Impacts			
	Individual	Equipment	Production	Environment
Overheating	Possible burns to personnel	Extreme temperatures that stress equipment components	Possible shutdown of compressor and loss of production	
Overpressurization	Leak or rupture venting process gas and causing injury to operating personnel Retraining	Leaks at seals, valves, etc. Damage to internal components	Possible shutdown of compressor causing loss of production	Leak or rupture venting process gas to the atmosphere
Cavitation		Damage to pump seal, impeller, bearings, and case	Loss of production time	
Leaks	Exposure to process gases or slipping if leak is liquid (such as oil)	Leakage and damage to component parts Fire or explosion if leak reaches hot surface	Possible shutdown of equipment causing loss of production	Harm to the atmosphere from both gas and liquid leaks
Vibration		Excessive wear on the pump seal, impeller, bearings, and case	Shutdown; loss of production time; cost of replacement parts	

Overheating

Pump overheating can be caused by improper lubrication. The moving parts, such as the shaft, bearings, and seals, these component parts can fail without lubrication and equipment damaged from generated friction and heat. This can cause mechanical failures, swelling, leakage, and decomposition of the process fluid.

Closing the discharge valve on a pump (deadheading) causes the generation of heat, allowing the liquid in the pump to be heated and vaporized by the mechanical energy of the motor. To prevent this situation, which can cause cavitation, recycle loops can be added on the discharge side of the pump. Recycle loops allow fluid flow circulation if the discharge valves or other downstream valves are closed.

Process technicians should always monitor rotating equipment for unusual sounds, unusual or excessive leaking at the seal, appearance of process fluid, and excessive heat, because operating equipment under these conditions can lead to permanent equipment damage or personal harm (e.g., burns).

Overpressurization

In Figure 16.9, a positive displacement pump, tank, and associated lines and valves are shown. Valve A is open and valve B is closed. This means the liquid is flowing through valve A and then back into the tank.

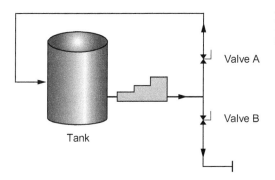

Figure 16.9 Improper valve operation (deadheading).

If a process technician wanted to redirect the flow of the liquid through valve B, that valve would be opened before valve A is closed.

If both valves are closed at the same time (deadheading), the liquid has no place to go. The result is back pressure or overpressurizaion that can damage the pump or cause serious personal injury.

Cavitation

Cavitation is a condition inside a pump in which the liquid being pumped partially vaporizes because of variables such as increased temperature and pressure drop. These variable changes cause vapor pockets (bubbles) to form and collapse (implode) inside a pump.

Cavitation occurs when the pressure at the suction eye of a pump impeller falls below the vapor pressure of the liquid being pumped. This is a very serious problem in dynamic pumps, especially centrifugal pumps. Cavitation is also a problem when the temperature of the liquid being pumped reaches its boiling point in the pump.

Key characteristics of cavitation include large pressure fluctuations, inconsistent flow rate, and severe vibration. Process technicians who identify cavitation should always try to eliminate it as quickly as possible. The resulting extreme vibration can cause excessive wear on the pump seal, impeller, bearings, and case.

To prevent cavitation, a pump should always be primed before it is started. **Priming** a pump involves filling the suction line and casing of a pump with liquid to remove vapors and to heat or chill the case metal to its service temperature. These steps eliminate the tendency of pumps to become vapor bound or lose suction.

Cavitation can occur with the pump running online if there are changes in the process stream. For example vapor bubbles can be created if the liquid becomes too hot. Cavitation can also be caused by the loss of **net positive suction head (NPSH)** in the suction line to the pump.

Pumps that have been damaged by cavitation frequently look as though the internal workings of the pump have been sandblasted or pitted, particularly on the outer edges of the impeller vanes.

Cavitation a condition inside a pump in which vapor bubbles develop in the liquid being pumped and then collapse, creating vibration and loss of flow.

Priming the process of filling the suction line and pump casing with liquid to remove vapors, and to heat or chill the case metal to its service temperature.

Net positive suction head (NPSH) pressure the amount of pressure needed at the suction of a pump to prevent cavitation.

Did You Know?

When a pump cavitates, it can sound as if rocks are being poured into the suction line. These bursts of liquid and gas create a "liquid hammer" that can seriously damage the inside of the pump and the impeller.

CREDIT: Andy Sears/Fotolia.

Leakage

Process technicians should always check pumps and drivers for sources of leaks, including process leakage, lubrication leakage, coolant leakage, steam leakage, or any associated equipment leakage. Leaks can introduce slipping hazards, exposure to harmful or hazardous substances, and process problems (e.g., inferior product produced as a result of improper feed supply).

Common places for leakage would be the pump seal or packing, cooling liquid applied to the pump, lubrication leakage at the supply or on the pump, steam leakage if the driver is a turbine, and process leakage in and around the pump equipment.

Vibration

Vibration can occur from various causes. Cavitation can cause vibration as a result of loss of suction to the pump. Damage to the pump seal or packing can create an imbalance in the pump that could cause vibration. Internal damage to pump components can also create vibration that can damage the pump. Changes in process flow to upstream and downstream equipment can create vibration and affect the pump.

Any condition that leads to vibration should be addressed quickly to avoid equipment damage and possible injury to personnel.

16.6 Monitoring and Maintenance Activities

When monitoring and maintaining pumps, process technicians must always remember to look, listen, and feel for the situations described in Table 16.2.

Failure to perform proper maintenance and monitoring could affect the process, result in equipment damage, and lead to personal injury.

Table 16.2 Monitoring and Maintenance Activities for Pumps

Look	Listen	Check (Using Correct Equipment)
■ Check oil levels to make sure they are satisfactory. ■ Check to make sure water is not collecting under the oil (water is not a lubricant, so it can cause bearing failure). ■ Check seals and flanges to make sure there are no leaks. ■ Check suction and discharge pressure gauges (note: technicians need to be aware that pressure gauges in vibrating areas can lose their calibration).	■ Listen for abnormal noises.	■ Check for excessive vibration. ■ Check for excessive heat.

Pump Symbols

In order to locate pumps on a piping and instrumentation diagram (P&ID) accurately, process technicians need to be familiar with the different types of pumps and their symbols. Figure 16.10 shows a few of the symbols for both centrifugal and positive displacement pumps.

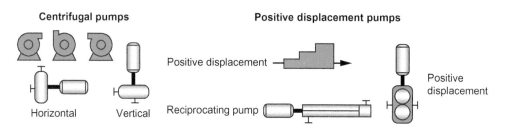

Figure 16.10 Common pump symbols.

Summary

Pumps are used to increase the pressure and move liquids from one location to another. Pumps are sized for capacity according to service needs. They are dynamic (centrifugal or axial) or positive displacement (reciprocating or rotary) in design.

Positive displacement pumps trap a specific amount of liquid within the housing and displace that volume of liquid at a higher pressure.

Centrifugal pumps operate on the principal of centrifugal force to increase the pressure of a liquid at its discharge.

Many hazards are associated with improper pump operation, including overpressurization, overheating, cavitation, vibration, and leakage. The process technician has a responsibility to check equipment in the operating area regularly for these hazards.

Checking Your Knowledge

1. Explain the purpose of pumps in the process industries.
2. Is the diagram below an example of a centrifugal pump or a positive displacement pump?

3. Which type of pump uses a piston to force liquids out of a chamber?
   a. axial
   b. centrifugal
   c. positive displacement
4. Label the following parts on the centrifugal pump.
   a. casing
   b. discharge nozzle
   c. eye of impeller
   d. shaft
   e. volute
   f. packing
   g. suction intake

   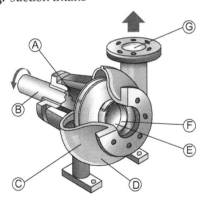

5. Label the following parts on the reciprocating pump.
   a. cylinder
   b. seal
   c. inlet valve
   d. outlet valve
   e. piston
   f. suction
   g. discharge

   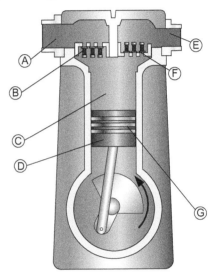

6. List four hazards associated with improper pump operation.
7. List five monitoring and maintenance activities that a process technician should perform when working with pumps.

NOTE: Answers to Checking Your Knowledge questions are in Appendix I.

Student Activities

1. Given a cutaway image of a centrifugal pump, identify the following components:
   a. casing
   b. discharge
   c. impeller
   d. shaft
   e. suction eye
   f. volute

2. Given a cutaway image of a reciprocating pump, identify the following components:
   a. casing
   b. connecting rods
   c. inlet valve
   d. outlet valve
   e. piston
   f. seal

3. Given a piping and instrumentation diagram (P&ID), identify all of the pumps as being centrifugal or positive displacement in type.
4. Divide into small groups. Each group will need a small pail and water. All students will need eye protection.
   a. Fill the small pail with water until it is half full.
   b. Locate an open area away from other individuals or obstructions.
   c. In the open area, have one person grasp the pail by its handle and swing it in a circular motion, arms fully extended, repeatedly raising it over the head and back down to knee level. After a minute, have the person gradually slow and stop the rotation.
   d. Discuss what happened. Did the water stay in the pail or did it spill out?
   e. What is the name for the principle at work in this experiment? Discuss how it is applicable to process technology.
5. Examine the suction and discharge actions of a reciprocating pump by doing the following:
   a. Obtain a syringe (without a needle) and a cup of water.
   b. Depress the syringe plunger all the way into the housing.
   c. Place the syringe in the cup of water and pull the plunger back until the syringe housing is full of water.
   d. Lift the syringe out of the water and then depress the plunger so the water is forced out of the syringe, into the cup.
   e. Repeat steps b through d several times. Each time, vary the amount of pressure applied to the plunger when forcing the water out of the syringe.
   f. Examine what happens. Did the amount of liquid discharged change as the pressure changed, or did it remain the same?

Chapter 17
Compressors

Objectives

Upon completion of this chapter, you will be able to:

17.1 Describe the purpose or function of compressors in the process industries. (NAPTA Compressors 1) p. 238

17.2 Explain the difference between a pump and a compressor in terms of what function each performs. (NAPTA Compressors 2) p. 238

17.3 Explain the difference between the two more common types of compressors used in the process industries: positive displacement and dynamic. (NAPTA Compressors 3) p. 239

17.4 Explain the differences between the rotary and reciprocating types of positive displacement compressors and their components. (NAPTA Compressors 6, 7, 8) p. 239

17.5 Identify the primary parts and operations of typical dynamic compressors: centrifugal and axial. (NAPTA Compressors 4, 5) p. 240

17.6 Discuss the hazards associated with the improper operation of both types of compressors. (NAPTA Compressors 9) p. 243

17.7 Describe the monitoring and maintenance activities associated with compressors. (NAPTA Compressors 10, 11) p. 245

Key Terms

Axial compressor—a dynamic-type compressor that uses a series of rotor blades with a set of stator blades between each rotating wheel and creates gas flow that is axial (parallel to) the compressor shaft, **p. 241**.

Blower—a mechanical device, either centrifugal or positive displacement in design, that has a lower ratio of pressure change from suction to discharge and is not considered to be a compressor; also can be termed a *fan*, **p. 242**.

Centrifugal compressor—a dynamic compressor in which the spinning impeller creates outward (centrifugal) force, increasing the velocity of the gas flowing from the inlet to the outer tip of the impeller blade; it converts velocity to pressure, **p. 241**.

Compression ratio—the ratio of discharge pressure (psia) to inlet pressure (psia), **p. 242**.

Compressor—mechanical device used to increase pressure of a gas or vapor, **p. 238**.

Discharge—in process technology, a term that refers to the outlet side of a pump, compressor, fan, or jet, **p. 238.**

Dynamic compressor—a category of compressor that adds kinetic energy/velocity (speed) to the gas and converts that speed to pressure at the discharge, **p. 240.**

Fan—see *blower*, **p. 242.**

Positive-displacement compressor—a device that uses screws, pistons, sliding vanes, lobes, gears, or diaphragms to trap a specific volume of gas, and reduce its volume, thereby increasing pressure at the discharge, **p. 239.**

Reciprocating compressor—a positive-displacement compressor that uses the inward stroke of a piston or diaphragm to draw gas into a chamber (intake) and then positively displace the gas using an outward stroke (discharge), **p. 240.**

Rotary compressor—a positive-displacement compressor that uses a rotating motion and internal elements of either screw, sliding vane, or lobe to pressurize and move the gas through the device, **p. 239.**

Suction—in process technology, a term that refers to the inlet side of a pump, compressor, fan, or jet, **p. 238.**

17.1 Introduction

Compressors are an important part of the process industries. The primary function of a compressor is to increase the pressure of gases for use in a facility.

Compressors are used in a wide variety of applications. They can be used to accelerate or compact gases (e.g., carbon dioxide, nitrogen, light hydrocarbons) and to provide the air that is used to control instruments or to operate equipment power tools.

The two most common compressor designs are positive displacement and dynamic. Positive-displacement compressors can be either reciprocating or rotary. Dynamic compressors use speed or kinetic energy to increase the pressure of gases. Both dynamic and positive-displacement compressors can be designed as single-stage or multistage, depending on process service requirements.

Dynamic compressors are more commonly used than positive-displacement compressors because they have a larger capacity and require less maintenance. They also can be less expensive and more efficient depending on the service application.

All compressors require a drive mechanism (e.g., electric motor, turbine) to operate and are rated according to capacity (referencing discharge pressure in psi and flow rate in cubic feet per minute). All industrial compressors require auxiliary components for cooling, lubrication, filtering, instrumentation, and control. Some compressors require a gearbox between unit and the driver to increase the speed of the compressor (Figure 17.1).

17.2 Differences Between Compressors and Pumps

Compressor mechanical device used to increase pressure of a gas or vapor.

Suction in process technology, a term that refers to the inlet side of a pump, compressor, fan, or jet.

Discharge in process technology, a term that refers to the outlet side of a pump, compressor, fan, or jet.

A **compressor** is a mechanical device used to increase pressure of a gas or vapor in a process system. Compressors and pumps are both commonly used in the process industries to move products. In addition, both use centrifugal or positive-displacement principles of operation. They both have similar operation and maintenance tasks. Factors that are important to both compressors and pumps are (1) the *line-up* (opening necessary valves in a piping system prior to placing a device in service), (2) **suction** (normally meaning the inlet side of a pump, compressor, fan, or jet), and (3) **discharge** (normally meaning the outlet side of a pump, compressor, fan, or jet).

The major difference between pumps and compressors is that pumps are used to move liquids from one location to another in a process system, and compressors are used to compress and move gases. Compressors cannot tolerate liquid, because liquids are not compressible. Compressors are used in gas service. Pumps are used in liquid and slurry service and are not designed for the entry of gas or vapor. Start-up procedures are different for compressors and pumps.

Figure 17.1 Flowchart of compressor types.

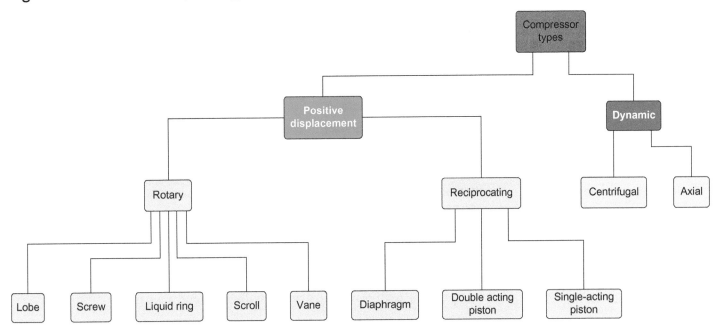

17.3 Common Types of Compressors

Positive-Displacement Compressors

Positive-displacement compressors use screws, pistons, sliding vanes, lobes, gears, or diaphragms to deliver a set volume of gas with each stroke. The two main categories of positive-displacement compressors are rotary and reciprocating.

17.4 ROTARY COMPRESSORS A **rotary compressor** is a positive-displacement compressor that uses a rotating motion to move gases. Rotary compressors move gases by rotating a screw, lobes, or a set of vanes. As these screws, lobes, or vanes rotate, gas is drawn into the compressor by negative pressure on one side and forced out of the compressor (discharged) through positive pressure on the other. Figure 17.2 shows an example of sliding vane, lobe, and rotary screw compressors.

Positive-displacement compressors a device that uses screws, pistons, sliding vanes, lobes, gears, or diaphragms to trap a specific volume of gas, and reduce its volume, thereby increasing pressure at the discharge.

Rotary compressor a positive-displacement compressor that uses a rotating motion and internal elements of either screw, sliding vane, or lobe to pressurize and move the gas through the device.

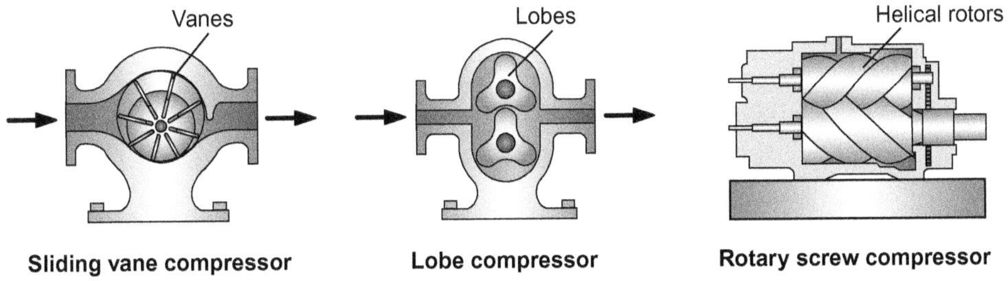

Figure 17.2 Examples of rotary compressors.

The diagram in Figure 17.3 shows the main components of a rotary compressor:

- Housing (casing)
- Shaft
- Inlet (suction)
- Outlet (discharge)
- Lobes, gears, or vanes

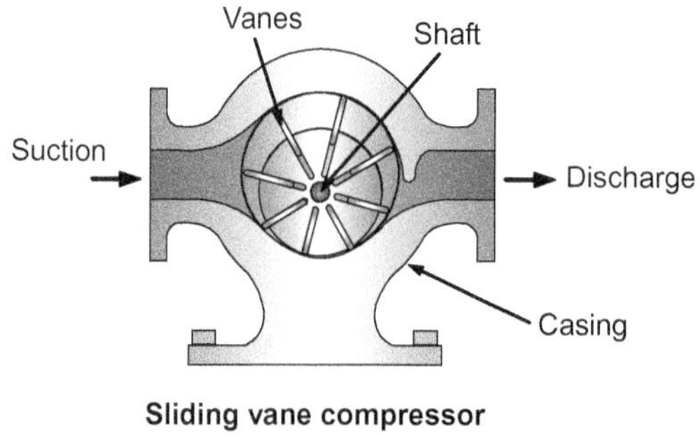

Figure 17.3 Rotary (sliding vane) compressor components.

Sliding vane compressor

Reciprocating compressor a positive-displacement compressor that uses the inward stroke of a piston or diaphragm to draw gas into a chamber (intake) and then positively displace the gas using an outward stroke (discharge).

RECIPROCATING COMPRESSORS A **reciprocating compressor** uses the inward stroke of a piston or diaphragm to draw gas into a chamber (intake) and then positively displace the gas using an outward stroke (discharge).

A good example of a reciprocating compressor is a manually operated bicycle pump. In this compressor, air is drawn into the chamber when the handle (attached to a piston) is pulled up and is forced out when the handle is pushed down. Figure 17.4 shows an example of a reciprocating compressor and how gas is drawn in and discharged.

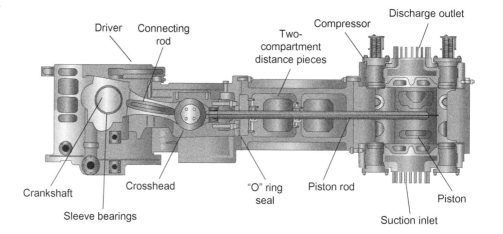

Figure 17.4 Piston-type reciprocating compressor and its components.

The main components of a reciprocating compressor are:

- Inlet (suction)
- Outlet (discharge)
- Pistons
- Piston rod or connecting rod
- Driver
- Counterweight
- Valves

17.5 Dynamic Compressors

Dynamic compressor a category of compressor that adds kinetic energy/velocity (speed) to the gas and converts that speed to pressure at the discharge.

A **dynamic compressor** is a continuous flow machine that uses speed (instead of pistons) to increase the pressure of gas at the discharge. Dynamic compressors are either centrifugal or axial, and they use principles such as centrifugal force or axial movement to increase the pressure of the gas.

CENTRIFUGAL COMPRESSORS A **centrifugal compressor** is a dynamic compressor in which the gas flows from the inlet located near the suction eye to the outer tip of the impeller blade. In a centrifugal compressor, the impeller spins, creating centrifugal force, which increases the velocity of the gas. The velocity is converted to pressure as the gas leaves the tips of the impeller, hits a diffuser, and enters either the next stage impeller suction eye or the discharge volute. As pressure increases, so does gas flow.

Centrifugal compressors can have a single stage or multiple stages, and the stages can be contained in one casing (case) or several different cases. Figure 17.5 shows examples of single-stage and multistage centrifugal compressors. Centrifugal compressors generally have split line cases that open in the midline and allow the top half to be removed; this avoids damage to the impellers.

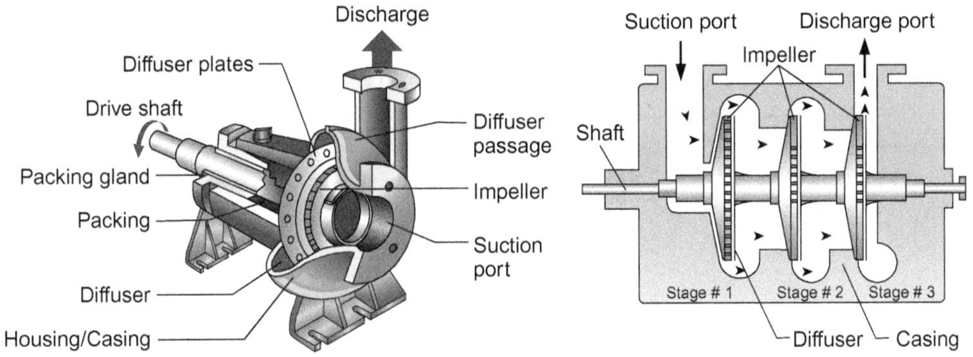

Centrifugal compressors a dynamic compressor in which the spinning impeller creates outward (centrifugal) force, increasing the velocity of the gas flowing from the inlet to the outer tip of the impeller blade; it converts velocity to pressure.

Figure 17.5 Centrifugal compressors. **A.** Single-stage; **B.** Multistage.

The main components of a centrifugal compressor are:

- Housing (casing)
- Shaft
- Suction eye (port or inlet)
- Discharge (outlet)
- Impeller
- Diffuser

Centrifugal compressors differ from positive-displacement compressors. The amount of gas centrifugal compressors deliver is dependent on the discharge pressure, not on the size of the chamber.

In centrifugal compressors, there are direct linear relationships among impeller speed, velocity, pressure, and flow. As the impeller speed increases, velocity increases. As velocity increases, pressure increases. As pressure increases, flow increases.

Centrifugal compressors are among the most common types of compressors in the process industries because they are relatively economical, deliver much higher flow rates than positive-displacement compressors, and take up less space.

Large multistage compressors can be extremely complex with many auxiliary systems, including bearing oil systems, seal oil systems, cooling systems, and extensive vibration detection systems.

AXIAL COMPRESSORS An **axial compressor** uses a series of rotor blades with a set of stator blades between each rotating wheel to move gases axially along the shaft. This is in contrast to centrifugal compressors, which force gases to the outer wall of the chamber.

With axial compressors, rotor blades are attached to the shaft, and stator blades are attached to the internal walls of the compressor casing. The stator blades decrease in size as the casing size decreases. Rotation of the shaft causes flow to be directed axially along the shaft, building higher pressure toward the discharge end of the machine. Figure 17.6 shows an example of an axial compressor.

Axial compressor a dynamic-type compressor that uses a series of rotor blades with a set of stator blades between each rotating wheel and creates gas flow that is axial (parallel to) the compressor shaft.

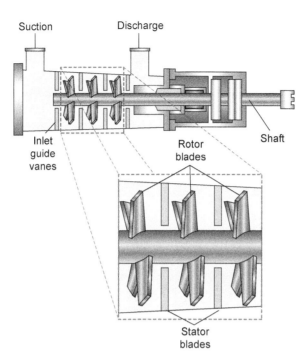

Figure 17.6 Axial compressor.

The main components of an axial compressor are:

- Suction (inlet)
- Inlet guide vanes
- Shaft
- Rotor blades
- Stator blades
- Discharge (outlet)

Multistage Compressors

Compression ratio the ratio of discharge pressure (psia) to inlet pressure (psia).

In compressors, the temperature of a gas increases as it is compressed. The amount of temperature increase is a function of the gas and the **compression ratio** (the ratio of discharge pressure [psia] to suction pressure [psia]). The compression ratio in a compressor usually is limited to 3:1 to 5:1 to avoid extremely high discharge temperatures.

However the desired discharge pressure may often be more than 10 times that of the inlet pressure. In such instances, a single-stage compressor cannot be used because of temperature limitations. Instead, multistage compressor with cooling after each stage is required. When the discharge gas from a compressor is cooled, liquids are frequently condensed. Because liquids are noncompressible, they could cause severe damage if they entered the compressor. So, these liquids are removed in "knockout drums" and are prevented from entering the compressor.

Blowers

Blower a mechanical device, either centrifugal or positive displacement in design, that has a lower ratio of pressure change from suction to discharge and is not considered to be a compressor; also can be termed a *fan*.

Fan see *blower*.

A **blower** (also called a **fan**) generally operates at a lower discharge pressure than a compressor does (Figure 17.7). The industry sector and process application determines whether the blower/fan type is centrifugal or positive-displacement. Centrifugal blowers are used to supply air for numerous applications, such as combustion air, air drying units, and movement of powders or pellets (via airveyors) from one point to another. Positive-displacement blowers also are used to move powders or pellets in wastewater aeration units, gas boosting, and other applications.

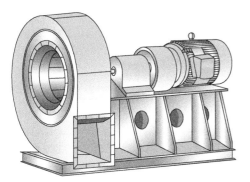

Figure 17.7 Industrial fan blower.

Did You Know?

Many top fuel dragster engines use a supercharger 14-71 Roots type blower, a positive-displacement lobe design.

The Roots blower was first developed by brothers Philander Higley Roots and Francis Marion Roots, who received a patent on the device in 1860.

Philip Pilosian/Shutterstock

17.6 Operational Hazards

When working with compressors, process technicians should always be aware of potential hazards, such as overpressurization, overheating, surging, and leakage (see Table 17.1).

Table 17.1 Hazards Associated with Use of Compressor

Improper Operation	Possible Impacts			
	Individual	Equipment	Production	Environment
Overpressurization	Retraining if operator, not equipment, is root cause	Leaks at seals, valves and other component parts Damage to compressor rotor and other internal components when valves are not aligned for correct flow path	Need to shut down the compressor, causing a loss of production	A leak, or rupture that sends process gas to the atmosphere, creating a hazard for operating personnel
Surging and operating at critical speeds	Retraining if operator, not equipment, is root cause	High vibration possibly leading to separation of equipment or damage to internal components	Possible loss of production if compressor is shut down	Possible leak to the air or ground surface if vibration causes rupture or leak
Rotating equipment	Possible injury to personnel standing close to an unguarded shaft; getting clothing, hair or other equipment caught in the rotating shaft	Possible damage to both the compressor and driver end equipment	Possible shutdown of the equipment causing loss of production	
Overheating	Possible burns to personnel	Extreme temperatures that stress metal in equipment components	Possible shutdown of compressor and loss of production	
Leaks	Possible exposure to process gases or slipping if leak is liquid (such as oil)	Leakage and possible damage to component part or compromise of a component such as a seal Possible fire or explosion if leak reaches hot equipment surface	Possible shutdown of equipment causing loss of production if leaks that cannot be repaired with equipment online	Harm to the atmosphere from both gas and liquid leaks; hazards for personnel and possibly near neighbors

Process technicians also should be aware of personal safety precautions they should take when working around high temperatures of hot metal, rotating equipment, leaks, pinch points, and tripping hazards.

Overpressurization

Because of the high pressures generated by a compressor, it is essential that valve alignment be reviewed both in a written procedure and by the process technician in the field when starting the machine. Overpressurization can occur if the valves associated with the compressor are incorrectly closed or blocked. In Figure 17.8 valves 1 through 7 would need to be open prior to starting the compressor.

Discharge block valves, recycle line block valves, and control valves must be in the designated open or closed position to avoid injury to personnel and damage to equipment.

Figure 17.8 Valve operation. This image shows a multistage compressor with stages in separate casings. In this image, valves 2 through 5 act as recycle/antisurge valves. Compressors with fewer stages might have one recycle/antisurge valve that returns flow from the last stage back to the first stage valve 2. In this compressor, all valves must be open before startup to avoid compressor overpressurization.

Based on MOGAS, https://www.mogas.com/fr-fr/industries/chemical-petrochemical/compressors

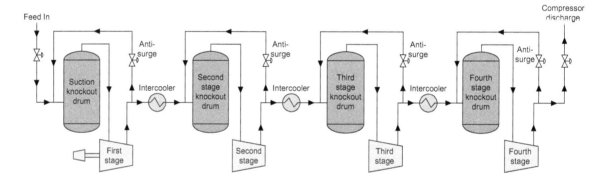

Overheating

Compressor overheating often is caused by improper lubrication. Without lubrication, bearings fail and equipment surfaces rub together, producing friction and generating heat.

Overheating also can occur, especially in multistage compressors, when compressor valves malfunction. This can result in excessive compression ratios and very high gas temperatures.

Process technicians should always monitor rotating equipment for unusual sounds and excessive heat, because operating equipment under these conditions can lead to permanent equipment damage or personal injury (e.g., burns).

Surging

Another operational hazard is surging. *Surging* is a temporary loss of flow to one or more impellers or stages of the compressor, which can be either single stage or multistage. Surging is typically associated with centrifugal compressors. It causes the compressor speed to fluctuate wildly and vibration to increase dramatically.

Leakage

Process technicians should always check compressors for leaks. Leaks can introduce harmful or hazardous substances into the atmosphere, create process problems, and create an unsafe work area for employees. Process gas leaks could occur at casing seals; steam leaks could occur on a turbine (compressor driver); oil leaks could occur at any point where lubrication is introduced on the machine. Each operating unit will have a set of procedures and checklists that guide the process technician through the equipment check. As a part of the daily work routine, process technicians make periodic rounds and take readings to ensure the compressor is working at design conditions.

17.7 Monitoring and Maintenance Activities

When working with compressors, process technicians should always conduct monitoring and maintenance activities to ensure the compressor is not overpressurizing, overheating, or leaking (Table 17.2).

Technicians must routinely check for excess vibration, correct oil flow, correct oil level, and check that temperatures and pressures are in range. They should also make sure that connectors, hoses, and pipes are in proper condition.

Periodically, technicians also should check overspeed trips and monitor for leaks at seals, packing, and flanges. Certain types of compressors might need to have the cylinders or housing checked for liquid.

Failure to perform proper maintenance and monitoring could affect the process, cause personal injury, and result in equipment damage.

Table 17.2 Monitoring and Maintenance Activities for Compressors

Look	Listen	Check
▪ Check oil levels to ensure they are satisfactory. ▪ Check seals and flanges to be sure there are no leaks. ▪ Check vibration monitors to ensure they are within operating range. ▪ Check liquid level in suction drum to be sure level is correct.	▪ Listen for abnormal noises.	▪ Use a handheld monitor or check the local panel board for excessive vibration. ▪ Use the appropriate tool such as a pyrometer to check for excessive heat.

Compressor Symbols

In order to locate compressors on a piping and instrumentation diagram (P&ID), process technicians must be familiar with the different types of compressors and their symbols. Figure 17.9 shows examples of symbols for both dynamic and positive-displacement compressors. However, these symbols can vary from facility to facility.

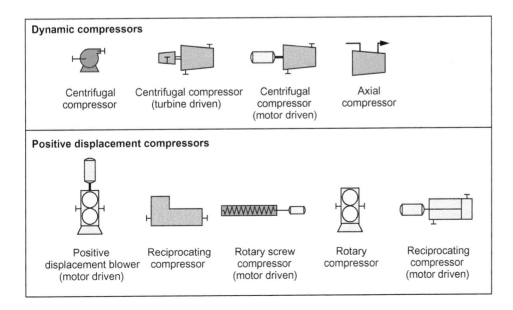

Figure 17.9 Compressor Symbols.

Summary

Compressors are used to increase the pressure of gases for use in a process facility. Compressors differ from pumps. Compressors are used with (compressible) gases, while pumps are used with (incompressible) liquids. Compressors can be single-stage or multistage and are sized in accordance with the amount of gas to be compressed. The two main types of compressors used in industry are positive-displacement compressors and dynamic compressors.

Positive-displacement compressors increase gas pressure by trapping specific amounts and reducing the volume at the discharge. They include rotary and reciprocating designs. In the rotary type, a screw, sliding vane, or lobe traps gas and reduces its volume. The reciprocating type uses the inward stroke of a piston or diaphragm to draw gas into a chamber (intake) and positively displace it using an outward stroke (discharge). Centrifugal compressors use a series of impellers in which the gas flows from the inlet located near the suction eye to the outer tip of the impeller blade. Axial compressors use a set of rotor and stator blades to move the gas parallel to the compressor shaft, increasing pressure as the casing and blades decrease in size.

Many hazards are associated with improper compressor operation, including overpressurization, overheating, surging, and leakage.

When monitoring and maintaining compressors, process technicians must always remember to look, listen, and check to ensure that oil levels are correct, that there are no leaks or abnormal noises, and that the equipment is not producing excessive heat or vibration.

Checking Your Knowledge

1. Explain the purpose of compressors in the process industries.
2. Describe the principle of operation of an axial compressor.
3. Which type of compressor uses vanes to force gases out of a chamber?

a. axial
b. centrifugal
c. positive-displacement
d. dynamic

4. Label the following parts on the single-stage centrifugal compressor.

a. casing
b. discharge
c. impeller
d. shaft
e. suction
f. packing
g. diffuser plates
h. packing gland

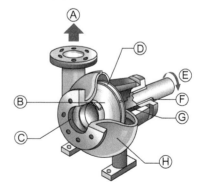

5. Label the following parts on the multistage compressor.
   a. discharge port
   b. casing
   c. diffuser
   d. suction port
   e. impeller
   f. shaft

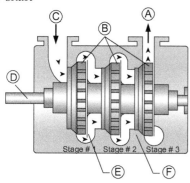

6. List three hazards associated with improper compressor operation.

7. List five monitoring and maintenance activities that a process technician should perform when working with compressors.

8. Draw piping and instrumentation diagram (P&ID) symbols for the following:
   a. centrifugal compressor
   b. axial compressor
   c. reciprocating compressor
   d. rotary compressor

NOTE: Answers to Checking Your Knowledge questions are in Appendix I.

Student Activities

1. Divide into groups. Have each group brainstorm the reasons:
   a. a top-fuel dragster needs a blower.
   b. a home AC unit needs a compressor.
   c. a home heating unit needs a blower.

   After 15 minutes, return to the large group to discuss these topics and explain them to the other students.

2. Given a compressor, perform the following maintenance and monitoring activities:
   a. Inspect for abnormal noise.
   b. Inspect for excessive heat.
   c. Check oil levels.
   d. Check for leaks around seals and flanges.
   e. Check for excessive vibration.

3. Given a piping and instrumentation diagram (P&ID), identify the category of all compressors as either dynamic (centrifugal) or positive-displacement design.

Chapter 18
Turbines

 Objectives

Upon completion of this chapter, you will be able to:

18.1 Describe the purpose and function of steam turbines in process industries. (NAPTA Turbines 1) p. 249

18.2 List and describe different types of turbines. (NAPTA Turbines 1) p. 249

18.3 Identify the primary parts and principles of operation of a typical noncondensing steam turbine:

casing
shaft
moving blades
fixed blades
governor
nozzle
inlet
outlet
throttle valve
trip valve. (NAPTA Turbines 2, 3) p. 252

18.4 Discuss the hazards associated with the improper operation of a steam turbine. (NAPTA Turbines 4) p. 254

18.5 Describe the monitoring and maintenance activities associated with a steam turbine. (NAPTA Turbines 5, 6) p. 255

Key Terms

Actuate—to put into action, **p. 252.**

Casing—the housing that contains the internal components of a turbine. The casing component of a steam turbine holds all moving parts, including the rotor, bearings, and seals, **p. 253.**

Controller—device that monitors the speed in RPMs of the turbine; also called a *governor*, **p. 253.**

Fixed blades—blades inside a steam turbine that remain stationary when steam is applied, **p. 253**.

Gas turbine—a device that consists of an air compressor, combustion chamber, and turbine. Hot gases produced in the combustion chamber are directed toward the turbine blades, causing the rotor to move. The rotation of the connecting shaft can be used to operate other equipment, **p. 250**.

Governor—a device used to control the speed of a piece of equipment such as a turbine, **p. 253**.

Hunting—a term used to describe the condition when a turbine's speed fluctuates while the governor or controller is searching for the correct operating speed, **p. 255**.

Hydraulic turbine—a turbine that is moved, operated, or affected by a fluid, usually water, **p. 251**.

Inlet—the point of entry for the fluid, **p. 253**.

Kinetic energy—energy associated with mass in motion, **p. 250**.

Mechanical energy—energy of motion that is used to perform work, **p. 250**.

Moving blades—rotor blades inside a steam turbine that move or rotate when steam is applied, **p. 253**.

Nozzle—the device in the turbine entry piping that converts pressure to velocity and directs the flow of steam to the turbine blades. There can be several nozzles in the turbine, depending on design and desired speed, **p. 253**.

Outlet—the point of exit for the fluid, **p. 253**.

Set point—the desired RPM indication where a controller is set for optimum operation, **p. 255**.

Shaft—a metal rod on which the rotor resides, suspended on each end by bearings; connects the driver to the shaft of the driven end or mechanical device, **p. 253**.

Steam turbine—a rotating device driven by the pressure of steam (potential energy) discharged at high velocity (kinetic energy) against the turbine blades; used as the motive force (mechanical energy) for equipment, **p. 250**.

Throttle valve—a device, or series of devices depending on turbine power, that can be opened or closed slowly (throttled) to control steam flow to the turbine, **p. 253**.

Trip valve—a device that can be closed quickly (tripped) to stop the flow of steam to a turbine, **p. 253**.

Turbines—devices that are used to produce power and to rotate shaft-driven equipment, **p. 249**.

Wind turbine—a mechanical device that converts wind energy into mechanical energy, **p. 251**.

18.1 Introduction

Turbines are devices that are used to produce power and to rotate shaft-driven equipment such as pumps, compressors, and electrical generators. Turbines are activated by the expansion of a fluid on a series of curved rotating blades. Some turbines also have stationary (stator) blades, attached to a central shaft.

Turbines can be powered by a variety of different fluids, including steam, gas, liquid, or air. Steam-powered turbines frequently are used as backups for electric motors in the event of a power outage.

Turbines devices that are used to produce power and to rotate shaft-driven equipment.

18.2 Types of Turbines

There are four main types of turbines—steam, gas, hydraulic, and wind—each of which is classified according to how it operates and the fluid that turns it (i.e., steam, gas, water, or air).

Although each of these turbine types is discussed, the most common turbine type used in industry is the steam turbine. For this reason, steam turbines will be the main focus of this chapter.

Steam turbine a rotating device driven by the pressure of steam (potential energy) discharged at high velocity (kinetic energy) against the turbine blades; used as the motive force (mechanical energy) for equipment.

Steam Turbines

A **steam turbine** is driven by the pressure of steam discharged at high velocity against the blades of the rotor assembly. There is a direct relationship between the steam pressure used to drive a turbine and the amount of speed and rotational force needed. Turbines can exhaust to lower pressure steam levels that are recycled into the system, or they can exhaust to condensate by pulling a vacuum on the exit side of the turbine.

Depending on the turbine size and workload, there can be several stages, with each successive stage operating under a lower pressure.

Turbines are used in industry to drive pumps, compressors, generators, and other rotating equipment. Steam lines and turbine casings may be insulated if steam turbines are located in areas of operation requiring monitoring by shift personnel, if they are located in closed areas of operation, or if they are exposed to the elements. Insulation helps protect personnel from burn injury and ensures that turbine speed is not affected by exposure to rain, snow, or other inclement weather elements.

Did You Know?

Steam can be a very powerful locomotive (moving) force.

It is so powerful, in fact, that it has been used to drive many types of large industrial equipment (e.g., trains, dredges, and shovels).

Mechanical energy energy of motion that is used to perform work.

Kinetic energy energy associated with mass in motion.

Mechanical energy involves motion used to perform work (as opposed to **kinetic energy**, which is associated with mass in motion).

Figure 18.1 shows an example of a steam turbine.

Steam turbines have many advantages over electrical equipment. For example, they are free from spark hazards, so they are good in areas where volatile substances are produced. Since they do not require electricity to run, they can be used during power outages. They are also suitable for damp environments that might cause electrical equipment to fail.

Figure 18.1 Steam turbine.
Courtesy of Bayport Technical

Gas turbine a device that consists of an air compressor, combustion chamber, and turbine. Hot gases produced in the combustion chamber are directed toward the turbine blades, causing the rotor to move. The rotation of the connecting shaft can be used to operate other equipment.

Did You Know?

Jet engines used in airplanes are actually gas turbines.

Gas Turbines

A **gas turbine**, also called a combustion turbine, consists of an air compressor, combustion chamber, and turbine. In a gas turbine, hot gases produced in the combustion chamber are directed toward the turbine blades, causing the rotor to move. As the rotor moves, the shaft turns. The rotation of the connecting shaft can be used to operate other equipment. Gas turbines are often coupled to a smaller steam turbine. The steam turbine spins the gas turbine and compressor to provide compressed air for startup.

Figure 18.2 shows an example of gas turbines.

In a gas turbine, an electric motor rotates the shaft to bring the air compressor up to speed and start the machine. After the turbine is started, it provides enough power to drive the

Figure 18.2 Gas turbines **A.** Illustration of a gas turbine. **B.** Sample gas turbine.

CREDIT: **A.** Fouad A. Saad/Shutterstock. **B.** pupikii/Fotolia

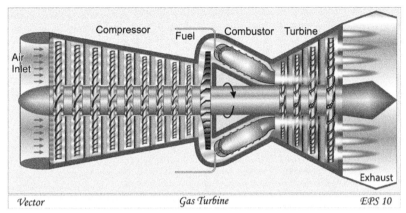
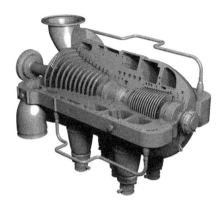

A. B.

air compressor to keep the gas turbine running and drive other connected equipment (e.g., pumps, compressors, and generators).

Gas turbines are most commonly used in the process industries to drive generators or compressors in remote locations (like on an offshore platform). Gas turbines are also used to power aircraft, some trains, some ships, and some tanks.

Hydraulic Turbines

A **hydraulic turbine** is one that is moved, operated, or affected by a fluid, usually water. These turbines operate on the principle of fluid mechanics.

In a hydraulic turbine a fluid, usually water, flows across the rotor blades, forcing them to move. The faster the fluid flows, the faster the turbine turns.

An example of a simple hydraulic turbine is a waterwheel. Another example would be where water from Lake Mead flows through the turbines at the bottom of Hoover Dam, generating power (Figure 18.3).

Hydraulic turbine a turbine that is moved, operated, or affected by a fluid, usually water.

A. B.

Figure 18.3 A. Waterwheel (hydraulic turbine). **B.** Power generation hydraulic turbines at Hoover Dam.

CREDIT: **A.** salajean/fotolia. **B.** CrackerClips/fotolia.

Wind Turbines

A **wind turbine** converts wind energy into mechanical energy. These types of turbines use air pressure to move a rotor. A windmill is an example of a wind turbine.

In a wind turbine, air currents move across fanlike blades, causing them to turn. As they turn, a shaft is rotated. The rotation of the shaft drives devices such as pumps or electrical generators.

Wind turbine a mechanical device that converts wind energy into mechanical energy.

Actuate to put into action.

Figure 18.4 shows an example of a simple wind turbine and how it can be used to **actuate** (put into action) a water pump. As people in the United States become more interested in renewable energy, wind farms, or wind parks, have been constructed to generate electrical power.

Figure 18.4 A. Wind turbine (windmill) actuating a pump. **B.** Wind farm generating electrical power.
CREDIT: B. Jim Parkin/Fotolia

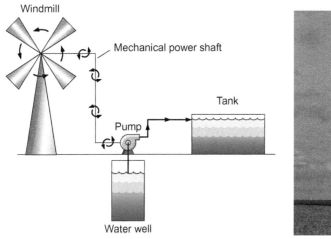

A. B.

Figure 18.5 Steam turbine components.

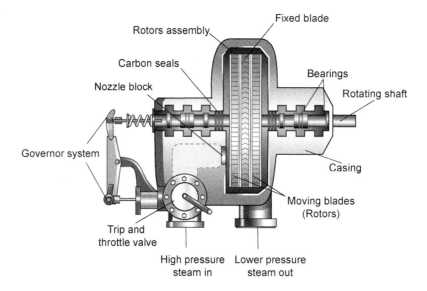

18.3 Steam Turbine Components

Turbine components vary depending on the type of turbine and its application. However, the main components of a steam turbine (shown in Figure 18.5) include:

- Casing
- Shaft
- Rotor assembly (moving blades and fixed blades on a shaft)
- Governor
- Nozzle
- Inlet
- Outlet

- Throttle valve
- Trip valve
- Bearings
- Seals

The **casing** is the housing around the internal components of a turbine. Within the casing are a shaft and a set of blades called the *rotor assembly*.

The **shaft** is a metal rod that is attached to the rotor assembly and is suspended on each side by bearings. The **moving blades** are rotor blades that move or rotate when steam is applied. The **fixed blades** remain stationary when steam is applied.

The **governor** is used to control the speed of the turbine as steam is channeled through the nozzle.

The **nozzle** is a device in the turbine entry piping that converts pressure to velocity and directs the flow of steam to the turbine blades. There might be several nozzles in the turbine, depending on design and desired speed.

Steam enters the nozzle from the steam piping **inlet** and exits the turbine through the steam piping **outlet**.

The **throttle valve** is a valve, or series of valves, that can be opened or closed slowly (called *throttling*) to control steam flow. This valve works in conjunction with the governor to control the speed of the turbine. The **trip valve** is a valve that can closed quickly (*tripped*) to stop or start the flow of steam to the turbine.

Principles of Operation

Turbines create rotational movement and are used to drive other equipment (e.g., pumps, compressors, and blowers). Although each of the different turbine types varies in how it works, all of them use the same basic principles. That is, they use some kind of force (e.g., water pressure, air pressure, steam pressure, or the pressure of combustion) to turn a rotor. As the rotor (or *driver*) is turned, another device (the *mover*) is moved or actuated. A pump is an example of a mover.

Modern steam turbines can have either reaction or impulse design. Impulse turbines have the capacity for quicker speed changes; reaction turbines generally offer more torque. Steam turbines can be single-stage or multistage, depending on speed and process service requirement. Steam turbines operate using steam supply pressures from low pressure (around 30 psi) to extremely high pressure (3300 psi).

One of the earliest types of turbines is the simple reaction turbine like the one shown in Figure 18.6. In this turbine (Hero's aeolipile), water is placed in a globe that contains two

Casing the housing that contains the internal components of a turbine. The casing component of a steam turbine holds all moving parts, including the rotor, bearings, and seals.

Shaft a metal rod on which the rotor resides, suspended on each end by bearings; connects the driver to the shaft of the driven end or mechanical device.

Moving blades rotor blades inside a steam turbine that move or rotate when steam is applied.

Fixed blades blades inside a steam turbine that remain stationary when steam is applied.

Governor a device used to control the speed of a piece of equipment such as a turbine.

Nozzle the device in the turbine entry piping that converts pressure to velocity and directs the flow of steam to the turbine blades. There can be several nozzles in the turbine, depending on design and desired speed.

Inlet the point of entry for the fluid.

Outlet the point of exit for the fluid.

Throttle valve a device, or series of devices depending on turbine power, that can be opened or closed slowly (throttled) to control steam flow to the turbine.

Trip valve a device that can be closed quickly (tripped) to stop the flow of steam to a turbine.

Figure 18.6 Reproduction of Hero's aeolipile, a simple turbine.

> **Did You Know?**
>
> Hero, an Egyptian scientist from Alexandria, developed the first "jet engine" about the year 100 BC. The simple reaction turbine, known as the *aeolipile*, consisted of a boiler and a sphere with two hollow bent tubes mounted on it. Hero is said to have used this device to open temple doors.

opposing nozzles. As the water is heated, steam and pressure are produced. The steam, being the lighter phase, is forced out of the nozzles. As the steam exits, propulsive (rotational) force is created. This force causes the globe to spin.

The other main type of steam turbine design is an impulse design. Figure 18.7 shows an example of an impulse turbine.

In both the reaction and impulse turbines, steam is channeled through a steam nozzle onto the turbine blades. As the steam passes through the nozzle, it is converted to velocity. This force causes the wheel to turn, thereby rotating the shaft and any equipment coupled to it on the driven end. Turbines can be designed with a radial or axial steam flow.

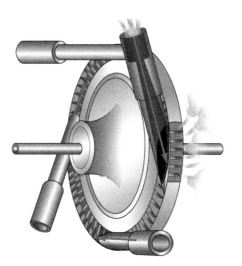

Figure 18.7 Impulse turbine concept.

An important operational hazard encountered with steam turbines and associated with the rotating shaft is vibration. During startup or shutdown of a turbine, there are areas where the rotational speed can equal the natural frequency of vibration, producing a *harmonic effect*. The harmonic effect creates dramatic vibration of the shaft, which affects the attached equipment of the system. Rotational speeds that cause harmonic effects are called *critical speeds*. Depending on many factors (such as size, metallurgy, and load), there can be more than one critical speed. It is very important for the turbine to be speeded up or slowed down rapidly through these critical speed points.

18.4 Operational Hazards

Many hazards are associated with steam turbines. Table 18.1 lists some of these hazards and their impacts.

Table 18.1 Hazards Associated with Improper Steam Turbine Operation

Improper Operation	Possible Impacts			
	Individual	Equipment	Production	Environment
Touching the external housing of a steam turbine that is full of steam	Burns caused by exposure to hot metal			
Loss of casing and/or line insulation	Possible burns to personnel from exposure to hot metal surfaces	Possible damage to hoses or other equipment left in contact with hot metal surfaces		Excess heat released into either an open or a closed area of operation
Allowing hot steam to enter the turbine without following the proper warm-up procedure	Burn injury from steam	Equipment damage because of thermal shock	Downtime because of repair, possible loss of product	

A turbine governor toggling the speed above or below normal operating **set point** (the point or place where the control index of a **controller** [device that monitors the speed in RPMs of the turbine] is set)	If overspeed trip does not activate, turbine could fly apart, causing injury	Insufficient pumping by driven equipment (e.g., a pump or compressor) if the turbine is running too slowly product can go off-spec because of improper pump speed **Hunting** possible (when a turbine's speed fluctuates while the governor or controller is searching for the correct operating speed)	Downtime because of repair, possible loss of product	
Failure to lubricate the linkage between the governor and the governor valve		Hunting as a result of valve sticking or binding		
Failure to maintain sufficient inlet steam pressure		Hunting because of insufficient steam pressure	Insufficient power to run the turbine	
Allowing "wet steam" or condensate into the turbine	Possible bodily injury if turbine explodes	Total failure of the turbine, explosion		
Allowing turbine to run at a critical speeds		Turbine damage as a result of harmonic vibrations if turbine is not moved through critical speeds as quickly as possible	Loss of production or process if turbine is compromised and needs shutdown	
High discharge steam pressure		Turbine slowdown and relief valve popping	Loss of production	Noise pollution
Steam leaks	Burn injuries	Possible need to shut down turbine for maintenance work	Possible loss of production	Noise hazard

18.5 Monitoring and Maintenance Activities

When monitoring and maintaining steam turbines, process technicians must always remember to look, listen, and feel for the items described in Table 18.2.

Failure to perform proper maintenance and monitoring could affect the process and result in equipment damage.

Set point the desired RPM indication where a controller is set for optimum operation.

Controller device that monitors the speed in RPMs of the turbine; also call a *governor*.

Table 18.2 Monitoring and Maintenance Activities for Turbines

Look	Listen	Check
■ Check oil levels to make sure they are satisfactory. ■ Check insulation to be in good condition. ■ Check bearing temperatures at gauges ■ Check seals and flanges to make sure there are no leaks. ■ Check for water in the oil reservoir. ■ Check steam pressure and temperature. ■ Observe governor for proper operation. ■ Check turbine RPMs to be in optimum operating range. This can be done using a portable tachometer or checked on a digital monitor. ■ Check valves and trip handles to be in the appropriate position.	■ Listen for abnormal noises.	■ Use appropriate instrument or review local panel reading to check for excessive vibration. ■ Use the appropriate instrument (pyrometer) to check for excessive heat.

Steam Turbine Symbol

In order to locate steam turbines on a piping and instrumentation diagram (P&ID) accurately, process technicians must be familiar with turbine symbols.

Figure 18.8 provides an example of a steam turbine symbol. However, this symbol can vary from facility to facility.

Hunting a term used to describe the condition when a turbine's speed fluctuates while the governor or controller is searching for the correct operating speed.

Figure 18.8 Steam turbine symbol.

Turbine driver

Summary

The steam turbine, the main type of turbine used in a process facility, is the focus of this chapter. Steam turbines are used to produce power and are the driving force for coupled equipment. They are activated by the expansion of steam on a series of curved vanes and blades attached to a central shaft.

Turbines are used to rotate shaft-driven equipment such as pumps, compressors, and electrical generators. Steam-powered turbines can be used as backups for electric motors.

There are four main types of turbines: steam, gas, hydraulic, and wind. Each of these turbine types is classified according to how it operates and the fluid that turns it. Gas turbines are most commonly used in the process industries to drive generators or compressors in remote locations (like on an offshore platform). Gas turbines are also used to power aircraft, some trains, some ships, and some tanks. Hydraulic and wind turbines are generally associated with the production of electrical power.

Improperly operating a steam turbine, or any other type of turbine, can lead to safety and process issues, including burns and equipment damage.

Technicians should always monitor turbines for abnormal noises, leaks, excessive heat, pressure drops, temperature drops, or other abnormal conditions, and conduct preventive maintenance to ensure that lubrication is sufficient for component parts.

Checking Your Knowledge

1. Define the following terms:
   a. casing
   b. fixed blades
   c. governor
   d. hunting
   e. inlet
   f. moving blades
   g. nozzle
   h. outlet
   i. shaft
   j. throttle valve
   k. trip valve
   l. rotor

2. True or false: A wind turbine is the only kind of turbine that uses a rotor.

3. A jet engine is an example of which type of turbine?
   a. gas turbine
   b. hydraulic turbine
   c. steam turbine
   d. wind turbine

4. A waterwheel is an example of which type of turbine?
   a. gas turbine
   b. hydraulic turbine
   c. steam turbine
   d. wind turbine

5. Which of the following are considered essential for safe turbine operation? Select all that apply.
   a. low steam inlet pressure
   b. running at critical speeds
   c. high discharge steam pressure
   d. running at set point
   e. appropriate lubrication
   f. loss of casing insulation

6. List three things a process technician should look, listen, and check when working with steam turbines.

NOTE: Answers to Checking Your Knowledge questions are in Appendix I.

Student Activities

1. Divide into teams of three. Prepare a short presentation for the class. Identify three hazards associated with steam turbines. For each hazard, describe any impacts these hazards might have on individuals, equipment, production, and the environment.

2. Given a piping and instrumentation diagram (P&ID), identify the turbines and type of equipment coupled to them.

3. Given a diagram of a steam turbine, label the following components and explain the function of each:
   a. casing
   b. shaft
   c. moving blades
   d. fixed blades
   e. governor
   f. nozzle
   g. inlet
   h. outlet
   i. throttle valve
   j. trip valve
   k. rotor

Chapter 19
Electricity and Motors

Objectives

Upon completion of this chapter, you will be able to:

19.1 Describe basic concepts about electricity and its use in process industries. (NAPTA Electricity & Motors 1) p. 259

19.2 Explain the difference between alternating current (AC) and direct current (DC) and their use in the process industries. (NAPTA Electricity & Motors 1, 2) p. 263

19.3 Describe the purpose or function of electric motors in the process industries. (NAPTA Electricity & Motors 4) p. 265

19.4 Identify the primary parts of a typical electric motor:

frame

shroud

rotor

shaft

stator

fan

bearings

power supply. (NAPTA Electricity & Motors 3, 5) p. 265

19.5 Discuss the hazards associated with the improper inspection and operation of an AC motor. (NAPTA Electricity & Motors 6) p. 267

19.6 Describe the monitoring and maintenance activities associated with an electric motor. (NAPTA Electricity & Motors 7, 8) p. 270

Key Terms

Ammeter—a device used to measure electrical current in amperes, **p. 261.**

Amperes (amps)—a unit of measure of the electrical current that flows in a wire; similar to "gallons of water" that flow in a pipe, **p. 260.**

Arc-rated—having protective characteristics in the fabric, as determined by a hazard/risk assessment, **p. 269.**

Circuit—a system of one or many electrical components connected together to accomplish a specified purpose, **p. 261**.

Conductor—a material that has electrons that can break free and flow more easily than other materials, **p. 260**.

Electromotive force (EMF)—the force that causes the movement of electrons through an electrical circuit, measured in volts, **p. 260**.

Ground fault circuit interrupter (GFCI)—a safety device that detects the flow of current to ground and opens the circuit to interrupt the flow, **p. 269**.

Grounding—the process of using the earth as a return conductor in a circuit, **p. 262**.

Insulator—a material that does not give up its electrons as easily; an insulator is a poor conductor of electricity, **p. 260**.

Intrinsically safe—condition in which devices such as motors are designed so as not to ignite a spark that could create a fire; a requirement for all equipment operating in areas where flammable gases and liquids are present, **p. 265**.

Motor—a mechanical driver with rotational output; usually electrically operated, **p. 265**.

Ohm—a measurement of resistance in electrical circuits; symbolized by the Greek letter omega: Ω, **p. 260**.

Resistance—the force that opposes the push of electrons, measured in ohms, **p. 260**.

Rotor—the rotating member of a motor that is connected to the shaft, **p. 266**.

Semiconductor—a material that is neither a conductor nor an insulator, **p. 260**.

Stator—a stationary part of the motor where the alternating current flows in and a large magnetic field is created using magnets and coiled wire, **p. 266**.

Transformer—a device that will raise or lower the voltage of alternating current of the original source, **p. 263**.

Volt—the electromotive force that will establish a current of one amp through a resistance of one ohm, **p. 260**.

Voltmeter—a device used to measure voltage, **p. 260**.

Watt—a unit of measure of electric power; the power consumed by a current of one ampere using an electromotive force of one volt, **p. 261**.

Introduction

Electricity and motors are important components in the process industry. Electricity allows us to power equipment and run electric lights. Electricity powers process instrumentation from point of measurement into the control room and back into the field, where it is typically converted into a pneumatic signal to power control valves. Motors allow us to operate rotating equipment such as pumps, compressors, and many other types of process equipment.

The electricity used by motors and other equipment can be either alternating (AC) or direct (DC) current. Alternating current, however, is the most common type used in process industries.

When working with electricity, process technicians should always wear proper protective gear, follow safety procedures, and be aware of potential hazards.

Process technicians also should be able to recognize the symbols for various pieces of electrical equipment on process drawings.

19.1 What is Electricity?

Electricity is a flow of electrons from one point to another along a pathway called a conductor. Process technicians should have a good understanding of electricity, because it is integral to the functioning of a plant and its systems.

Did You Know?

Ben Franklin's famous experiment with flying a kite in a thunderstorm demonstrated the true nature of lightning—that it is static electricity.

CREDIT: Victor Brave/Shutterstock.

Conductor a material that has electrons that can break free and flow more easily than other materials.

Insulator a material that does not give up its electrons as easily; an insulator is a poor conductor of electricity.

Semiconductor a material that is neither a conductor nor an insulator.

Electromotive force (EMF) the force that causes the movement of electrons through an electrical circuit, measured in volts.

Volt the electromotive force that will establish a current of one amp through a resistance of one ohm.

Voltmeter a device used to measure voltage.

Resistance the force that opposes the push of electrons, measured in ohms.

Ohm a measurement of resistance in electrical circuits; symbolized by the Greek letter omega (Ω).

Amperes (amps) a unit of measure of the electrical current that flows in a wire; similar to "gallons of water" that flow in a pipe.

Chapter 11 described an atom as the smallest particle of an element that still retains the properties and characteristics of that element. Atoms contain positively and neutrally charged particles in the nucleus and negatively charged particles around the nucleus. The negatively charged particles are called electrons.

When electrons flow from one atom to another, an electrical current is created. Free electrons flow along a path like a river. This pathway is called a conductor.

A **conductor** is a material that has electrons that can break free and flow more easily than other materials. Metals, as well as some types of hot gases (plasmas) and liquids, are good conductors.

A material that does not give up its electrons as easily is called an **insulator**. Insulators, including air, rubber, and glass, are poor conductors of electricity.

A material that is neither a conductor nor an insulator is called a **semiconductor**. Semiconductors are most commonly used in the electronics industry and are being used increasingly in power applications.

A unique type of electricity is static electricity. Static electricity occurs when a number of electrons build up on the surface of a material but have no positive charge nearby to attract them and cause them to flow. When the negatively charged surface comes into contact with (or comes near) a positively charged surface, current flows until the charges on each surface become equalized, sometimes creating a spark. Lightning and the shock that occurs from touching a doorknob after shuffling across carpeting are both good examples of static electricity.

To better understand electricity, process technicians need to be familiar with its principles and terms—volts, ohms, watts, and amps.

Volts

Using the water analogy again, electric current flows like a river down a slope (the path of least resistance). The greater the angle of a slope, the faster the water flows. Electric current behaves in a similar way. If the difference between positive and negative charges is low, electrons flow with little force. Increase the difference, and the electrons will flow with greater force. The force that makes electrons flow is called voltage or **electromotive force (EMF)**, and it is measured in units called volts (or V).

Volts are a measurement of the potential energy required to push electrons from one point to another. A **volt** is the electromotive force that will establish a current of one amp through a resistance of one ohm.

A **voltmeter**, which is a device used to measure voltage, can be connected to a circuit to determine the actual voltage present.

Resistance (Ohms)

The force that opposes the push of electrons is called **resistance** (or R). Resistance in electrical equipment is measured in ohms. One **ohm** (Ω) is the amount of resistance at which one volt produces a flow of one amp. Conductors provide little resistance to electricity; insulators provide high resistance.

Amps

Electrical current is measured using **amperes (amps)**. Amps (or I) are units of measure of the electrical current flow in a wire (similar to "gallons of water" that flow in a pipe). Amps describe how many electrons are flowing at a given time.

The amount of work produced by a circuit is measured in amps. In other words, amps show the capacity of a battery or other source of electricity to produce electrons. An

ammeter, which is a device used to measure electrical current, must be connected in series to an electrical circuit to display the actual amps.

Ammeter a device used to measure electrical current in amperes.

Ohm's Law

Ohm's law describes how volts, ohms, and amps act upon each other. This law states that the amount of steady current through a conductor is proportional to the voltage across that conductor. This means a conductor with one ohm of resistance has a current of one amp, under the potential of one volt. Simply put, volts equals amps times ohms (V = IR).

Voltage (Ohm's Law) The formula for Ohm's law is:

$$V = IR$$

(V = volts, I = current flow in amps, R = resistance to flow in ohms)

When working with Ohm's law, if you know the value of two units, you can always figure out the third using one of the following calculations:

$$I = V/R \text{ (Current = Volts ÷ Resistance) or}$$
$$R = V/I \text{ (Resistance = Volts ÷ Current)}$$

Watts

Electrical power is measured in watts. A **watt** is a unit of measure of electric power or the power consumed by a current of one amp using an electromotive force of one volt.

(watts = volts × amps)

Watt a unit of measure of electric power; the power consumed by a current of one ampere using an electromotive force of one volt.

Wattage The formula for determining how many watts a circuit uses, if you know the volts and amps, is:

$$W = VI$$

(W = watts, V = potential difference in volts, I = current flow in amps)

The formula for determining how many watts a circuit uses, if you know the amps and resistance, is:

$$W = I^2R$$

(W = watts, I = current flow in amps squared, R = resistance to current flow in ohms)

Circuits

A **circuit** is a system of one or many electrical components connected together to accomplish a specified purpose. Circuits combine conductors with a power supply and usually some kind of electrical component (such as a switch or light) in a continuously conducting path. In a circuit, electrons flow along the path, uninterrupted, and return to the power supply to complete the circuit.

Circuit a system of one or many electrical components connected together to accomplish a specified purpose.

Circuits fall into one of two types: series and parallel.

- Series—Electrons have only one path to flow along (Figure 19.1). An example is a string of lights. In a series type, if one bulb burns out, then the circuit (path) is interrupted and none of the bulbs will light.

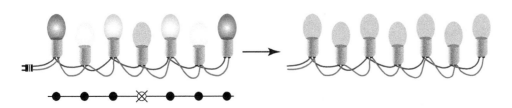

Figure 19.1 Series circuit example. When the center light on the left goes out, the whole string goes out.

- Parallel—Electrons are given a choice of paths to flow along (Figure 19.2). Some strings of lights use a parallel type of circuit, so that if one bulb burns out, the others continue to shine. Houses are wired using parallel circuits.

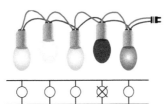

Figure 19.2 Parallel circuit example. The light that is out does not affect the rest of the string of lights.

Grounding

All energized conductors supplying current to equipment are kept insulated from each other, from the ground (earth), and from the equipment user. Many types of equipment have exposed conductive parts, such as metal covers, that are routinely touched during normal operation. If these surfaces become energized, a difference in electrical potential will exist between the equipment and ground. The process technician could complete the circuit and receive an electrical shock. An energized equipment case can present a shock hazard.

Because of the shock hazard, noncurrent-carrying conductive materials should be used to enclose electrical conductors. The equipment should be properly grounded to limit the voltage to ground on these materials. This **grounding** is typically accomplished by a separate conductor specifically designed for the purpose (Figure 19.3). With this conductor in place, if the equipment case becomes energized, a low-resistance path for the flow of ground current back to the source is already in place that reduces or eliminates the shock hazard to the operator.

Grounding the process of using the earth as a return conductor in a circuit.

Figure 19.3 Grounding wire.
CREDIT: Leonid Eremeychuk/Fotolia.

Electrical Transmission

Electricity must come from a power source, such as from batteries or generators in a power station. Although electricity is a form of energy, it is not an energy source and must be manufactured. This section describes how electricity is moved from a power station to homes and businesses.

Generators convert mechanical energy into electrical energy. For a power station, the most common way to manufacture electricity is by burning fuels that make turbines rotate magnetic fields inside generators to create electric current. Other methods include hydroelectric and nuclear power generation.

Electricity flows in a continuous current from a high-potential point (the power source) to a point of lower potential (a house or a plant), through a conductor (usually wire). High-voltage electricity is transmitted from the power plant to the power grid through a system of wires (Figure 19.4).

Figure 19.4 Electrical power lines.
CREDIT: MEzairi/Shutterstock.

High-voltage electricity is routed to a substation that steps down the electricity to a lower, safer voltage. The substation distributes the electricity through feeder wires to a **transformer**, a device that will raise or lower the voltage of alternating current. Transformers are used to step up or step down AC voltages. Transformers do not convert AC to DC or DC to AC (Figure 19.5).

The electricity then is used as energy to do work, such as lighting a bulb or operating a motor to run a pump. Safety devices such as fuses, protective relays, ground fault detectors, and others are used throughout the power transmission process to make the system as safe as possible.

Transformer a device that will raise or lower the voltage of alternating current of the original source.

Figure 19.5 Power transformer.

19.2 Understanding Alternating (AC) and Direct (DC) Current

The two types of electrical currents are alternating current (AC) and direct current (DC):

- AC—electrical current that uses a back-and-forth movement of electrons in a conductor. The movement of alternating current is similar to water sloshing backward and forward in a pipe. When a negative charge is at one end of a conductor and a positive charge is at the other end, the electrons move away from the negative charge. But if the charges (polarity) at the end of the conductors are reversed, the electrons switch directions. In the United States, the AC power supply changes direction 120 times per second; this is called frequency or cycles per second, written as 60 cycle AC or 60 Hertz AC.

- DC—electrical current that results in a direct flow of electrons through a conductor. Direct current flows like water moving in one direction through a pipe. With a battery as a power source, electricity flows in only one direction through a circuit.

> **Did You Know?**
>
> In 1882, a street in New York was the first to be illuminated by electric lighting, using Thomas Edison's direct current generator.

CREDIT: spyrakot/fotolia

In the late 1800s, Edison power stations (created by Thomas Edison) supplied direct current electricity to customers scattered across the United States. DC had a limiting factor, though—It could only be sent economically over a short distance (about a mile) before the electricity began to lose power.

To relieve these distance limitations, George Westinghouse introduced an alternating current power system, designed by Nikola Tesla, as an alternative to DC power. AC voltages can be easily increased (stepped up) or decreased (stepped down) using transformers. By employing transformers to boost voltage levels, AC systems economically distribute electricity for hundreds of miles.

Transformers take low-voltage current and make it high voltage and vice versa. Power stations send out high-voltage electricity that is stepped down using a transformer so homes and businesses can use the electricity.

AC and DC produce the same amount of heat, which is proportional to the product of the current (amps) squared times the resistance (ohms) and written using the formula I^2R. However, it is more difficult to measure a circuit using AC than DC, because AC voltage cycles from zero, to positive, to zero, to negative, to zero during every cycle.

Table 19.1 summarizes the similarities and differences between alternating and direct current.

Table 19.1 Differences and Similarities Between AC and DC Currents

AC	DC
Polarity is switched constantly	Polarity is fixed
Voltage varies during cycles	Voltage remains constant
It can be varied for power distribution (transformers can amplify or reduce AC)	It cannot be varied (a steady value is produced that transformers do not affect)
Measurements are more complex than DC	Measurements are easier than AC
Voltage can be stepped up or down by a transformer	Voltage cannot be stepped up or down by a transformer
Heating effect is the same as that of DC	Heating effect is the same as that of AC

Which Type of Current is Used Most in the Process Industries?

Alternating current is the most common type of electrical current used in the process industries. AC provides most of a plant's electrical power requirements.

> **Did You Know?**
>
> An English scientist, Michael Faraday, is credited with generating electrical current on the first practical scale.

CREDIT: happystock/fotolia

Plants typically receive high-voltage electricity from the electric company. Transformers then step down this electricity so it can be distributed throughout the plant.

When AC power reaches a unit, it is stepped down to 4,160 V for use with large motors and is then stepped down again to 480 V (sometimes referred to as 440 V), which is a common motor voltage used around a plant.

Finally, the power can be stepped down once again to somewhere between 120 V and 240 V for use with air conditioning, heating, lighting, and other applications.

Direct current is used primarily in the form of batteries, which can provide electrical power for portable equipment or in emergency situations involving the loss of electric power. For example, DC can power flashlights and other types of lighting, basic power tools, carts, and other small motors designed for DC current.

To maintain power to critical equipment and instrumentation during a power outage or emergency, a device called an uninterruptible power supply (UPS) is used. A typical UPS first converts AC to DC and then remanufactures AC from DC for use by plant instrumentation and other critical loads. A battery bank is connected to, and kept charged by, the AC current that was converted to DC current with a rectifier for the UPS. If the plant AC power should fail, the UPS uses DC from the battery which is converted to AC power for the critical loads using an inverter. It is crucial to maintain power to plant instrumentation. When a total loss of power occurs, control loop instrumentation (specifically the control valve) is designed to move toward its fail position at either 100% open or closed. This could create unsafe conditions in the process and for personnel. More information covering the UPS will be given in the instrumentation course and textbook. It is important to note that a UPS system is a temporary source of power that is running on batteries. Without being recharged, the batteries will expend their power in a short period of time.

Plants also can generate their own electricity in some cases. For example, if power fails, emergency diesel generators can provide electricity to critical equipment and instrumentation. Depending on geographical location, plants can be designed to incorporate a cogeneration plant, which provides both electricity and steam for plant usage.

19.3 Uses of Electric Motors in the Process Industries

A **motor** is a mechanical driver (usually electrically operated) that converts one form of energy into mechanical energy. Electric motors turn electrical energy into useful mechanical energy and provide power for a variety of rotating equipment including:

Motor a mechanical driver with rotational output; usually electrically operated.

- Pumps
- Compressors
- Mixers
- Fans
- Feeders
- Valves
- Blowers

Did You Know?

Nikola Tesla, a Serbian engineer who invented the first AC motor, discovered a way to broadcast radio waves before Guglielmo Marconi, who is generally credited with inventing radio.

Tesla also created the world's first radio-controlled boat.

CREDIT: RS89/Shutterstock

The rotation of a motor affects the speed of rotating equipment. Different types of motors are used depending on the application; examples include a two-phase (240/220V) or a three-phase (generally 480/460V) motor. Of the many types of motors used in industrial applications, induction motors are the simplest, most rugged, and most common. The motor consists of a stator (coils of wire attached to magnets in the casing) and a rotor (coils of wire attached to magnets that are attached to the shaft). AC current flows to the stator creating a magnetic field. Like magnetic fields repel each other; the rotor (shaft) magnets create rotation when repelled by the stator magnets. As long as power is supplied, continuous repulsion causes the shaft to rotate, driving the driven-end equipment.

All motors in process plants that are located in the area of flammable gases or liquids are required to be intrinsically safe. **Intrinsically safe** means that the operation of these devices will not ignite a spark that could ignite the gases or liquids as a fuel source.

Intrinsically safe condition in which devices such as motors are designed so as not to ignite a spark that could create a fire; a requirement for all equipment operating in areas where flammable gases and liquids are present.

19.4 Primary Components of a Typical AC Induction Motor

The primary parts of a typical AC motor include a frame, shroud, rotor, shaft, stator, fan, bearings, and AC power source.

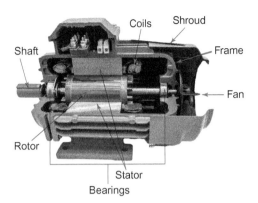

Figure 19.6 AC motor components.
CREDIT: wi6995/fotolia

The following is a description of each of the AC motor components (Figure 19.6):

- **Frame**—a structure that holds the internal components of a motor and motor mounts.
- **Shroud**—a casing over the motor that allows air to flow into and around the motor. The air flow keeps the temperature of the motor cool.
- **Rotor**—the rotating member of a motor that is connected to the shaft.
- **Shaft**—a round metal tube that holds the rotor and is coupled to the driven end equipment.
- **Stator**—a stationary part of the motor where the alternating current flows in and a large magnetic field is created using magnets and coiled wire.
- **Fan**—rotating blades that cool the motor by pulling air in through the shroud.
- **Bearings**—rotating machine parts that support the load of the shaft and its components and minimize friction.
- **AC power source**—a device that supplies current to the stator.

Rotor the rotating member of a motor that is connected to the shaft.

Stator a stationary part of the motor where the alternating current flows in, and a large magnetic field is created using magnets and coiled wire.

Principles of Operation of AC Electric Motors

Early pioneers in electricity discovered the principle of electromagnetism. When current is run through a coil of insulated wire that is wrapped around a soft iron bar, a strong magnetic field is created. When the current is removed, the magnetic field diminishes. This is electromagnetism.

Electromagnetism plays an important role in electric motors. Electric motors, whether AC or DC, operate on the same three electromagnetic principles:

- Electric current generates a magnetic field.
- Like magnetic poles repel each other (positive to positive or negative to negative), while opposite poles attract each other (positive to negative).
- The direction of the electrical current determines the magnetic polarity.

A motor consists of two main parts: a stationary member, called a stator, and a rotating member, called a rotor. The stator includes field magnets and field coils, through which AC current runs. This generates a rotating magnetic field. The rotor features an iron core and copper bars (two highly conductive metals).

The magnetic field of the stator induces an electric current in the rotor, generating a second magnetic field. When the two magnetic fields interact, it causes the rotor to turn. The rotor then supplies mechanical energy to another device, such as a pump or compressor. So, motors convert electrical energy into mechanical energy that can be used to power other equipment (Figure 19.7).

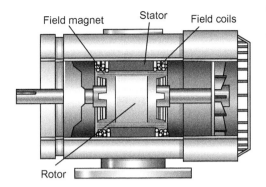

Figure 19.7 AC motor field magnet, stator, field coils, and rotor.

19.5 Operational Hazards

All personnel working with electrical equipment *must* be properly trained and authorized for the particular type and level of work being performed.

OSHA considers electricity to be a hazardous energy source. OSHA Standards 29 CFR 1910.147 and 29 CFR 1910.333 cover the procedures and practices that are mandated to be completed by personnel to ensure the energy source is disabled prior to work on any piece of equipment. Industry calls this set of procedures and practices Lockout-Tagout (LOTO) of hazardous energy. Process technicians and operating personnel can be specifically trained to complete the entire procedure in some instances. The main responsibility of the process technician is to prepare the equipment by shutting off the electrical source to the motor at the switch and by preparing the process end of the equipment for the procedure. The process is also considered an energy source. Documentation is then created to substantiate that the LOTO process is ready for completion. In many industries, personnel from the electrical instrumentation department, or other specifically trained group of workers, then isolate the energy source at the breaker in an electrical room or substation.

The most challenging electrical control and maintenance activities are likely to be reserved for personnel with specialized training in this area. However, process technicians might be required to perform some electrically related tasks or assist an electrician.

Electricity can be dangerous. Electrical current passing through a body causes a shock, which can result in:

- Serious bodily injuries
- Burns
- Death

Shock can make muscles tighten (e.g., chest muscles and diaphragm), which can restrict breathing. Shock also can interrupt the rhythm of the heart by interfering with the natural electrical impulses that control it.

Shock and other electrical hazards occur when a person contacts a conductor carrying electricity while also touching the ground or an object that has a conductive path to the ground. The person completes the circuit as the current passes through the body.

The amount of current and the contact point or path determine the amount of damage to the body. For example, current passing from a finger to an elbow causes less damage than current passing from a hand to a foot.

Electrical shocks can occur when a person:

- Comes in contact with a bare wire (either bare on purpose or as a result of cracked or worn insulation)
- Uses improperly grounded electrical equipment
- Works with electrical equipment in a wet or damp environment, or when the person is sweating heavily
- Works on electrical equipment without checking that the power source has been turned off
- Uses long metal equipment, such as cranes or ladders, that can come into contact with a power source

Did You Know?

Approximately 15% of the OSHA general industry citations relate to electrical hazards.

CREDIT: Luciano Cosmo/Shutterstock.

Static electricity discharges and lightning strikes are two other ways a person can be shocked.

Short circuits are a common cause of electrical hazards. A short circuit occurs when electrons in a current find a path of least resistance which is outside of the normal circuit and flow to it. For example, if the insulation is cracked on two wires that are near each other, the electrons jump between the wires and create a "short" circuit (a shortcut outside the intended circuit).

Water is another potential hazard when dealing with electricity. Water decreases the resistivity of materials to electricity. For example, dry skin provides resistance of up to 100,000 ohms. Wet skin, however, reduces resistivity to as low as 1,000 ohms (per NIOSH). Even sweat on the skin can decrease resistivity.

When energized conductors are close together or touching, electricity can arc through the air from one to the other and complete the circuit. Also, static electricity can discharge, causing a spark. Arcs or sparks can ignite nearby hydrocarbons or other flammable materials, causing a fire or explosion that can harm workers, the unit or plant, and nearby communities.

Along with electrical hazards, motors present mechanical hazards to workers. Motors have moving parts that can pull, tear, or rip clothing or skin if not properly de-energized before working on them. Motors can also affect production (slowing or halting it) if they become inefficient or seize up.

All motors should have controllers that incorporate protection against instantaneous overload, such as a short circuit, as well as thermal overload from working beyond design limits. Motors also will have the means to disconnect and isolate the circuitry for maintenance work. Typically, the motor's overload, starting and stopping mechanisms, and disconnecting device are built together in one motor controller. A motor control center (MCC) is a grouping of these motor controllers. Motors must be "locked out" at the equipment panel breaker using the OSHA Lockout/Tagout procedure. Breakers are generally located in an electric room associated with equipment found in the same general area.

Many accidents are the result of loss of concentration, the routine repetitive work of the process, or rushing to complete a task. Equipment must be operated and prepared for maintenance in a manner to avoid electrical injury. If proper safety procedures are followed when working on electrical circuits and/or electrical devices, then the risk of any hazards occurring are minimized.

The following are some general, yet essential, tips for reducing electrical hazards (Box 19.1).

Box 19.1 Tips for Reducing the Risk of Electrical Hazards

- Consider all equipment and electrical systems to be energized until it is verified that they are not.
- Always inspect and test safety equipment.
- Plan your work before you start.
- Familiarize yourself with the electrical equipment you are using and the electrical circuit on which you will work.
- Do not wear conductive metal (such as jewelry).
- If trained in the appropriate procedures, and as governed by facility policies, disconnect and lock the electrical circuit or system yourself; do not depend on someone else to do it. Depending on the process industry, the process technician might rely on someone else to perform this task.
- After a circuit or system is disconnected, test the circuit to ensure it was properly disconnected and is not still energized or does not contain residual energy.
- Use equipment and tools only as they are intended.
- Use a buddy system when working on electrical circuits.
- Never distract or startle a coworker.
- For electrical fires, use only an extinguisher approved for such a fire; never use water.
- Make sure you are trained in CPR and First Aid.

Precautions that should be taken to minimize the risks associated with specific electrical hazards include the following:

- Wear proper personal protective equipment (PPE) when engaging circuit breakers of any type and rating. This includes appropriate **arc-rated** clothing, which is clothing with protective characteristics in the fabric, as determined by a hazard/risk assessment. Failure to do this could cause serious bodily injury or death. It also could lead to an equipment breaker explosion and flash fire.

- Use a **ground fault circuit interrupter (GFCI)**. A GFCI works on the principle that the amount of current fed from the source to a device, such as a hand tool, equals the amount of current returned from the device to the source. So, simultaneous measurement of the current going to and from any device should always be equal in a properly working system. The GFCI continuously monitors and verifies that the two currents are, in fact, equal. The GFCI assumes that any imbalance is most likely because the current is returning to the source through a human body and ground, rather than through the circuit conductors. If the GFCI measures an imbalance of 5 milliamps (ma) (0.005 ampere) or greater, the GFCI will automatically shut off the circuit to prevent possible electrocution. The typical threshold of feeling for most people is five milliamps, well below the lethal level.

- Use properly grounded 110 V lighting with a GFCI when working inside a vessel. Unlike circuit breakers, which sense only excessive current, GFCIs detect the flow of current to ground and open the circuit to interrupt the flow of current, GFCIs can prevent serious bodily injury, burns, or death, as well as a short in electrical equipment or energized vessel floors and walls.

- Use properly grounded electrical tools (with a GFCI) and make sure electrical tools are working properly (e.g., housings are not accidentally energized) and do not have frayed cords. Failure to do so could lead to serious bodily injury, burns, or death.

- If trained in the appropriate procedures, and as governed by facility policies (Figure 19.8), turn off and lock out breakers using the proper methods (lockout or tagout) to prevent injury to you and your coworkers. The actual opening/closing of the breaker in the electrical room is usually the responsibility of department personnel specifically trained in that work process.

- Inspect wiring for corrosion or fraying and insulation for cracks, burns, and degradation.

Arc-rated having protective characteristics in the fabric, as determined by a hazard/risk assessment.

Ground fault circuit interrupter (GFCI) a safety device that detects the flow of current to ground and opens the circuit to interrupt the flow.

Figure 19.8 A lockout/tagout.
CREDIT: digitalreflect/fotolia

19.6 Monitoring and Maintenance Activities

Scheduled maintenance activities are an essential part of keeping equipment in good condition and preventing equipment failure.

Table 19.2 describes typical, basic maintenance routines necessary for the upkeep of motors.

Table 19.2 Monitoring and Maintenance Activities for Electrical Motors

Look	Listen	Check
■ Inspect wires to ensure they are insulated and the insulation is not cracked or worn. ■ Check for loose covers and shrouds. ■ Visually inspect bearings for wear. ■ Prepare equipment for maintenance personnel to measure bearings and bushing to check for excessive wear. ■ Look for signs of corrosion. ■ Lubricate bearings when scheduled or noisy. ■ Check the motor shaft for wear and verify that it is not bent. ■ Inspect oil passages to make sure they are not plugged. ■ Ensure oil wells or holes are clean.	■ Listen for abnormal noise. ■ Listen to and check the cooling fan. ■ Listen to the bearings.	■ Using the appropriate tool, check for excessive heat. ■ Check for excessive vibration. ■ Check the bearings for heat.

Symbols Associated with Electricity

Several types of equipment are associated with electricity. Each type has a unique symbol that is used to identify it on process drawings. Figure 19.9 provides some examples of these symbols.

Figure 19.9 Common electrical symbols.

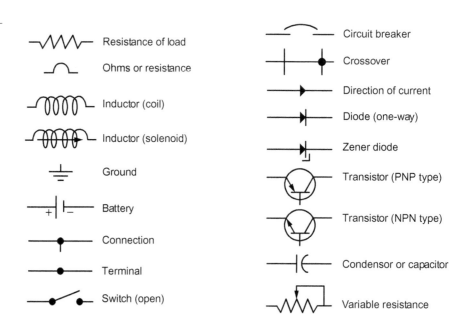

Summary

Electricity is a flow of electrons from one point to another along a pathway, called a conductor. Conductors are materials that have electrons that can break free and flow more easily than other materials. An insulator is a material that does not give up its electrons as easily; an insulator is a poor conductor. A semiconductor is neither a conductor nor an insulator.

Electricity travels through circuits, a system of one or many electrical components connected together to accomplish a specified purpose. An ohm is a measurement of resistance in electrical circuits. An amp is a unit of measure of the electrical current flow in a wire; similar to "gallons of water" flow in a pipe. A volt is the electromotive force that will establish a current of one amp through a resistance of one ohm. A watt is a unit of measure of electric power or the power consumed by a current of one amp using an electromotive force of one volt.

Electrical energy can flow as either alternating current or direct current. AC uses a back-and-forth movement of electrons in a conductor (similar to water sloshing backward and forward in a pipe). DC results in a direct flow of electrons through a conductor (such as water moving in one direction through a pipe). DC is most commonly used in batteries. AC is the most commonly used power type in industry.

AC and DC power both can be used to turn motors. Motors are mechanical drivers that turn electrical energy into useful mechanical energy and provide power for a variety of rotating equipment (e.g., pumps, compressors, and fans). The main components of an electric motor include a frame, shroud, rotor, shaft, stator, fan, bearings, and power supply.

Both AC and DC motors work off the same three principles of electromagnetism: (1) Electric current generates a magnetic field. (2) Like magnetic poles repel each other (positive to positive or negative to negative), while opposite poles attract each other (positive to negative). (3) The direction of the electrical current determines the magnetic polarity.

Because of the hazards associated with electric shock, all personnel working with electrical equipment MUST be properly trained and authorized for the particular type and level of work being performed.

Electricity is dangerous and should be respected. Electrical current passing through a body causes a shock, which can result in serious bodily injuries, burns, or death. For this reason, process technicians should always wear proper protective gear, follow safety procedures, and be aware of potential hazards (e.g., frayed cords and cracked insulation).

Process technicians also should inspect and perform scheduled maintenance on equipment in order to keep it in good condition and prevent equipment failure. Among other tasks, they should lubricate bearings, inspect wires to make sure they are not cracked or worn, listen for abnormal noises, and check for excessive heat or vibration.

Checking Your Knowledge

1. Define the following terms:
   a. ampere
   b. circuit
   c. ground fault circuit interrupter (GFCI)
   d. grounding
   e. insulator
   f. motor
   g. ohm
   h. transformer
   i. volt
   j. watt

2. *(True or False)* DC current is the most common current used in the process industries.

3. What type of equipment converts mechanical energy to electrical energy?

4. What type of equipment converts electrical energy to mechanical energy?

5. The _____ is a casing over an electric motor that allows air to flow into and around the motor.
   a. stator
   b. shell
   c. shroud
   d. frame

6. What is the stator on a motor?

7. Which of the following is NOT a component of a motor?
   a. stator
   b. bearings
   c. pump
   d. shroud
   e. fan

8. Which of the following hazards can result in death?
   a. engaging 480 V and above breakers without PPE
   b. using faulty electrical equipment in the field without a GFCI
   c. using 110 V lighting inside a vessel without a GFCI
   d. all of the above

9. Most accidents involving electric motors:
   a. are caused by loss of attention
   b. cannot be avoided
   c. are someone else's fault
   d. are minor enough to be ignored

10. Electrical shock occurs when your _____ becomes part of the circuit.

11. What is the resistivity of wet skin versus dry skin?

12. Name three items to inspect visually on an electric motor.

13. What does the following symbol represent? What does this device do?

NOTE: Answers to Checking Your Knowledge questions are in Appendix I.

Student Activities

1. Describe the basic principles of electricity, including the difference between AC and DC and which type is commonly used in the process industry.

2. Write a one-page paper explaining how an electric motor works.

3. Given a picture or a cutaway of an AC motor, identify the following components:
   a. frame
   b. shroud
   c. rotor
   d. shaft
   e. stator
   f. fan
   g. bearings
   h. power supply

4. List five causes of electrical shock and explain what you could do to prevent them.

5. Given two strings of lights (one wired in series and the other wired parallel), examine what happens when a "good" bulb (one that lights properly) is replaced with a "bad" bulb (one that no longer lights). Did the entire string stay lit? If not, explain what happened.

6. Using the formula for Ohm's law:
   a. Determine the voltage (V) if current (I) = 0.2 A and resistance (R) = 1,000 ohms.
   b. Determine the current (I) if voltage (V) = 110 V and resistance (R) = 22,000 ohms.
   c. Determine the resistance (R) if voltage (V) = 220 V and current (I) − 5 A.

7. Determine the watts required if the voltage is 480 and the amps are 5.

Chapter 20
Heat Exchangers

 Objectives

Upon completion of this chapter, you will be able to:

20.1 Describe the purpose and major types of heat exchangers in the process industries. (NAPTA Heat Exchangers I & II, 1) p. 274

20.2 Identify the primary parts of a typical shell and tube heat exchanger. (NAPTA Heat Exchangers I, 3; II, 2) p. 277

20.3 Describe the principles of operation of a typical shell and tube heat exchanger. (NAPTA Heat Exchangers I, 2 & 4) p. 277

20.4 Describe the different applications of typical heat exchangers. (NAPTA Heat Exchangers I, 5) p. 280

20.5 Discuss the hazards associated with the improper operation of a heat exchanger. (NAPTA Heat Exchangers I, 6) p. 281

20.6 Describe the monitoring and maintenance activities associated with a heat exchanger. (NAPTA Heat Exchangers I, 7, 8) p. 281

Key Terms

Aftercooler—a heat exchanger, located on the discharge side of equipment, with the function of removing excess heat from the system generated by the process, **p. 280**.

Air-rumble—the process of introducing bursts of compressed air in the cooling water to remove contaminants, **p. 282**.

Back-flush—the process of removing contaminants by reversing the normal flow; typically associated with the use of shell and tube exchangers or plate and frame exchangers that employ cooling water, **p. 282**.

Baffles—partitions located inside a shell and tube heat exchanger that direct flow and increase turbulent flow to help reduce hot spots, **p. 277**.

Channel head—(also called the *exchanger head*); a device located at the end of a heat exchanger that directs the flow of the fluids into and out of the tubes, **p. 277**.

Chiller—a device used to cool a fluid to a temperature below ambient temperature; chillers generally use a refrigerant as a coolant, **p. 280**.

Condenser—a heat exchanger that is used to change the phase of a material by condensing vapor to liquid, **p. 280**.

Distillation—the separation of the components of a liquid mixture by their boiling points, completed by partial vaporization of the mixture and separate recovery of vapor and residue, **p. 280.**

Fin fan—an air-cooled heat exchanger in which air moves across wide fins containing fluid to release heat, **p. 276.**

Heat exchanger—a device used to move heat from one fluid to another without direct physical contact of the fluids, **p. 274.**

Interchanger—(also called a *cross exchanger* or *intercooler*) one of the process-to-process heat exchangers, **p. 280.**

Laminar flow—a condition in which fluid flow is smooth and unbroken; viewed as a series of laminations or thin cylinders of fluid slipping past one another inside a tube, **p. 279.**

Preheater—a heat exchanger used to warm liquids before they enter a distillation tower or other part of the process, **p. 280.**

Reboiler—a tubular heat exchanger, placed at the bottom of a distillation column or stripper, used to supply the necessary column heat, **p. 280.**

Shell—the outer housing of a heat exchanger that covers the tube bundle, **p. 277.**

Shell inlet and outlet—openings that allow process fluids to flow into and out of the shell side of a shell and tube heat exchanger, **p. 277.**

Spacer rods—the rods that hold the tubes in a tube bundle apart so they do not touch one another, **p. 277.**

Tube bundle—a group of fixed or parallel tubes, such as are used in a heat exchanger, through which process fluids are circulated; the tube bundle includes the tube sheets with the tubes, the baffles, and the spacer rods, **p. 277.**

Tube inlet and outlet—openings that allow process fluids to flow into and out of the tube bundle in a shell and tube heat exchanger, **p. 277.**

Tube sheet—a flat plate to which the tubes in a heat exchanger are fixed, **p. 277.**

Turbulent flow—a condition in which the fluid flow pattern is disturbed so there is considerable mixing, **p. 279.**

Introduction

In process industries, heat is created through various mechanisms. Temperatures must be maintained within specific ranges for processes to occur efficiently. Heat exchangers are heat-transferring devices that enable industry processes to run more smoothly and cost effectively.

Many different types of heat exchangers are in use today. Although their designs differ, all have similar components and use the same principles of heat transfer.

20.1 Purpose of Heat Exchangers

A **heat exchanger** is a device used to move heat from one substance to another without direct physical contact of the fluids. By using heat exchangers, many processes can be executed in a more energy-efficient manner. Although inline heaters and cooling towers can enable processes to be done, heat exchangers allow them to be run more efficiently.

Heat exchanger a device used to move heat from one fluid to another without direct physical contact of the fluids.

Types of Heat Exchangers

Heat exchangers come in a variety of types (e.g., fin fan or air-cooled, plate and frame, double-pipe, reboiler, and shell and tube). Most common are the shell and tube exchangers, which have differing functional designs, such as: single-pass, multipass, floating head, and U-tube.

The simplest form of the shell and tube heat exchanger is the single-pass shell and tube heat exchanger (Figure 20.1A). As suggested by its name, fluid moves through the tubes just once in the single-pass exchanger. In contrast, the multipass heat exchanger (Figure 20.1B) has channel heads on the ends of the heat exchange, with the result that fluid in the tubes makes multiple passes through the heat exchanger's shell. Both of these types accommodate the lowest temperature differential between the hottest fluid and the coldest fluid. They also tend to be the least expensive.

Figure 20.1 Common types of heat exchangers. **A.** Single-pass shell and tube heat exchanger. **B.** Multipass shell and tube heat exchanger.

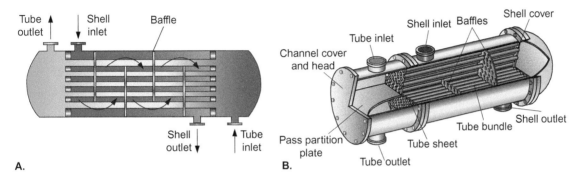

A shell and tube heat exchanger that can accommodate a slightly larger heat differential is a floating head heat exchanger (Figure 20.2). On one end of the tube bundle, a rigid tube sheet is mounted to the shell body. On the other end of the tubes is a floating head with an inner channel head. This floating head allows the tubes to expand at a different rate than the shell. The floating head heat exchanger falls in the middle range of cost.

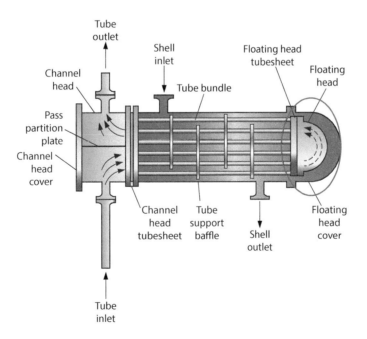

Figure 20.2 The floating head heat exchanger is also a shell and tube exchanger.

For the largest temperature differentials, the U-tube heat exchanger is used (Figure 20.3). It has both ends of the tube attached to the same tube sheet and is designed so that the tubes can expand without putting stress on the tube sheet. The channel head directs the flow of the fluid into one side of the U-tubes and then out of the channel head when it exits the tubes. This type of heat exchanger is the costliest type of shell and tube heat exchanger to manufacture.

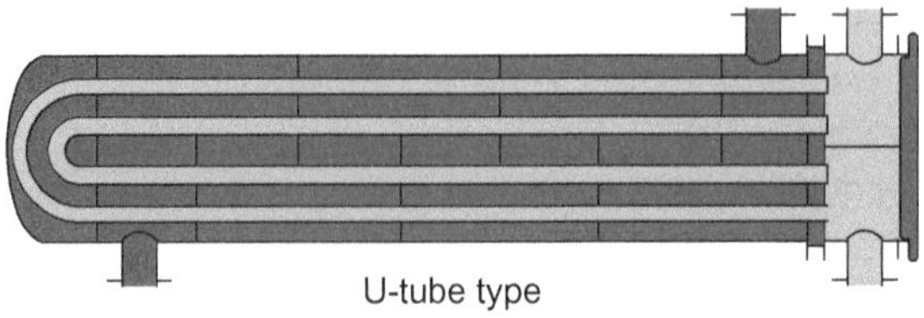

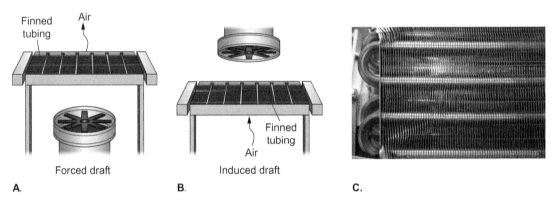

Figure 20.3 U-tube heat exchanger. This shell and tube heat exchanger can tolerate the greatest temperature differentials.

Fin fan an air-cooled heat exchanger in which air moves across wide fins containing fluid to release heat.

The **fin fan** is an air-cooled heat exchanger (Figure 20.4). It works like a radiator in a car. The fluid to be cooled is in the fin-covered tubes. The fins help disperse more heat by adding surface area. The design of a fin fan cooler can be forced draft (in which the fan pushes air into the fin tubes) or induced draft (in which the fan pulls the air through the fin tubes). The fin fan heat exchanger does not recycle heat, but instead discards the heat from the process.

Figure 20.4 Fin fan (air cooled) heat exchangers. **A.** Forced draft. **B.** Induced draft. **C.** Close-up of fin fan construction.

CREDIT: **C.** Aleksander Tasevski/Shutterstock.

The plate and frame style of heat exchanger (Figure 20.5) has a series of thin plates separating the two fluids. In this design, the hot and cold fluids alternate between plates. Plate and frame heat exchangers tend to be very efficient, but have low pressure capacity.

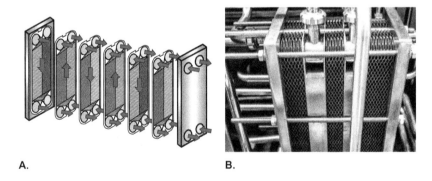

Figure 20.5 Plate and frame heat exchanger.

CREDIT: **B.** Mrs_ya/Shutterstock.

The simplest type of heat exchanger is the double pipe (Figure 20.6). This type of heat exchanger consists of a small pipe inside a larger pipe.

Figure 20.6 Double pipe heat exchanger. This is the simplest form of heat exchanger.

The reboiler (or *kettle reboiler*) (Figure 20.7) is a special type of heat exchanger in which heat is added to a liquid for the purpose of boiling out a vapor entrained in the liquid. This type of heat exchange has a path for the vapor to exit.

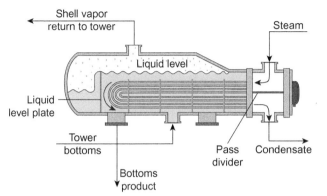

Figure 20.7 Reboiler (kettle reboiler) heat exchanger.

20.2 Components of a Typical Shell and Tube Heat Exchanger

The components of a heat exchanger can vary based on the design and purpose of the exchanger. However, there are some commonalities among exchangers.

The components of a typical shell and tube heat exchanger include (see Figure 20.1B):

- Tube bundle
- Tube sheet
- Baffles
- Tube inlet and outlet
- Shell cover
- Shell inlet and outlet
- Channel cover and head
- Pass partition plate(s)

The **tube bundle** is a group of fixed or parallel tubes through which process fluids are circulated. Tube bundles include the **tube sheet**, a flat plate to which the tubes in a heat exchanger are fixed; **baffles**, partitions located inside a shell and tube heat exchanger that increase turbulent flow and reduce hot spots; and **spacer rods** (not shown in Figure 20.1B). Spacer rods hold the tubes in a tube bundle apart so they do not touch one another.

Fluids move into the tube bundle through the **tube inlet** and out through the **tube outlet**. Covering the tube bundle is an outer housing called a **shell**. Fluid moves into the shell through the **shell inlet** and out through the **shell outlet**.

The **channel head** (also called *exchanger head*) is located at the end of a heat exchanger, and the pass partition plate directs the flow of fluids into and out of the tubes. The channel head has a removable cover to allow access to the tubes for maintenance and cleaning.

20.3 Principles of Operation of Heat Exchangers

Heat exchangers facilitate the transfer of heat, primarily through two heat transfer methods: conduction and convection (Figure 20.8).

As discussed in the physics chapter, several methods of heat transfer exist. *Conduction* is flow of heat through a solid (e.g., a frying pan transferring heat to an egg). *Convection* is

Tube bundle   a group of fixed or parallel tubes, such as are used in a heat exchanger, through which process fluids are circulated; the tube bundle includes the tube sheets with the tubes, the baffles, and the spacer rods.

Tube sheet   a flat plate to which the tubes in a heat exchanger are fixed.

Baffles   partitions located inside a shell and tube heat exchanger that direct flow and increase turbulent flow to help reduce hot spots.

Spacer rods   the rods that hold the tubes in a tube bundle apart so they do not touch one another.

Tube inlet and outlet   openings that allow process fluids to flow into and out of the tube bundle in a shell and tube heat exchanger.

Shell   the outer housing of a heat exchanger that covers the tube bundle.

Shell inlet and outlet   openings that allow process fluids to flow into and out of the shell side of a shell and tube heat exchanger.

Channel head   (also called the *exchanger head*); a device located at the end of a heat exchanger that directs the flow of the fluids into and out of the tubes.

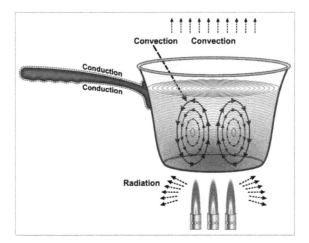

Figure 20.8 Examples of heat transfer through convection, conduction, and radiation.

CREDIT: Fouad A. Saad/Shutterstock.

Did You Know?

Heat always travels from hot to cold.
 This means that when ice melts it is actually absorbing heat and giving up cold.

CREDIT: zentilia/fotolia.

transfer of heat through a fluid medium (e.g., warm air circulated by a hair dryer). *Radiation* is the transfer of heat through space (e.g., warmth emitted from the sun).

The main methods of heat transfer in a shell and tube heat exchanger are conduction (through the tube wall and tube surface, to shell fluid) and convection (fluid movement within the shell and the tubes).

In shell and tube exchangers (see Figures 20.1 to 20.3), liquids or gases of varying temperatures are pumped, or pressured, into the shell and tubes in order to facilitate heat exchange.

For example, hot process fluids can be pumped through the tubes while cool water is pumped through the shell. As the cool water circulates around the tubes, heat is transferred from the tubes to the water in the shell.

A practical example of an air-cooled (fin fan) heat exchanger (see Figure 20.5) is a car radiator. In this type of exchanger, hot fluids flow through the radiator tubes. As the car moves forward, cool air is drawn into the radiator grill and over the radiator tubes by the fan. As air passes over the tubes, heat from the fluid is transferred to the air via conduction and convection.

Flow pattern designs in heat exchangers must take into consideration the process material composition, the amount of heat that needs to be added or removed, and the size of the exchanger (Figure 20.9). Flow patterns can be:

Parallel—both materials flow in the same direction (see Figure 20.9A)

Counter flow—the two materials enter the exchanger at separate ends and flow against each other (see Figure 20.9B)

Cross flow—the two materials flow through the exchanger perpendicular to each other (see Figure 20.9C)

Figure 20.9 Examples of flow pattern designs. **A.** Parallel flow. **B.** Counter flow. **C.** Cross flow.

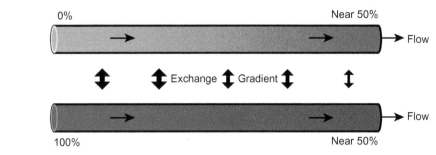

A.

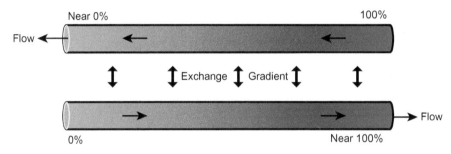

B.

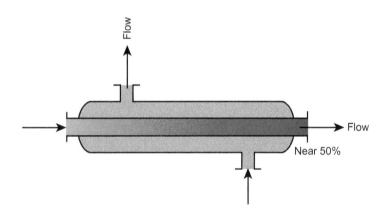

C.

As fluids move through a pipe or a heat exchanger, the flow can be either **laminar** (a condition in which the fluid flow is smooth and unbroken, such as a series of laminations or thin cylinders of fluid slipping past one another inside a tube) or **turbulent** (a condition in which the fluid flow pattern is disturbed so there is considerable mixing); see Figure 20.10.

Laminar flow a condition in which fluid flow is smooth and unbroken; viewed as a series of laminations or thin cylinders of fluid slipping past one another inside a tube.

Turbulent flow a condition in which the fluid flow pattern is disturbed so there is considerable mixing.

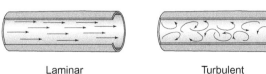

Figure 20.10 Examples of laminar and turbulent flow.

> **Did You Know?**
>
> Turbulent flow provides more fluid mixing than laminar flow. This makes it the desired flow type in most heat exchangers.

The ideal flow type in a heat exchanger is turbulent flow because it provides more mixing and better heat transfer. The appropriate flow rate through the tubes will cause turbulent flow in the tubes, and baffles will promote a turbulent flow on the shell side. Baffles support the tubes, increase turbulent flow and heat transfer rates, and reduce hot spots.

Shell and tube exchangers may be designed as single-pass, meaning the process to be cooled or heated enters the exchanger on one side and leaves the exchanger on the opposite end. Or, exchangers may be multipass in design, meaning the process enters one end of the exchanger, turns at the header, and returns through another pass to exit the exchanger at the point where it entered. The passes are separated by a divider plate or *pass partition*. Review Figure 20.1B for an example of a multipass exchanger. Process enters the tube end, moves through and enters the head, is redirected through the second tube pass, and exits the exchanger.

20.4 Heat Exchanger Applications

The most common applications for heat exchangers are reboilers, preheaters, aftercoolers, condensers, chillers, and interchangers.

Reboiler (see Figure 20.7) is the term for a heat exchanger, usually placed at the bottom of a distillation column or stripper, which is used to supply the column heat required for the process operation. The main purpose of a reboiler is to supply heat to help vaporize the liquid. The heating medium for a reboiler can be steam or hot process fluids from other parts of the plant.

Preheater is the term for a heat exchanger used to warm fluids before they enter a distillation tower or other equipment in the process.

Aftercooler is the term for a heat exchanger, located on the discharge side of equipment, whose function is to remove from the system excess heat that is generated by the process.

Condenser is the term for a heat exchanger that is used to change vapor to liquid. The design of a condenser can be the same as that of a preheater (i.e., a typical shell and tube exchanger), but condensers can also have air fins or be air cooled, depending on the process application. The difference between condensers and preheaters is the temperature of the fluid being used for heat exchange (i.e., warm vs. cool).

Chiller is the term for a device used to cool a fluid to a temperature below ambient temperature. Chillers generally use a refrigerant as a coolant.

Interchanger (*cross exchanger*) is a process-to-process heat exchanger. Interchangers are used on columns or towers where the hot fluids leaving the column are pumped through the tube side of an exchanger, and the feed to the column is put through the shell side. This action preheats the feed to the column and cools the hot material leaving the column. Other applications are possible in a process plant and are used as an effective cost-saving measure.

Relationships Among Different Types of Heat Exchangers

To illustrate the various applications of heat exchangers, consider the distillation process.

Distillation is the separation of components of a liquid mixture by their boiling points, completed by partial vaporization of the mixture and separate recovery of vapor and residue. In order for vaporization to occur, heat must be added to the process fluids. This can be accomplished through preheaters and reboilers.

As mentioned, preheaters warm the fluid before it enters the column. Reboilers add additional heat to fluids that are already in the column so they will vaporize. The overhead vapors are condensed in the condenser and returned to a liquid state.

Figure 20.11 illustrates the relationship between a preheater, a reboiler, and a condenser in the distillation process.

Reboiler a tubular heat exchanger, placed at the bottom of a distillation column or stripper, used to supply the necessary column heat.

Preheater a heat exchanger used to warm liquids before they enter a distillation tower or other part of the process.

Aftercooler a heat exchanger, located on the discharge side of equipment, with the function of removing excess heat from the system generated by the process.

Condenser a heat exchanger that is used to change the phase of a material by condensing vapor to liquid.

Chiller a device used to cool a fluid to a temperature below ambient temperature; chillers generally use a refrigerant as a coolant.

Interchanger (also called a *cross exchanger* or *intercooler*) one of the process-to-process heat exchangers.

Distillation the separation of the components of a liquid mixture by their boiling points, completed by partial vaporization of the mixture and separate recovery of vapor and residue.

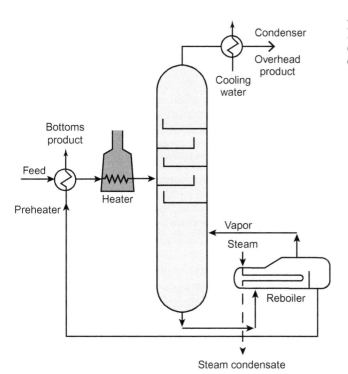

Figure 20.11 Heat exchangers associated with a distillation column.

20.5 Operational Hazards

Many hazards are associated with improper heat exchanger operation. Table 20.1 lists some of these hazards and their impacts.

Table 20.1 Hazards Associated with Improper Heat Exchanger Operation

Improper Operation	Possible Impacts			
	Individual	Equipment	Production	Environment
Putting the exchanger online with the bleeder valve open on process side	Exposure to chemicals or steam			Spill to the environment; possible fire or explosion
Applying heat to the exchanger without following the proper warm-up procedure	Exposure to chemicals or steam if the tubes rupture	Metal stress from thermal shock; possible tube or shell rupture	Downtime because of repair	Spill to the environment if the tubes rupture
Misalignment: cooling, heating, or not opening inlet and outlet		Overheating, tube fouling, or melting	Downtime because of repair	
Operating a heat exchanger with ruptured tubes	Exposure to chemicals or steam	Possible damage if the safety valve is not working properly	Product ruined by contamination; downtime because of repair	Spill to the environment
Opening a cold fluid to a hot heat exchanger		Overpressurizing the heat exchanger; potential for thermal shock		

20.6 Monitoring and Maintenance Activities

When working with heat exchangers, process technicians need to be aware of problems that can have an impact on the exchanger and be able to perform preventive maintenance. Typically, cooler fluid should be flowing first when putting a heat exchanger into service, and warmer fluid should be put in service second. When taking a heat exchanger out of service, hot fluid should be removed first, then cooler fluid.

Table 20.2 describes some of the things a process technician should monitor when working with heat exchangers.

Table 20.2 Monitoring and Maintenance Activities for Heat Exchangers

Look	Listen	Check
■ Check for external leaks. ■ Check for internal tube leaks by collecting and analyzing samples. ■ Look for abnormal pressure changes (could indicate tube plugging). ■ Look for temperature changes. ■ Look for hot spots or uneven temperatures. ■ Inspect for leaky gaskets. ■ Inspect insulation. ■ Monitor inlet and outlet temperature gauges. ■ Monitor inlet and outlet pressure gauges. ■ Monitor inlet and outlet flow rate. ■ Monitor inlet and outlet samples.	■ Listen for abnormal noises (e.g., rattling or whistling). Be extremely careful around high pressure leaks; they are often invisible but very audible.	■ Check for excessive vibration (vibration can loosen the tubes in the tube sheet). ■ Check inlet and outlet for heat and coolness. ■ Check for excessive temperature.

When the heat exchanger is disassembled:

- Inspect for leaky tubes.
- Inspect tube sheet.
- Inspect divider plate (dual pass).
- Inspect baffles for proper fit.
- Inspect for corrosion.
- Inspect for distorted tubes.
- Inspect vents to ensure they are working properly.

For startup or shutdown, monitor for temperature gradients/shock.

In addition to the tasks listed above, process technicians may choose to air-rumble or back-flush the cooling water side of a heat exchanger as needed to reduce fouling. If a heat exchanger that has cooling water in the tubes is not cooling the process fluid (on the shell side of the heat exchanger), it is a sign that the heat exchange needs to be cleaned. Not all heat exchangers can be shut down for a proper cleaning. Air-rumble and back-flushing are procedures that can be used while a heat exchanger is in service.

Air-rumble introduces a two-phase flow. It is designed to remove fouling in cooling water with bursts of compressed air. On the inlet of the cooling water, a quarter-turn valve is added to a utility air header. On the cooling water outlet, a bleed valve is added that goes into a container, such as a bucket. After the bleed valve is opened, the cooling water flow back to the cooling tower is blocked for the duration of the procedure. With cooling water flowing through the heat exchanger, bursts of compressed air are applied, and the cooling water exiting the heat exchanger into the bucket is monitored. Airbursts are repeated as long as sediment is being removed. After the procedure is complete, the flow of cool water is returned to the cooling tower and the air line is removed. (See more about cooling towers in Chapter 21.)

Back-flushing (back-washing) is washing by reversing the normal flow. During back-flushing, water is run backward through a heat exchanger to remove deposits and reduce fouling. Heat exchangers are equipped with special valves so that back-flushing can be performed. Back-flushing an in-service exchanger can create serious process disturbance, so it is performed after careful planning and communication with all personnel and with supervision in the affected area. Any such maintenance activity should be monitored closely at all times.

Failure to perform proper maintenance and monitoring could affect the process and result in personnel or equipment damage.

Air-rumble the process of introducing bursts of compressed air in the cooling water to remove contaminants.

Back-flush the process of removing contaminants by reversing the normal flow; typically associated with the use of shell and tube exchangers or plate and frame exchangers that employ cooling water.

Heat Exchanger Symbols

Figure 20.12 shows examples of heat exchanger symbols a process technician might encounter on a piping and instrumentation diagram (P&ID).

Figure 20.12 Heat exchanger symbols.

Heat exchanger symbols				
U-tube HEX	Floating head HEX	S&T multi-pass HEX	Kettle reboiler	Plate exchanger
Plate and frame heat exchanger	Heat exchanger	Heat exchanger 2	Heat exchanger 3	Heater
Exchanger	Condenser	Shell and tube heat	Shell and tube heat 2	Shell and tube heat 3
Cooler	Air-blown cooler	Induced flow air cooler	Fin fan cooler	Straight tubes heat exchanger
Coil tubes heat exchanger	Finned tubes heat exchanger	Floating head heat exchanger	Plate heat exchanger	Kettle heat exchanger
Reboiler heat exchanger	Reboiler	Single pass heat exchanger	Single pass heat exchanger	Fin-tube exchanger
Hairpin exchanger	Spiral heat exchanger	Spiral heat exchanger 2	Air cooled exchanger	U-tube heat exchanger
U-tube heat exchanger	Condenser			

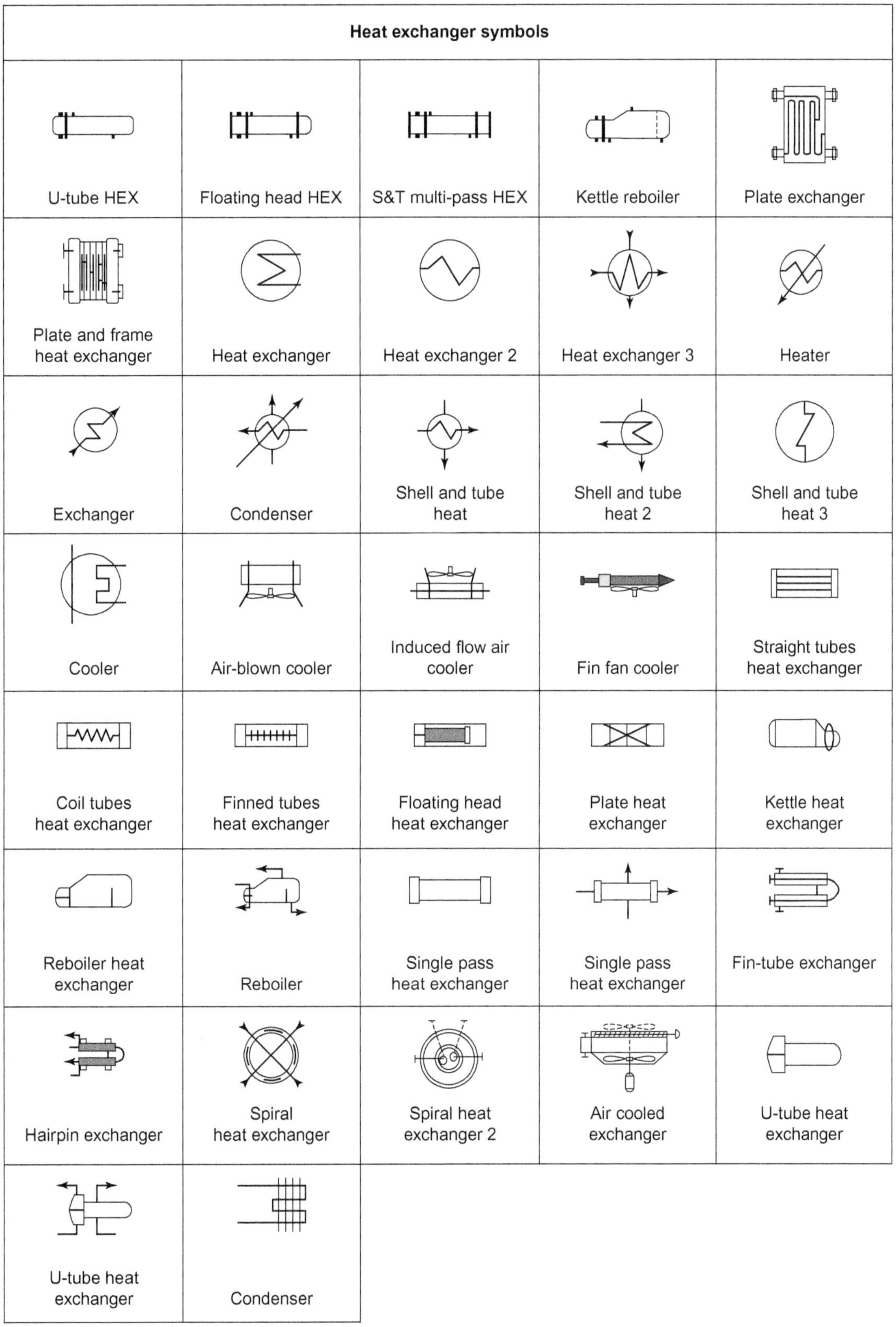

Summary

Heat exchangers are devices that use conduction, convection, and radiation to heat or cool process fluids.

The most common type of heat exchanger in process industries is the shell and tube design. The primary components of this type of exchanger are a tube bundle, tube sheet, baffles, spacer rods, tube inlet, tube outlet, shell, shell inlet, shell outlet, and an exchanger (channel) head.

In a shell and tube exchanger, liquids or gases of varying temperatures are pumped into the shell and tubes. As the fluids move through the shell, baffles cause turbulence (mixing) and direct the flow of the fluid. Turbulence is desired because it facilitates heat transfer and creates more surface area for heating and cooling.

Air-cooled (fin fan) heat exchangers remove heat by the movement of air over a wide surface area containing the fluids. They do not recycle heat but release it.

Heat exchangers can be used in a variety of applications, including reboilers, pre-heaters, aftercoolers, condensers, chillers, or interchangers.

Improperly operating a heat exchanger can lead to safety and process issues, including possible exposure to chemicals or hot fluids, or equipment damage because of thermal shock. Generally, when putting a heat exchanger into service, the cooler fluid should be flowing first; the warmer fluid should be put in service second. When taking the heat exchanger out of service, the hot fluid should be removed first, and then the cooler fluid should be removed.

Technicians should always monitor heat exchangers for abnormal noises, leaks, excessive heat, pressure drops, temperature drops, or other abnormal conditions, and conduct preventive maintenance as needed.

Checking Your Knowledge

1. Define the following terms:
   a. baffle
   b. condenser
   c. distillation
   d. heat exchanger
   e. laminar flow
   f. preheater
   g. reboiler
   h. shell
   i. tube bundle
   j. turbulent flow

2. Which of the following heat exchanger components is used to create turbulence in the process fluids and facilitate heat exchange?
   a. shell
   b. baffle
   c. exchanger head
   d. tube inlet

3. *(True or False)* Heat transfer always moves from cold to hot.

4. The transfer of heat through a fluid medium is called:
   a. convection
   b. conduction
   c. radiation
   d. evaporation

5. List three applications of heat exchangers.

6. List at least three things a process technician should look, listen, and check for when monitoring a heat exchanger.

7. The type of heat exchanger that is most commonly used in process industries is:
   a. fin fan
   b. double pipe
   c. shell and tube
   d. reboiler

8. The heat exchanger that can handle the greatest temperature differentials is:
   a. multipass shell and tube
   b. U-tube
   c. plate and frame
   d. floating head

NOTE: Answers to Checking Your Knowledge questions are in Appendix I.

Student Activities

1. Given a cutaway or a picture of a heat exchanger, identify the following components:
   a. tube bundle
   b. tube sheet
   c. baffles
   d. tube inlet and outlet
   e. shell
   f. shell inlet and outlet
   g. exchanger (channel) head

2. Sketch a simple diagram of distillation column and identify where the following components might be found:
   - preheater
   - condenser
   - reboiler

3. List five hazards associated with improper heat exchanger operations, and identify the possible impacts these hazards could have on individuals, equipment, production, and/or the environment.

4. Given a piping and instrumentation diagram (P&ID), correctly identify all of the heat exchangers and describe their application to the process.

Chapter 21
Cooling Towers

Objectives

Upon completion of this chapter, you will be able to:

21.1 Describe the purpose and types of cooling towers in the process industries. (NAPTA Cooling Towers 1) p. 287

21.2 Identify the primary parts and support systems of a typical open cooling tower. (NAPTA Cooling Towers 2) p. 289

21.3 Describe the principles of operation and applications of a cooling tower. (NAPTA Cooling Towers 3) p. 290

21.4 Discuss the hazards associated with the improper operation of a cooling tower. (NAPTA Cooling Towers 4) p. 292

21.5 Describe the monitoring and maintenance activities associated with a cooling tower. (NAPTA Cooling Towers 5, 6) p. 293

Key Terms

Basin—a compartment at the base of a cooling tower that is used to store cooled water until it is pumped back into the process, **p. 290**.

Blowdown—the process of taking basin water out of a cooling tower to reduce the concentration level of impurities, **p. 291**.

Chemical treatment—chemicals added to cooling water that are used to control algae, sludge, scale buildup, and fouling of exchangers and cooling equipment, **p. 291**.

Closed circuit cooling tower—a cooling tower that uses a closed coil over which water is sprayed and a fan provides air flow, and in which there is no direct contact between the water being cooled and the water and air used to cool it, **p. 288**.

Cooler—a heat exchanger that uses a cooling medium to lower the temperature of a process material, **p. 292**.

Cooling tower—a structure designed to lower the temperature of a cooling water stream by evaporating part of the stream (latent heat of evaporation); these towers are usually made of fiber-reinforced plastic (FRP), galvanized metal, wood, or stainless steel and are designed to promote maximum contact between water and air, **p. 287**.

Drift eliminators—devices that prevent water from being blown out of the cooling tower; the main purpose of a drift eliminator is to minimize water loss, **p. 290**.

Evaporation—a process in which a liquid is changed into a gas through the latent heat process, **p. 290**.

Fan—a device used to force or draw air, **p. 290**.

Fill—the material that breaks water into smaller droplets as it falls inside the cooling tower, **p. 290**.
Forced draft tower—a cooling tower that has fans or blowers at the bottom of the tower; the fans or blowers force air through the tower components, **p. 288**.
Induced draft tower—a cooling tower that has fans at the top of the tower; these fans pull air through the tower components, **p. 288**.
Makeup water—water brought in from an external source such as a lake or river that is treated and used to replace the water lost during the blowdown and/or evaporation process, **p. 290**.
Natural draft tower—a cooling tower that uses temperature differences inside and outside the stack to facilitate air movement, **p. 288**.
Open circuit cooling tower—a cooling tower design that cools cooling water by evaporation and has the cooling water exposed to air, **p. 288**.
Suction screens—devices located on the inlet side of equipment used to filter out debris; they are most often associated with pump suction, **p. 290**.
Water distribution header—a device that distributes water evenly across the top of the cooling tower through the use of distribution nozzles; the header helps create more effective cooling and allows the water to flow evenly into the tower fill, **p. 290**.

Introduction

Cooling towers play an important role in many industry processes. Like heat exchangers and other types of equipment, cooling towers enable change of temperature of process fluids.

Many different types of cooling towers are in use today. Cooling towers are classified as either the wet type, which is open design, or the dry type, which is closed design. The open design is the most common type used in process industries. Although their designs differ, all cooling towers have similar components and use the same principles of heat transfer.

21.1 Purpose of Cooling Towers

Cooling towers are structures designed to lower the temperature of a water stream by evaporating part of the stream (latent heat of evaporation) and having a small amount of the heat transfer take place via mixing of fluids (convection).

Process industries use water from local waterways such as lakes, streams, and rivers to circulate through plant equipment as a means of cooling processes. Cooling towers are found in industries from petrochemical plants to HVAC systems to thermal power plants.

The main purpose of open circuit cooling towers is to remove excess heat from process cooling water used in heat exchangers so that the water can be recycled and recirculated through the plant. The open (wet) circuit cooling tower is generally the most commonly used in industrial process applications.

Closed (dry) circuit cooling towers tend to be used in refrigeration and other HVAC systems.

Cooling tower a structure designed to lower the temperature of a cooling water stream by evaporating part of the stream (latent heat of evaporation); these towers are usually made of fiber-reinforced plastic (FRP), galvanized metal, wood, or stainless steel and are designed to promote maximum contact between water and air

Types of Cooling Towers

Cooling towers have many different design structures that reflect the required cooling of the process load. They may be evaporative (wet), dry (closed), or a combination of evaporative and dry. In a wet design tower, the water to be cooled comes directly into contact with the air flow. Open (wet) circuit cooling towers are usually made of plastic, FRP fiber, galvanized metal, wood, or stainless steel. The component design works to break the water into tiny droplets and promote maximum direct water-to-air contact. In a dry design-type tower, the fluid to be cooled is circulated through a heat coil and does not directly contact the water or air flow.

Did You Know?

Many homes use evaporative cooling units to cool the air. These units lower the temperature of the air by filtering it through water.

Because humidity has an impact on the process of evaporation, this type of unit works best in drier climates.

CREDIT: Constantine Pankin/Shutterstock.

Natural draft tower a cooling tower that uses temperature differences inside and outside the stack to facilitate air movement

Forced draft tower a cooling tower that has fans or blowers at the bottom of the tower; the fans or blowers force air through the tower components

Induced draft tower a cooling tower that has fans at the top of the tower; these fans pull air through the tower components

Cooling towers may utilize a natural or mechanical draft application. Natural draft cooling towers are suitable in areas of high wind and low humidity. However, the most common draft type is mechanical, which uses a fan system. Air flows in mechanical draft applications may be either cross flow or counter flow. Figure 21.1 shows examples of the different types of cooling towers.

- **Natural draft towers**—use temperature differences inside and outside the stack to facilitate air movement (see Figure 21.1A). These towers are more dependent on wind currents than other cooling tower designs.
- **Forced draft towers**—have fans or blowers at the bottom of the tower that force air through the equipment (see Figure 21.1B).
- **Induced draft towers**—have fans at the top of the tower that pull air through the equipment components (see Figure 21.1C).

Figure 21.1 Types of drafts for open cooling towers. **A.** Natural draft tower. **B.** Forced draft tower. **C.** Induced draft tower.
CREDIT: A. martinlisner/fotolia **C.** chatchawal/fotolia.

A.

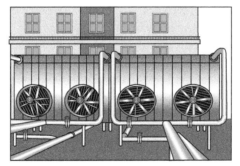

B.

C.

Closed circuit cooling tower a cooling tower that uses a closed coil over which water is sprayed and a fan provides air flow, and in which there is no direct contact between the water being cooled and the water and air used to cool it

Open circuit cooling tower a cooling tower design that cools cooling water by evaporation and has the cooling water exposed to air

- **Closed circuit cooling towers**—use a heat exchange coil through which the working fluid passes and over which water is sprayed. A fan draws air through the water spray, pulling heat away from the coil by the process of convection. There is no direct contact between the water being cooled and the sprayed water. Figure 21.2 illustrates a closed circuit cooling tower.
- **Open circuit cooling tower**—a cooling tower design that cools cooling water by evaporation and has the cooling water exposed to air. This can be a forced draft, natural draft, or induced draft type with countercurrent or cross current design.

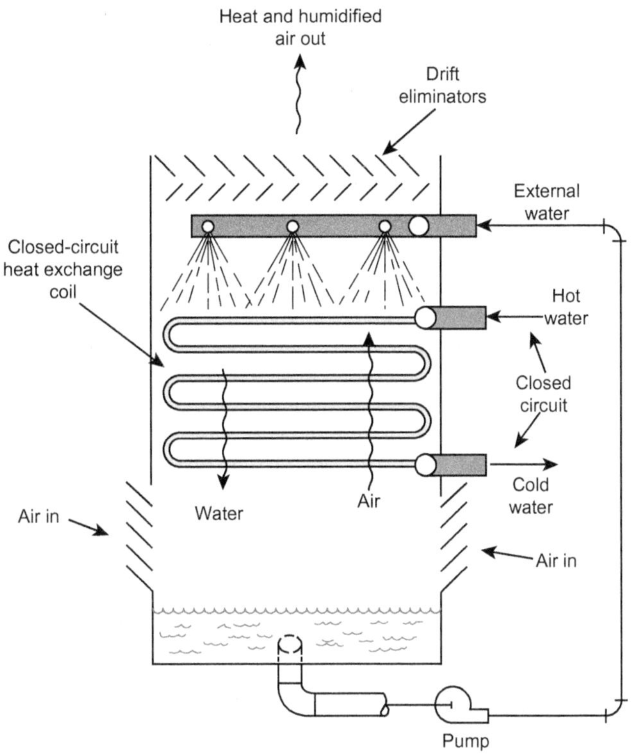

Figure 21.2 Closed circuit cooling tower.

21.2 Component Parts of an Open Cooling Tower

Although cooling tower designs can vary, most cooling towers are made of fiber-reinforced plastic (FRP), galvanized metal, wood, or stainless steel and have similar components. These components include (Figure 21.3):

- Water distribution header and cells—distribute the warm water
- Fill (splash bars)—redirects flow of air and water
- Fan (induced or forced draft)—moves air through the tower
- Louvers (slats)—allow air intake
- Drift eliminators—minimize water loss
- Makeup water—replaces lost fluid
- Basin—stores water
- Suction screens—filter out debris

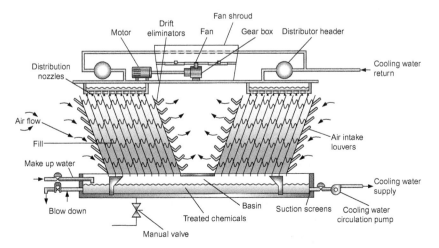

Figure 21.3 Parts of an induced draft, cross flow cooling tower.

Water distribution header a device that distributes water evenly across the top of the cooling tower through the use of distribution nozzles; the header helps create more effective cooling and allows the water to flow evenly into the tower fill

Fill the material that breaks water into smaller droplets as it falls inside the cooling tower

Fan a device used to force or draw air

In a cooling tower, the **water distribution header** distributes the incoming water evenly across the top of the cooling tower through the use of distribution nozzles. It allows the water to flow through the louvers evenly into the tower. As the water leaves the header, it falls down into the cells, which may contain splash bars called "fill" or "packing." **Fill** is the material that breaks water into smaller droplets as it falls inside the cooling tower, creating greater surface exposure area. Cooling towers may be designed with one or with many cells, depending on the process application and needed workload. One complete cell consists of all components found in Figure 21.3.

In induced or forced draft cooling towers, **fans** are used to force or draw air through the cooling tower. As the water falls down, the air contacts the water, thereby reducing the temperature of the water by evaporation. Air flow patterns vary by design. The two main air flow patterns are counter flow and cross flow (Figure 21.4). In counter flow, water falls downward and the air rises upward through the water. In cross flow, water falls downward and air flow, entering through louvers (slats), flows perpendicular to (across) the water flow.

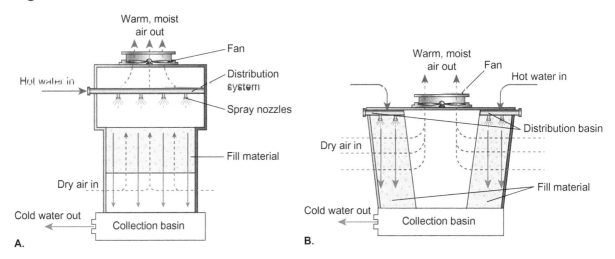

Figure 21.4 Illustration of types of flow in a cooling tower. **A.** Counter flow. **B.** Cross flow.

Drift eliminators devices that prevent water from being blown out of the cooling tower; the main purpose of a drift eliminator is to minimize water loss

Makeup water water brought in from an external source such as a lake or river that is treated and used to replace the water lost during the blowdown and/or evaporation process

Basin a compartment at the base of a cooling tower that is used to store cooled water until it is pumped back into the process

Suction screens devices located on the inlet side of equipment used to filter out debris; they are most often associated with pump suction

Evaporation a process in which a liquid is changed into a gas through the latent heat process

Drift eliminators minimize water loss and prevent water from being blown out of the cooling tower with the evaporated water up through the fan. However, some water is always lost despite these measures. In order to compensate for this water loss, makeup water is added to the basin.

Makeup water is water that is treated and used to replace the water lost during blowdown and the process of evaporation. The **basin** is a compartment, located at the base of a cooling tower, in which water is stored until it is pumped back into the process. Makeup water is taken from a lake, river, or other waterway located close to the cooling tower.

Debris in the basin can damage or destroy the cooling water circulation pumps. In order to prevent this damage, **suction screens** (usually cone-shaped or flat metal strainers) are used to filter out debris (e.g., wood fibers and trash). Periodic maintenance is conducted by shutting down the tower, or an individual cell in a tower with multiple cells, to clean the fill and basin of build-up or debris.

21.3 Principles of Operation of Open Circuit Cooling Towers

Cooling tower operation is based on the process of **evaporation**, an endothermic process in which a liquid is changed into a gas through the latent heat process. Through the process of evaporation, hot water entering the cooling tower from the process is cooled. This cool water is then sent back to the plant for reuse by the equipment (Figure 21.5).

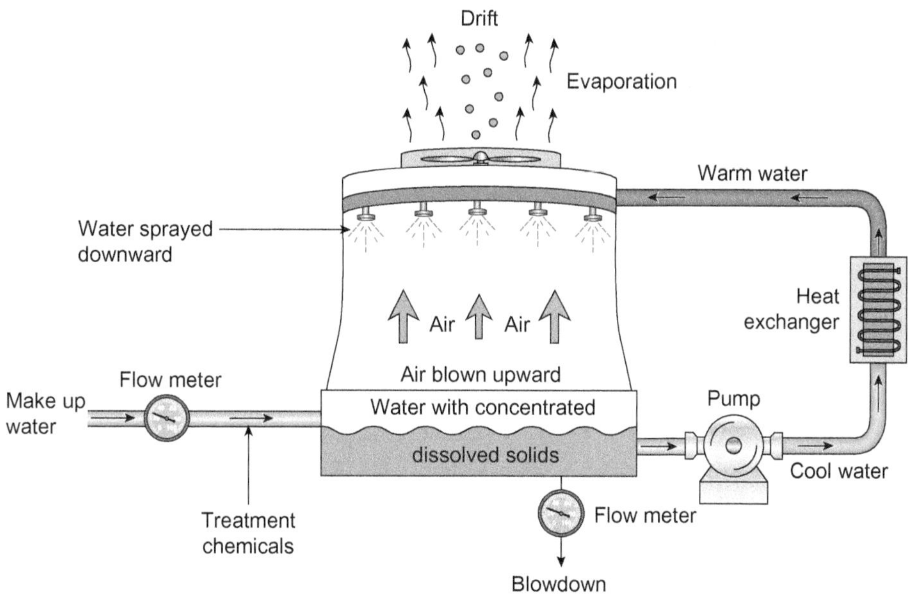

Figure 21.5 Overview of process cooling.

The following is an overview of the cooling process:

1. Water from the cooling tower is pumped to the process plant equipment.
2. Heat from the process equipment is transferred to the cool water.
3. Hot water from the equipment is returned to the cooling tower distribution header.
4. The distribution header sprays the hot water downward onto the fill.
5. The fill breaks the water into smaller droplets, thereby increasing the surface area and facilitating heat exchange.
6. As water falls through the tower, it is exposed to air that removes heat through latent heat of evaporation, dropping the water temperature for reuse in the plant.
7. When the cool water reaches the bottom of the tower, it is collected in the basin.
8. Pumps carry water out of the basin and return it to the cooling water header that supplies process plant equipment for reuse.
9. Steps 2–8 are repeated as a continuous process.

During the cooling process, some water is lost because of evaporation, splashing, leakage, or blowdown. This water is replaced with makeup water, often using a simple level control to maintain a certain water level in the basin.

All cooling water, including makeup water, must undergo **chemical treatment**. Chemicals must be added to the cooling water in order to prevent algae growth, sludge, and scale buildup, which cause fouling of the exchangers and cooling equipment. They also control foaming, which can act as a blanket and diminish the cooling process.

In addition to chemical treatment, blowdown must be performed on cooling tower water. **Blowdown** is a process in which water is removed from a cooling tower basin to maintain the correct water chemical balance and to reduce the concentration level of impurities. The blowdown process can occur continuously or periodically.

Chemical treatment chemicals added to cooling water that are used to control algae, sludge, scale buildup, and fouling of exchangers and cooling equipment

Blowdown process of taking basin water out of a cooling tower to reduce the concentration level of impurities

Factors that Affect Cooling Tower Performance

Many factors have an impact on cooling tower performance. These include:

- Ambient air temperature
- Relative humidity of ambient air
- Wind velocity and direction
- Water contamination
- Cooling tower design

Table 21.1 lists each of these variables and explains how they affect the performance of the cooling tower.

Table 21.1 Factors that Affect Cooling Tower Performance

Variable	Impacts on Cooling Tower Performance
Temperature	↑ambient temperature = ↑cooling tower demands
Humidity	↑humidity = ↑water vapor in the air
	↑water vapor = ↓cooling
	(Note: If humidity reaches 100%, evaporation will not occur because the air is completely saturated with water vapor.)
Wind velocity and direction	↑wind velocity = ↑airflow
	↑airflow = ↑cooling
	Wind direction opposite of cooling tower orientation = decreased evaporation
Water contamination	↑contamination (e.g., algae buildup) = ↓cooling (e.g., because of fouling)
Tower design	↑airflow = ↑cooling
	(Note: Induced or forced draft towers have greater airflow than natural draft towers and are not dependent on wind currents.)

Cooling Tower Applications

Cooling towers are used in many industrial processes, including petrochemical and refining plants, food processing plants, power generation plants, and industrial HVAC systems.

Within the process industries, exchangers constitute the heaviest usage of cooling water. These exchangers can be categorized as coolers, condensers, and chillers, each with the same responsibility of cooling the process material to a desired temperature for further processing.

Cooler, condenser, and chiller are terms given to any exchanger that lowers the temperature of the process fluid. The placement of the exchanger is related to its work in the process plant and the particular name used to describe it. In lubricating systems (found in rotating equipment), coolers are used to cool lubrication oils on steam turbines and other large pieces of equipment.

Cooler a heat exchanger that uses a cooling medium to lower the temperature of a process material

21.4 Operational Hazards

Several hazards are associated with improper cooling tower operation. Table 21.2 lists some of these hazards and their possible impacts.

Table 21.2 Hazards Associated with Improper Cooling Tower Operation

Improper Operation	Possible Impacts			
	Individual	Equipment	Production	Environment
Improper/incorrect treatment of cooling water chemically	Exposure to harmful microorganisms (e.g., *Legionella* bacteria, which causes Legionnaires' disease)	Algae growth or sludge buildup that can foul process equipment	Off-spec products created because of improper heat exchange. Foaming can be created by too much chemical addition	Foaming in the cooling tower can allow the air current to blow the foam into nearby areas exposing personnel and equipment to possible harm
Improper valve operation	Scalding or burns from exposure to hot process cooling water	Pump damage because of "deadheading" or improper feed-water supply	Off-spec products created because of improper heat exchange	
Circulating cooling water through a heat exchanger with ruptured tubes	Exposure to chemicals, steam, or bacteria	Tower contamination; depletion of treatment chemicals	Product ruined by contamination	Hazardous chemicals spilled to the environment
Failure to use caution on wet, icy, or slippery surfaces, or when ice fog has formed	Injury because of slipping and falling			Injury/motor vehicle crashes due very slippery conditions (e.g., on adjacent highways)

Improper operation of the chlorine system	Severe injury or death from chlorine exposure (e.g., don't climb a tower when shock chlorinating unless wearing the correct PPE for protection)	Possible algae growth or sludge buildup that fouls the exchanger (if chlorine levels are not high enough)	Off-spec products created because of improper heat exchange (e.g., if the exchanger fouls because chlorine levels are too low)
Ingesting or exposing skin to chemically treated cooling water	Injury or death because of water treatment chemical exposure		
Formation of ice-fog when operating in cold climates	Injury due to very slippery conditions		Injury/motor vehicle crashes due very slippery conditions (e.g., on adjacent highways)

21.5 Monitoring and Maintenance Activities

When working with cooling towers, process technicians should be aware of the correct operating parameters for each season of the year to ensure they recognize problems that can have an impact on them, and they should be able to perform preventive maintenance.

Table 21.3 describes some of the things a process technician should monitor when working with cooling towers.

In addition to the tasks listed above, process technicians should run only the number of fans required to produce the proper cooling water supply temperature. Fan usage can be affected by both the season of the year and the time of day.

Failure to perform proper maintenance and monitoring could affect the process and result in personal or equipment damage.

Table 21.3 Monitoring and Maintenance Activities for Cooling Towers

Look	Listen	Check
■ Look for leaks. ■ Check basin water levels to make sure they are adequate. ■ Check chemical balance (pH and hardness). ■ Observe filter screens for plugging. ■ Monitor temperature differentials. ■ Look for broken fill materials to fix at next turnaround. ■ Look for ice buildup or ice fog in cold climates. ■ Watch for foaming because of possible improper chemical treatment or incorrect blowdown stream. ■ Check for proper water distribution on top of the tower.	■ Listen for abnormal noises (e.g., grinding sounds associated with pump cavitation, or high-pitched sounds associated with improperly lubricated fan bearings).	■ Check for excessive heat or vibration in fans and pumps.

Cooling Tower Symbols

In order to locate cooling towers on a piping and instrumentation diagram (P&ID) accurately, process technicians need to be familiar with the different cooling tower symbols. Figure 21.6 shows some of the symbols used to indicate cooling towers.

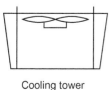
Cooling tower

Induced draft cross-flow

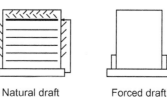
Natural draft counter-flow   Forced draft cooling tower

Figure 21.6 Cooling tower symbols.

Summary

Cooling towers are structures designed to lower the temperature of a water stream by evaporating part of the stream (latent heat of evaporation) and having a small amount of the heat transfer take place because of the mixing of fluids (convection). The main purpose of cooling towers is to remove heat from process cooling water so it can be recycled and recirculated through the process. Cooling towers may be of open circuit or closed circuit operational design. Open design is the type generally seen in process industries.

Cooling towers are used in conjunction with many types of process equipment. Heat exchangers are the heaviest users of cooling water. Heat exchangers remove heat from process fluids. Cooling towers remove heat from the water used in heat exchangers. The heating and cooling relationship between exchangers and towers is a continuous process.

Cooling towers are designed in accordance with process workload and may have different design structures. The three main airflow types are natural draft (which relies on temperature differences), forced draft (which uses fans to push air through the cooling tower), and induced draft (which uses fans to pull air through the cooling tower).

The main components of an open cooling tower include a water distribution header, fill, louvers, drift eliminators, basin, makeup water, fan(s), and suction screens.

Many factors affect cooling tower performance, including ambient air temperature, humidity, wind velocity, water contamination, and tower design.

Cooling tower water can be used in a variety of applications, including petrochemical and refining plants, power generation plants, food processing plants, and industrial HVAC systems.

Monitoring and maintenance of cooling towers is critical. Many entire plant systems rely on cooling water as the main source of cooling for process streams.

Checking Your Knowledge

1. Define the following terms:
   a. basin
   b. blowdown
   c. cooler
   d. cooling tower
   e. drift eliminators
   f. louvers (slats)
   g. evaporation
   h. fill
   i. makeup water
   j. suction screens
   k. water distribution header

2. What is the purpose of a cooling tower?

3. List three different types of draft in cooling towers.

4. List five factors that affect cooling tower performance.

5. List at least three monitoring and maintenance activities associated with cooling towers

6. A sound associated with pump cavitation is:
   a. humming
   b. grinding
   c. high-pitched whistling
   d. rumbling

NOTE: Answers to Checking Your Knowledge questions are in Appendix I.

Student Activities

1. On a model or diagram of a cooling tower, label the following components and discuss the purpose of each:
   a. water distribution header
   b. fill (splash bars)
   c. fan
   d. drift eliminators
   e. makeup water
   f. basin
   g. louvers

2. List the steps associated with the cooling process (i.e., explain how a cooling tower works) and explain why cooling water must be chemically treated.

3. Given illustrations of the three most common types of cooling towers, identify them by name and place arrows in the correct place for flow direction. Indicate:
   a. type of cooling tower
   b. flow path of the water
   c. flow path of the air

Chapter 22
Furnaces

Objectives

Upon completion of this chapter, you will be able to:

22.1 Describe the purpose or function of furnaces in the process industries. (NAPTA Furnaces 1) p. 296

22.2 Identify the primary parts of a typical furnace. (NAPTA Furnaces 3) p. 297

22.3 Describe the operation of a furnace, including types of fuel and feedstock. (NAPTA Furnaces 2) p. 299

22.4 Describe the different furnace designs and draft types. (NAPTA Furnaces 4, 5) p. 299

22.5 Discuss the hazards associated with the improper operation of a furnace. (NAPTA Furnaces 7) p. 302

22.6 Describe the monitoring and maintenance activities associated with furnaces. (NAPTA Furnaces 6, 8) p. 302

Key Terms

Air register—also called the *burner damper*, a device located on a burner that is used to adjust the primary and secondary airflow to the burner; air registers provide the main source of air entry to the furnace, **p. 299**.

Balanced draft furnace—a furnace that uses fans to facilitate airflow; fan(s) inducing flow out of the firebox, through the stack (induced draft); and fan(s) providing positive pressure to the burners (forced draft), **p. 301**.

Breech—area below the stack where flue gases are collected past the last convection tubes, **p. 299**.

Burner—a device with open flame used to provide heat to the firebox; there may be one or many depending on the furnace's size and capacity, **p. 298**.

Convection section—the upper and cooler section of a furnace where heat is transferred and where convection tubes absorb heat, primarily through the method of convection; the part of the furnace where process feed and/or steam enters, **p. 299**.

Convection tubes—furnace tubes located above the shock bank tubes that receive heat through convection, **p. 299**.

Damper—a valve, movable plate, or adjustable louver used to regulate the flow of air or draft in a furnace, **p. 299**.

Draft gauges—devices calibrated in inches of water that measure the firebox pressure, airflow, and differential pressure between the outside of the furnace and the flue gas inside, **p. 299**.

Firebox—the combustion area and hottest section of the furnace where burners are located and radiant heat transfer occurs, **p. 298**.

Flame impingement—a condition in which the flames from a burner touch the tubes in the furnace, **p. 302**.

Forced draft furnace—a furnace that uses fans or blowers located at the base of the furnace to force the air required for combustion into the burner's air registers, **p. 301**.

Fuel gas valve—a valve that controls the flow of fuel gas entering the furnace burners, **p. 298**.

Furnace—a type of process equipment in which heat is liberated and transferred directly or indirectly to a fluid mass for the purpose of increasing the temperature of a process fluid, **p. 296**.

Furnace purge—a method of using steam or air to remove combustibles from a furnace firebox before lighting the burner and startup, **p. 299**.

Hot spots—areas within a radiant section furnace tube, or areas within a furnace firebox, that have been overheated due to flame impingement or some other cause, **p. 302**.

Induced draft furnace—a furnace that uses fans, located in the stack, to pull air up through the furnace and induce airflow by creating a lower pressure in the firebox, **p. 301**.

Natural draft furnace—a furnace that has no mechanical draft or fans; instead, the heat in the furnace causes draft, **p. 299**.

Pilot—an initiating device used to ignite the burner fuel, **p. 298**.

Radiant section—the lower portion of a furnace (firebox) where heat transfer occurs, primarily through radiation, **p. 298**.

Radiant tubes—tubes located in the firebox that receive heat primarily through radiant heat transfer, **p. 298**.

Refractory lining—a form of insulation used inside high temperature boilers, incinerators, heaters, reactors, and furnaces; a common type is called fire brick, **p. 298**.

Shock bank (shock tubes)—also called the *shield section*; a row of tubes located directly above the firebox in a furnace that receives both radiant and convective heat; it protects the convection section from exposure to the radiant heat of the firebox, **p. 299**.

Stack—a cylindrical outlet, located at the top of a furnace, through which flue (combustion) gas is removed from the furnace, **p. 299**.

Introduction

Furnaces play an important role in many types of processes, and they have a wide range of applications in process industries. Many different types of furnaces are in use today. Although their designs and applications differ, they have similar components and use the same principles of combustion and heat transfer.

When working with furnaces, process technicians should be aware of the operational aspects of the furnace and the factors that could affect furnace performance and safety.

22.1 Purpose of Furnaces

A **furnace** is an apparatus in which heat is liberated and transferred directly or indirectly to a fluid mass for the purpose of increasing the temperature of the process stream. It is also referred to as a process heater or fired heater.

Furnace a type of process equipment in which heat is liberated and transferred directly or indirectly to a fluid mass for the purpose of increasing the temperature of a process fluid.

Furnaces are used for many applications. They can be used to heat water, superheat steam, incinerate waste products, heat a process feed stream, remove metals from ore (smelting), and facilitate chemical reactions. For example, large furnaces in the refining and petrochemical industries are used to raise the temperature of feedstock to ready it for the distillation process. These furnaces also separate the components of crude oil in order to break heavier hydrocarbon molecules into lighter hydrocarbon molecules (Figure 22.1). In a process commonly called *cracking*, feedstocks such as butanes, butylene, propane, and ethane are separated so that products like gasoline and plastic can be formed.

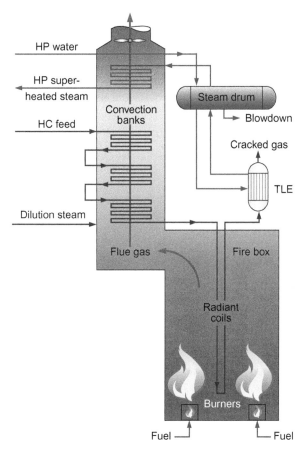

Figure 22.1 Cracking furnace layout.

22.2 Components of a Furnace

Furnaces contain different sections and component parts. The most common components are:

- Radiant section
- Firebox
- Fan (induced or forced draft)
- Refractory lining
- Radiant tubes
- Burner (gas or oil)
- Fuel gas valve
- Pilot
- Air register
- Draft gauges
- Furnace purging (steam or air)
- Convection section

Did You Know?

The *Titanic* had 159 coal-fired furnaces that used as much as 825 tons of coal per day!

These furnaces were used to generate the steam for turbines that were attached to and turned the propellers.

CREDIT: Christopher Krohn/Fotolia

- Convection tubes
- Shock bank
- Stack
- Damper

Figure 22.2 shows an example a cabin furnace and its components. Cabin furnaces are called this because their shape resembles a cabin with a chimney.

Furnace internals are divided into two sections: radiant and convection. The radiant section contains the burners. The convection section contains the shield or shock bank and the breech area. The stack provides the outlet for discharge of flue gas from the furnace.

Figure 22.2 Cabin furnace components.

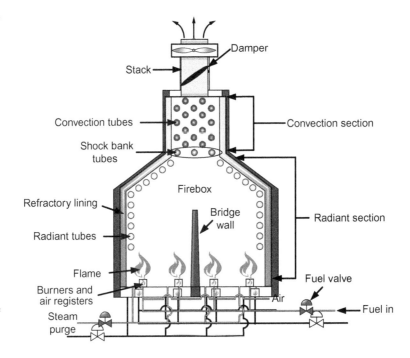

Radiant section the lower portion of a furnace (firebox) where heat transfer occurs, primarily through radiation.

Firebox the combustion area and hottest section of the furnace where burners are located and radiant heat transfer occurs.

Refractory lining a form of insulation used inside high temperature boilers, incinerators, heaters, reactors, and furnaces; a common type is called fire brick.

Radiant tubes tubes located in the firebox that receive heat primarily through radiant heat transfer.

Burner a device with open flame used to provide heat to the firebox; there may be one or many depending on the furnace's size and capacity.

Fuel gas valve a valve that controls the flow of fuel gas entering the furnace burners.

Pilot an initiating device used to ignite the burner fuel.

Radiant Section

The **radiant section** is the lower portion of a furnace where heat transfer takes place primarily through radiation. It is the second area of exposure of the process to heat. The radiant section contains the firebox, radiant tubes, and burners.

The **firebox** is the combustion area. It is the hottest section of the furnace, where the burners are located and radiant heat is transferred. (The terms firebox and radiant section are sometimes used interchangeably.) Because so much heat is generated in this section, the firebox must be lined with a special refractory lining.

The **refractory lining**, sometimes called fire brick, is a bricklike form of insulation used inside high temperature applications. This lining may be made of other materials, such as ceramic fiber or castable (poured like concrete). The purpose of this lining is to reflect heat back into the firebox and to protect the external walls and floor of the furnace.

Radiant tubes are tubes located in the firebox that receive heat primarily through radiant heat transfer from the burners.

A **burner** is a device used to burn an air/fuel mixture in the firebox section of a furnace. Common furnace fuels include natural gas, fuel oil, process oil, process gas, and fuel gas.

The **fuel gas valve** is a valve that controls the flow of fuel gas entering the furnace burners. The **pilot** is an initiating device used to ignite the burner fuel.

An **air register**, also called the *burner damper*, provides the main source of air to the furnace. Air registers are used to adjust the primary and secondary airflow to the burner. They control excess oxygen and help achieve the most efficient burn of the fuel.

Draft gauges are devices calibrated in inches of water that measure the firebox pressure, airflow, and differential pressure between the outside of the furnace and the flue gas inside.

A **furnace purge** is the process of adding steam or air flow to the firebox in order to remove, or sweep, combustibles from the furnace before lighting the burner and startup.

Convection Section

The **convection section** is the cooler upper section of a furnace where heat transfer primarily occurs by convection. The convection section contains the shield section, or shock bank, and the breech area. Within this section are convection tubes (also called *preheat tubes*), where the convection tubes absorb heat. Some of the convection tubes can preheat fluids like process feedstock, water, or steam.

Convection tubes are furnace tubes containing process feedstock, water, or steam. These tubes transfer heat through the process of convection. Below the convection tubes, and above the firebox, is a row of tubes called the shield section or **shock bank**. The shock bank is the tube section that protects the convection section and stack from high heat temperatures. Shock bank tubes receive both radiant and convective heat.

Above the convection tubes is the breech. The **breech** is the transition area between the convection section and stack where flue gases are collected. Most of the heat from the flue gas should be recovered by the time it reaches this area.

The **stack** is a cylindrical outlet, located at the top of a furnace through which flue (combustion) gas is removed from the furnace. Within the stack is a valve or movable plate (louver) called the **damper**. The purpose of the damper is to regulate the flow of air or draft in the furnace. Emission measurements are taken here as an issue or factor of compliance with federal regulation.

22.3 Principles of Operation of Furnaces

Furnaces burn fuel through the use of burners inside the firebox to heat the process fluid to a specified temperature. The temperature is determined by process feedstock effluent analysis. In the heated firebox, process feedstock fluids are pumped through tubes that are exposed to the heat of a direct flame. Heat is transferred through the walls of the tubes to the process fluid via conduction. With the aid of fans from the firebox, the product of combustion (flue gas) flows up through the convection section and into the stack, where it is released into the atmosphere.

22.4 Furnace Designs and Draft Types

Furnace Designs

Furnaces are structurally designed to accommodate different process applications. Design depends on factors such as tube and burner arrangement and whether the furnace is part of a larger process unit. The most common process furnace structures (Figure 22.3) are box, cylindrical, and cabin designs.

Furnace Draft Types

Furnace design and process load affect the type of draft system used in the unit. Types of draft systems include natural draft, forced draft, induced draft, and balanced draft. The different draft-system applications are utilized according to furnace design.

NATURAL DRAFT A **natural draft furnace** has no fans or mechanically produced draft. Furnace applications for using a natural draft depend on the difference in density between the warm air rising from the burner(s) and the cool air that enters the furnace. As hot air rises

Figure 22.3 Box, cylindrical, and cabin furnace designs.

CREDIT: *(Bottom row left to right)* eakkaluktemwanich/Shutterstock; tamapapat/Shutterstock; meepoohfoto/Shutterstock.

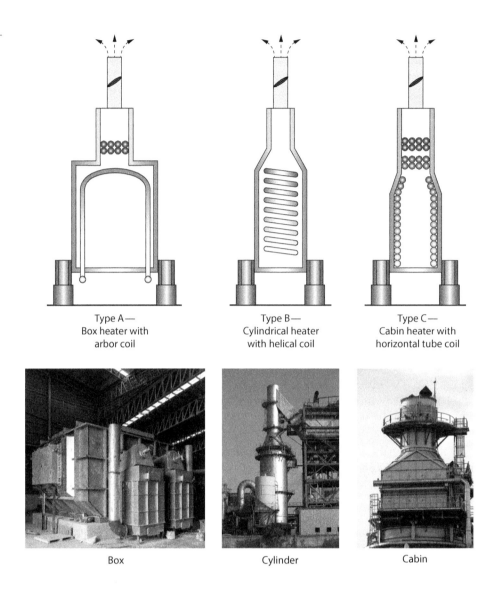

Type A—
Box heater with arbor coil

Type B—
Cylindrical heater with helical coil

Type C—
Cabin heater with horizontal tube coil

Box

Cylinder

Cabin

through the stack, pressure is created inside the firebox burner's air registers. Circulation is created as warm air leaves the furnace, allowing cooler air to enter. This type of furnace is not suitable for heavy process load applications. Figure 22.4 shows an example of a natural draft furnace.

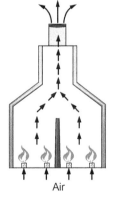

Figure 22.4 Natural draft furnace (cabin-type).

In order to achieve proper draft, the stacks on natural draft furnaces must be taller than those using other draft methods. In general, natural draft furnaces are not highly efficient in operation and are not suitable for use in many process plant applications.

FORCED DRAFT A **forced draft furnace** uses fans or blowers located at the base of the furnace to force air into the air registers. Figure 22.5 shows an example of a forced draft furnace.

Forced draft furnace a furnace that uses fans or blowers located at the base of the furnace to force the air required for combustion into the burner's air registers.

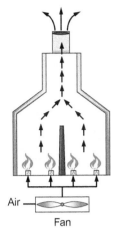

Figure 22.5 Forced draft furnace (cabin-type).

During combustion in a forced draft furnace, positive pressure is created by the fan or blower, which forces air into the burner's air registers as the hotter flue gases leave the firebox.

INDUCED DRAFT An **induced draft furnace** uses fans, located at the top of the furnace, to pull air up through the furnace. This induces airflow by creating a lower pressure in the firebox. Figure 22.6 shows an example of an induced draft furnace.

Induced draft furnace a furnace that uses fans, located in the stack, to pull air up through the furnace and induce airflow by creating a lower pressure in the firebox.

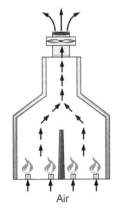

Figure 22.6 Induced draft furnace (cabin-type).

BALANCED DRAFT A **balanced draft furnace** uses fans to facilitate airflow: fan(s) providing positive pressure to the burners (forced draft), and fan(s) pulling flow out of the firebox through the stack (induced draft). Figure 22.7 shows an example of a balanced draft furnace.

Balanced draft furnace a furnace that uses fans to facilitate airflow, with fan(s) pulling flow out of the firebox, through the stack (induced draft); and fan(s) providing positive pressure to the burners (forced draft).

Figure 22.7 Balanced draft furnace (cabin-type).

22.5 Operational Hazards

There are several hazards associated with improper furnace operation. Table 22.1 lists some of these hazards and their possible impacts.

Table 22.1 Operational Hazards Associated with Furnaces

Improper Operation	Possible Impacts			
	Individual	Equipment	Production	Environment
Opening inspection ports when the firebox has a positive pressure or pressure greater than atmospheric	Burns, injuries, or death due to hot flames or combustion gases forced out of the inspection port			
Failure to follow flame safety procedure (e.g., lighting off burners without purging firebox)	Burns, injuries, or death	Explosion in firebox or *flashback* (a backfire that occurs inside the fuel management system)	Lost production due to downtime for repairs	Exceeding EPA opacity limits
Poor control of excess air and draft control	Burns, injuries, or death	Flame impingement and/or tube rupture and explosion	Lost production due to downtime for repairs	Exceeding EPA opacity limits
Bypassing safety interlocks	Burns, injuries, or death	Explosion in firebox	Lost production due to downtime for repairs	
Failure to wear proper protective equipment (e.g., gloves and face shield) when opening furnace inspection ports	Burns, injuries, or death if exposed to hot gases and/or fire			

22.6 Monitoring and Maintenance Activities

Flame impingement a condition in which the flames from a burner touch the tubes in the furnace.

Process technicians, working at the control board and in the field, are responsible for monitoring and maintenance of furnaces. They must always remember to look, listen, and check for the items listed in Table 22.2.

Table 22.2 Monitoring and Maintenance Activities for Furnaces

Look	Listen	Check
■ Check for secondary combustion in the convection section. ■ Check for **flame impingement** on tubes (when the flames from a burner touch the tubes in a furnace). ■ Visually check the firebox for tube leaks that can cause impingement on other tubes, furnace floors, or walls ■ Visually check tubes to be mounted on hangers ■ Check for **hot spots** on furnace tubes and walls that have been overheated due to flame impingement or some other cause. ■ Check draft balance (pressures). ■ Check temperature gradient. ■ Check firing efficiency (CO_2 and O_2 in the stack). ■ Check burner balance (fuel-to-air) ratios. ■ Check all burners to be lit. ■ Check controlling instruments (fuel flow, feed, flow and pressure, and temperature). ■ Check for proper flame pattern (important if all burners are not being lit in the furnace). ■ Observe flame color in firebox. ■ Check for fireballs inside firebox (an indication of incorrect fuel and air ratio to burners).	■ Listen for abnormal noise (fans, burners, and leaks).	■ Using the appropriate tool, check for excessive vibration (fans and burners).

Hot spots areas within a radiant section furnace tube, or areas within a furnace firebox, that have been overheated due to flame impingement or some other cause.

Failure to perform proper maintenance and monitoring could impact the process and result in equipment damage and possible injury to personnel. There are written procedures for furnace shutdown and cleaning that vary according to furnace design and are not addressed in this chapter.

Furnace Symbol

In order to locate furnaces on a piping and instrumentation diagram (P&ID) accurately, process technicians need to be familiar with the different types of furnaces and their symbols.

Figure 22.8 shows an example of a furnace symbol. However, this symbol may vary from facility to facility.

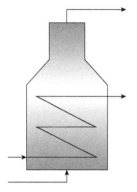

Figure 22.8 Furnace symbol.

Summary

Furnaces are devices used to heat and increase the temperature of various process fluids. Furnaces are divided into several sections: the radiant, convection, shield, breech, and the stack sections. The radiant section includes components such as the firebox, burners, radiant tubes, and refractory lining. The convection section includes components such as the convection tubes for water, steam, and process-fluid preheat. The shield section protects the convection section tubes. The breech is the transition area to the stack. The stack is the component that directs the flue gas flow out of the furnace.

Radiant furnaces work by burning fuel inside a containment area called the firebox. Within the firebox are tubes that receive heat and transfer it to process fluids in the tubes through the process of conduction. Common furnace fuels include natural gas, fuel oil, process oil, process gas, and fuel gas.

Furnaces utilize many designs and draft types. Common designs are box, cylindrical, and cabin. Natural draft furnaces have no mechanical draft or fans. Instead, they use the difference in density between hot combustion gases and cold outside air to cause draft. Forced draft furnaces use fans or blowers, located at the base of the furnace, to force airflow through the furnace via air registers. Induced draft furnaces use fans, located in the stack, to induce airflow through the furnace via the air registers. Balanced draft furnaces use both forced and induced draft fans to facilitate airflow: one at the top to induce airflow, and one at the bottom to force airflow through the furnace.

When working with furnaces, process technicians should be aware of the hazards associated with improper operation, and the impact those hazards could have on individuals, equipment, production, and the environment. They should always wear proper protective equipment and look, listen, and check for potential problems or hazards with the equipment in the field location.

Checking Your Knowledge

1. Define the following terms:
   a. balanced draft furnace
   b. burner
   c. convection section
   d. firebox
   e. forced draft furnace
   f. furnace purge
   g. induced draft furnace
   h. natural draft furnace
   i. pilot
   j. radiant section
   k. refractory
   l. stack

2. The purpose of a furnace includes which of the following? Select all that apply.
   a. separating components of crude oil
   b. superheating steam
   c. removing metal from ore
   d. cleaning fluid for use in a process
   e. raising the temperature of a process stream

3. Which of the following are types of furnace drafts? Select all that apply.
   a. reduced draft
   b. balanced draft
   c. forced draft
   d. cross draft
   e. natural draft

4. How is heat transferred to process feedstock fluids in a furnace?
   a. convection
   b. conduction
   c. radiation
   d. evaporation

5. Which of the following does not belong in a list of feedstock for a furnace?
   a. octane
   b. butane
   c. propane
   d. butylene

6. Which of the following is a common type of furnace design?
   a. U-tube
   b. box
   c. lodge
   d. heater

7. List at least three hazards associated with improper furnace operation.

8. List at least three things a process technician should look, listen, and check for during normal monitoring and maintenance of furnaces.

NOTE: Answers to Checking Your Knowledge questions are in Appendix I.

Student Activities

1. Given a picture of a furnace, identify the following components:
   a. radiant section
   b. firebox
   c. refractory lining
   d. radiant tubes
   e. burner
   f. convection section
   g. convection tubes
   h. shock bank
   i. stack
   j. fans (if any are present)

2. Given a picture of a furnace, tell whether the furnace is natural draft, forced draft, induced draft, or balanced draft. Explain the reason for your answer.

3. Compare and contrast the operational differences between natural draft, forced draft, induced draft, and balanced draft furnaces.

Chapter 23
Boilers

 Objectives

Upon completion of this chapter, you will be able to:

23.1 Describe the principles and purpose of boiler operation in the process industries. (NAPTA Boilers 1, 2) p. 306

23.2 Identify the primary parts and support systems of a typical fuel-fired boiler. (NAPTA Boilers 3) p. 306

23.3 Describe the operating principles and types of fuels used in a boiler. (NAPTA Boilers 4) p. 308

23.4 Describe the different types of boilers (NAPTA Boilers 5, 6) p. 309

23.5 Discuss the hazards associated with improper operation of a boiler. (NAPTA Boilers 8) p. 310

23.6 Describe the monitoring and maintenance activities associated with operating boilers. (NAPTA Boilers 7, 9) p. 311

Key Terms

Boiler—a closed pressure vessel in which water is boiled and converted into steam under controlled conditions, **p. 306.**

Desuperheater—a system that controls the temperature of steam leaving a boiler by using water injection through a control valve, **p. 308.**

Downcomers—tubes exposed to the firebox that transfer water from the steam drum to the mud drum, **p. 308.**

Economizer—the section of a boiler used to preheat feed water before it enters the main boiler system, **p. 307.**

Mud drum—(1) the lower drum of a boiler that is used as a junction area for boiler tubes. (2) A low place in a boiler where heavy particles in the water will settle out so that they can be blown down, **p. 307.**

Risers—tubes that go into the side or top of the steam drum, allowing heated water and steam to flow up from the mud drum, **p. 308.**

Steam drum—the top drum of a boiler where all of the generated steam gathers before entering the separating equipment, **p. 307.**

Superheated steam—steam that has been heated to a very high temperature so that a majority of the moisture content has been removed (also called dry steam), **p. 308.**

Superheater—a set of tubes, located near the boiler outlet, that increases (superheats) the temperature of the steam, **p. 308.**

Introduction

Boilers are vessels that are used to create steam. Steam has many applications in process industries. For example, it is used to heat and cool process fluids, drive equipment, fight fires, and purge equipment. Steam provides the energy required to drive many types of equipment and reactions. Without boilers, many processes would operate less efficiently.

23.1 Purpose of Boilers

A **boiler** (also referred to as a steam generator) is a closed pressure vessel in which water is boiled and converted into steam under controlled conditions.

The steam produced by boilers provides energy to drive equipment such as turbines, compressors, and pumps. It is also used to provide the heat energy required to aid distillation and induce other physical and chemical reactions.

Boilers may function as stand-alone units or as a component operated in conjunction with a furnace or bank of furnaces. Boilers that are components of furnace operation use the excess heat of the flue gas before it is released to the atmosphere.

Boiler a closed pressure vessel in which water is boiled and converted into steam under controlled conditions.

23.2 Parts of a Boiler

Industrial boilers are complex and contain many parts (Figure 23.1). The most common components of a water tube boiler include the:

- Firebox
- Stack
- Damper
- Air register
- Burner (gas or oil)
- Raw gas burners
- Premix burners
- Combination burners
- Pilot
- Downcomer tubes
- Riser tubes
- Fan
- Mud (lower) drum
- Economizer
- Superheater
- Steam (upper) drum
- Safety valves

Boilers contain a firebox. The firebox is the section of the boiler where the burners are located and where radiant heat transfer occurs. In order to contain and redirect heat back into the firebox, a special refractory lining (a bricklike form of insulation) is used.

Radiant tubes within the firebox receive heat, primarily through radiant heat transfer. These tubes contain water (called boiler feed water) that is heated and turned into steam.

Burners are devices used to mix and burn an air/fuel mixture in the firebox section (e.g., natural gas or fuel oil). Burner types include raw gas, premix, or a combination. A pilot is a device used to light a burner.

Air registers control the flow of air to the firebox section of a boiler. They are used to adjust the primary and secondary airflow to the burner. Air registers maintain a proper air/

Figure 23.1 A. Components of a water tube boiler. B. Water tube boiler in a plant.
CREDIT: B. Melnikofd/Fotolia.

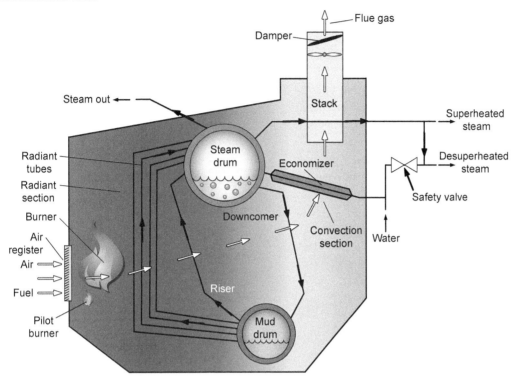

A.

B.

fuel ratio to mix gas and air efficiently and to prevent smoke or soot. Fans provide the main source of air in the boiler firebox. They are used to force airflow through the firebox.

The stack is located at the outlet of the boiler, which removes flue gas from the boiler. The stack may contain a damper (a movable plate), which is used to regulate the flow of air or draft. Emission detectors and analyzers are located in the stack.

The **steam drum** is the upper drum section of the boiler where the generated steam gathers before entering the separating equipment. Water enters the steam drum from the **economizer**, the section of a boiler used to preheat feed water before it enters the main boiler system.

The steam drum is connected to a lower drum, commonly called the **mud drum**. The mud drum is connected to the steam drum through a set of downcomer and riser tubes.

Steam drum the top drum of a boiler where all of the generated steam gathers before entering the separating equipment.

Economizer the section of a boiler used to preheat feed water before it enters the main boiler system.

Mud drum (1) the lower drum of a boiler that is used as a junction area for boiler tubes. (2) A low place in a boiler where heavy particles in the water will settle out so that they can be blown down.

Downcomers tubes exposed to the firebox that transfer water from the steam drum to the mud drum.

Risers tubes that go into the side or top of the steam drum, allowing heated water and steam to flow up from the mud drum.

Superheater a set of tubes, located near the boiler outlet, that increases (superheats) the temperature of the steam.

The mud drum is used as a junction area for the boiler tubes. Boiler sediments, or solids, are removed from the mud drum through the process of controlled blowdown, which is dictated by water analysis.

Downcomers are tubes located in the firebox that transfer water from the steam drum to the mud drum. The tubes contain cooler water descending from the steam drum. As water flows through the downcomers, it picks up heat from the firebox and replenishes the water supply to the mud drum.

Risers are tubes that go into the side or top of the steam drum and allow heated water and steam to flow up from the mud drum.

The **superheater** is a set of tubes, located near the boiler outlet, which increase (superheat) the temperature of the steam. The steam drum is usually connected to the superheater through a coil or pipe. The superheater coil passes back through the boiler firebox where it is exposed to the hot flue gas effluent leaving the boiler. **Superheated steam** is steam that has been heated to a very high temperature so that most of the moisture content has been removed. It is also called dry steam.

A **desuperheater** is a system designed to control the temperature of steam leaving a boiler by using water injection through a control valve.

23.3 How a Boiler Works

Boilers use a combination of heat and pressure to convert water to steam. To illustrate how boilers work, consider a simple boiler like the one shown in Figure 23.2.

This simple boiler consists of a heat source, a water drum, a water inlet, and a steam outlet. In this type of boiler, the water drum is partially filled with water, and then heat is applied.

Did You Know?

American inventors George Herman Babcock and Stephen Wilcox patented the water tube boiler in 1867.

This type of boiler became the standard for all large boilers because it allowed for higher pressures than earlier boilers.

CREDIT: Science & Society Picture Library/Contributor/Getty Images.

Figure 23.2 Simple boiler.

Superheated steam steam that has been heated to a very high temperature so that a majority of the moisture content has been removed (also called dry steam).

Desuperheater a system that controls the temperature of steam leaving a boiler by using water injection through a control valve.

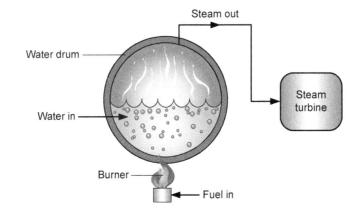

Once the water is sufficiently heated, steam forms. As the steam leaves the vessel, it is captured and sent to other parts of the plant for use in applications, such as turning a steam turbine or heating a process fluid in a heat exchanger. Makeup water is then added to the drum to compensate for the fluid lost during the production and removal of steam.

Boilers use the principle of differential density to assist fluid circulation. Hot water is less dense than cold water, so it tends to rise. Colder, denser water tends to sink. The difference in density creates fluid circulation without the need for a mechanical pump. This results in passive heat exchange, called the *thermosiphon effect*.

In order for boilers to work properly, they must have adequate amounts of heat and water flow. Factors that affect boiler operation include pressure, temperature, water level, and differences in water density.

As fluid is heated, molecules expand and become less dense. When cooler, denser water is added to hot water, convective currents are created that facilitate water circulation and mixing.

Fuels Used in Boilers

Boiler design dictates the type of fuel used to supply the burners. The burners are ignited by a pilot flame supplied by natural gas or methane.

Burners may use a fossil fuel in the form of fluidized pulverized (crushed) coal, which ignites at the pilot burner. Burners can use a mix of fuels, including oil sprayed into the burner throat in a fine mist, or combination fuels such as methane, off-gas, or natural gas in vapor form. Burner placement also affects the type of fuel used as the heat source.

23.4 Boiler Types

Boilers are designed in two major types: water tube and fire tube. Water tube boilers work at higher operating pressures, while fire tube boilers operate under limited pressure.

The boiler service usage dictates the type of draft system. For example, a water tube boiler being used as a waste heat boiler for a furnace or bank of furnaces could have an induced draft system to accommodate the flue gas exit from the furnaces. A separate boiler unit might have a forced draft system to ensure positive pressure in the firebox. Unit design and process economics affect the type of draft system used in boiler applications.

Fire Tube Boiler

Fire tube boilers are heat exchanger type devices that pass hot combustion gases through the tubes to heat the water on the shell side of the boiler. Figure 23.3 shows an example of a fire tube boiler.

As the water begins to boil, steam is formed. This steam is directed out of the boiler to other parts of the process. Makeup water is added to compensate for the fluid loss.

In this type of system, the water level within the shell must always be maintained so that the tubes are covered. Otherwise, the tubes could overheat and become damaged.

Figure 23.3 Fire tube boiler in a locomotive.

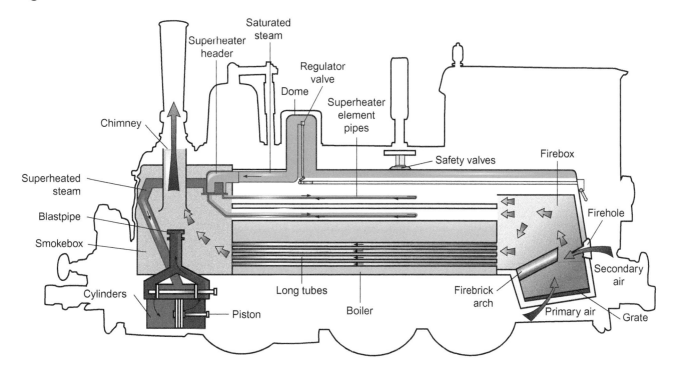

Fire tube boilers operate at lower pressures than water tube boilers. They are used in industrial processes such as laundry and cleaning, food processing, making molded brick, and brewing beer. Fire tube boilers are commonly used to power steam locomotives.

Water Tube Boiler

Water tube boilers are one of the most common types of boilers. In a water tube boiler, water flows through the tubes and is heated by combustion gases flowing over the outside of the tubes. See Figure 23.4 for an example of an industrial water tube boiler.

Figure 23.4 Water tube boiler burning coal dust and methane.

CREDIT: Nutthapat Matphongtavorn/Shutterstock.

As mentioned, water tube boilers have an upper (steam) and lower (mud) drum connected by tubes. Proper water level in a water tube boiler is controlled in the steam drum. This water level must be maintained for safety and operating purposes. If the water level gets too low, then the boiler could be damaged. If the water level gets too high, then it can carry over into the steam exiting the boiler and result in unit upsets and equipment damage.

Chemicals are used in boilers to prevent fouling, control conductivity, and prevent corrosion. Combustion occurs at the burners located in the firebox, which heats the water in the tubes. Water tube boilers are more suitable for process industry applications in the petrochemical and refining, oil and gas, and power generation sectors.

23.5 Operational Hazards

As with other pressured vessels, boilers must meet strict code enforcement regarding implementation of safety devices, such as safety valves and other fail-safe measures. Boilers are governed by state agencies that receive direction from the federal government.

There are many hazards associated with boilers. Table 23.1 lists some of these hazards and their potential impact.

Table 23.1 Operational Hazards Associated with Boilers

Improper Operation	Possible Impacts			
	Individual	Equipment	Production	Environment
Opening pressured header drains and vents	Burns and eye injuries			
Failing to purge firebox (start-up)	Burns or injuries if firebox ruptures	Explosion in firebox; damage to boiler internals	Facility upset, lost steam production, and downtime for repairs.	Exceeding EPA opacity limits; fines
Poor control of excess air and draft control		Flame impingement	Decreased boiler efficiency	Exceeding EPA opacity limits; fines
Loss of boiler feed water		Tube rupture, loss of downstream equipment use	Lost production due to downtime for repairs	
Loss of fuel gas or oil		Loss of downstream equipment use	Lost steam production	

23.6 Monitoring and Maintenance Activities

When monitoring and maintaining boilers, process technicians must always remember to look, listen, and check for the items shown in Table 23.2.

Failure to perform proper maintenance and monitoring could affect the process and result in equipment damage.

Table 23.2 Monitoring and Maintenance Activities for Boilers

Look	Listen	Check
■ Check firebox for flame impingement on tubes. ■ Check burner flame color. ■ Check for wall hot spots (external and internal). ■ Check draft balance (pressures). ■ Check temperature gradient. ■ Check firing efficiency (CO_2 and O_2 in the stack). ■ Check burner balance (fuel and air). ■ Check controlling instruments (water level, fuel flow, feed water flow, pressure, steam pressure, and temperature).	■ Check for abnormal noise (fans, burners, water leaks, steam leaks, or external alarms).	■ Check for excessive vibration (fans and burners).

Boiler Symbols

In order to locate boilers on a piping and instrumentation diagram (P&ID), process technicians need to be familiar with the different types of boilers and their symbols.

Figure 23.5 shows an example of a boiler symbol. Note that symbols may vary from facility to facility.

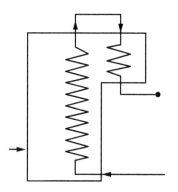

Figure 23.5 Boiler symbol.

Summary

Boilers are vessels that are used to create steam. Steam has many applications in process industries. It is used to heat and cool process fluids, fight fires, purge and drive equipment, and facilitate reactions.

Boilers contain many components. The main components include a firebox, stack, burners, steam (upper) drum, mud (lower) drum, economizer, and superheater. The two types of boilers are fire tube and water tube.

Boilers use the principle of differential density to create a thermosiphon effect for water circulation. In order for boilers to work properly, they must have adequate amounts of heat and water flow. Factors that affect boiler operation include pressure, temperature, water level, airflow, and differences in water density.

Improperly operating a boiler can lead to safety and process issues, injury to personnel, and equipment damage.

Technicians should always monitor boilers for flame impingement, hot spots, pressure drops, excessive temperatures, abnormal noises, excessive vibrations, or other abnormal conditions, and conduct preventive maintenance as needed.

Checking Your Knowledge

1. Define the following terms:
   a. downcomers
   b. economizer
   c. mud drum
   d. risers
   e. steam drum
   f. superheater

2. A fire tube boiler is called this because:
   a. the tubes are painted to look like flames.
   b. the tubes are in direct contact with the burner.
   c. the tubes are surrounded by hot gas.
   d. the heat source is located on the tube side.

3. In a water tube boiler, the steam drum is connected to the mud drum through tubes called _____ _____ and _____.

4. List three uses for steam in the process industries.

5. List two common types of fuel used to fire boilers.

6. List at least three hazards associated with improper boiler operation.

7. List at least three things a process technician should look, listen, and check for during normal boiler monitoring and maintenance.

NOTE: Answers to Checking Your Knowledge questions are in Appendix I.

Student Activities

1. Given a picture or cutaway of a water tube boiler, identify the following components:
   a. firebox
   b. stack
   c. damper
   d. burner (gas or oil)
   e. downcomer(s)
   f. riser(s)
   g. fan
   h. mud drum
   i. economizer
   j. superheater
   k. steam drum

2. Explain the differences between fire tube and water tube boilers.

3. Given a piping and instrumentation diagram (P&ID), identify the major parts of the boiler, such as the steam drum, mud drum, stack, and burners.

Chapter 24
Distillation

Objectives

Upon completion of this chapter, you will be able to:

24.1 Describe the purpose or function of a distillation column or tower in the process industries. (NAPTA Distillation 1) p. 314

24.2 Identify the primary parts and support systems of a typical tray-type distillation column. (NAPTA Distillation 2) p. 315

24.3 Describe the principles of operation and methods involved in the distillation process. (NAPTA Distillation 3) p. 318

24.4 Describe the use of packing as it pertains to distillation. (NAPTA Distillation 4) p. 319

24.5 Discuss the hazards associated with the improper operation of a distillation column. (NAPTA Distillation 5) p. 319

24.6 Describe the monitoring and maintenance activities associated with distillation column operations. (NAPTA Distillation 6, 7) p. 320

Key Terms

Bottoms—see *Heavy ends*, **p. 316.**

Column—also called *tower*, the distillation component where separation of materials takes place, **p. 314.**

Distillation—a method for separating a liquid mixture using the different boiling points of each component, **p. 314.**

Feed tray—the tray located immediately below the feed line in a distillation tower; this is where liquid enters the column, **p. 318.**

Flash zone—the section located between the rectifying and stripping sections in a distillation column, **p. 316.**

Fraction—one component of the distillate that has a particular boiling point that is different from the boiling point of the other fractions in the column, **p. 318.**

Heavy ends—materials in a distillation column that boil at the highest temperature, found at the bottom of a distillation column, **p. 316.**

Light ends—also called *distillate, overhead*, or *overhead takeoff*, product or material stream in a distillation column that boils at the lowest temperature and that comes off the top of a fractionation tower, **p. 316.**

Overhead—see *Light ends*, **p. 316.**

Packed tower—a distillation column that is filled with specialized packing material instead of trays, **p. 319.**

Packing—material used in a distillation column to maximize the contact area between the liquid and gas to effect separating, **p. 319.**

Rectifying (fractionating) section—also called *the enriching section*, the section of a fractionating tower between the feed tray and the top of the tower where the overhead product is purified, **p. 316.**

Reflux—cooled and condensed overhead vapor that is returned as a liquid to the top of the tower to provide cooling and to help with more efficient column operation, **p. 316.**

Stripping section—the section of a distillation tower below the feed tray where the heavier components with higher boiling points are located, **p. 316.**

Tower—*see* Column, **p. 314.**

Tray downcomer—the component of a tower tray located below the outlet weir that directs flow to the next succeeding tray, **p. 316.**

Trays—metal plates with openings (valve, bubble cap, or sieve design), placed within a distillation column, instead of packing, to increase contact between vapors and liquid in order to facilitate separation of components, **p. 316.**

Weir—flat or notched dams or barriers used to maintain a given depth of liquid on a tray, **p. 316.**

24.1 Introduction

Distillation is a commonly used technique in the process industries. The process of distillation requires the use of heat, pressure, and boiling point to separate liquid mixtures into their different components.

Column also called *tower*, the distillation component where separation of materials takes place.

Tower *see* Column.

A distillation **column** or **tower** is the distillation component where the separation of materials takes place. (Note that the terms **column** and **tower** are used interchangeably.) Column height is determined by the process being separated. The specific process also dictates whether the tower contains trays or packing. Associated equipment includes preheaters, condensers, and reboilers, which are all required as part of the process of separation and recovery of products.

Distillation processes can occur either as a batch or as a continuous operation. In a batch operation, a set amount of product is produced, the system is shut down, and then the system is restarted with the next batch of materials. In a continuous operation, materials are fed to the system and the operation and production continue day and night for long periods of time.

The distillation process can be achieved with systems that operate under varying pressures: vacuum, atmospheric pressure, or greater than atmospheric pressure. The material or materials being separated determine the column pressure setting. For example, a material that is heat sensitive might be separated in a column that operates under a pressure less than atmospheric (vacuum) to help protect the integrity of the product. In contrast, a material that would exist as a vapor at atmospheric pressure would need to be distilled under pressure in order to complete the separation process and ensure product quality.

Purpose of Distillation

Distillation a method for separating a liquid mixture using the different boiling points of each component.

Distillation is a method for separating a liquid mixture using the different boiling points of each component. During the distillation process, a liquid mixture is separated into two or more components through partial vaporization and the separate recovery of vapors and residue (usually through the process of condensation).

The distillation process has a wide array of uses throughout the process industries. For example, it is used in water treatment (e.g., removing contaminants from water), food and beverage manufacturing (e.g., creating alcoholic beverages), oil refining (e.g., creating gasoline and other products from crude oil), chemical manufacturing (e.g., making specialty chemicals and additives), paper and pulp manufacturing, and manufacturing of pharmaceuticals.

24.2 Distillation System Components

Distillation columns are designed with different heights and diameters, depending on the process material being separated. This chapter focuses primarily on the most commonly used type of column in process industries: the trayed distillation column (Figure 24.1). Common components of this type of system include:

- Tower/column
- Preheater
- Trays
- Condenser
- Reboiler.

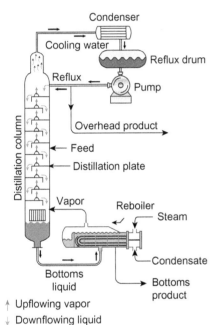

Figure 24.1 Distillation column with associated components.

The separation process takes place within the tower or column, often with preheated liquid. The preheater is a heat exchanger that warms the liquid before it enters the tower. Often, preheating materials provides a cost savings and helps achieve the best separation with the least energy usage. The preheating process helps the tower maintain a more constant temperature and also reduces the amount of time required for the substance to reach its boiling point once it enters the tower.

Fluids are inventoried into the column, and a reboiler is used to apply additional heat to the base to begin boiling the bottoms liquid. Low boilers will begin to vaporize and rise through the column, contacting the liquid lying on the trays. The base heat addition helps the column maintain a proper temperature and pressure profile. Liquids not vaporized as low boilers are considered the high boiler material and are removed from the base of the column.

After low boiler fluids have completed the vaporization process and moved to the top of the column, they are routed to a condenser. The condenser removes heat and changes the phase from vapor to liquid. Some of these liquids are pumped back into the distillation

Reflux cooled and condensed overhead vapor that is returned as a liquid to the top of the tower to provide cooling and to help with more efficient column operation.

column as **reflux** (cooled and condensed overhead vapor that is returned as a liquid to the top of the tower to provide cooling and to help with more efficient column operation). Reflux improves separation of rising vapors. The remaining liquid is sent to other parts of the process or to product storage.

Column Sections

Distillation columns have three sections: a flash zone, a rectifying (fractionating) section, and a stripping section Figure 24.2 provides examples of column sections.

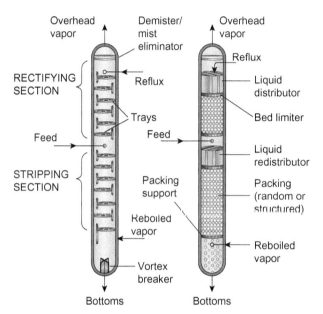

Figure 24.2 Rectifying and stripping sections of distillation columns.

Flash zone the section located between the rectifying and stripping sections in a distillation column.

Rectifying (fractionating) section also called *the enriching section*, the section of a fractionating tower between the feed tray and the top of the tower where the overhead product is purified.

Light ends also called *distillate*, *overhead*, or *overhead takeoff*; materials in a distillation column that boil at the lowest temperature and that come off the top of a fractionation tower.

Stripping section the section of a distillation tower below the feed tray where the heavier components with higher boiling points are located.

Heavy ends materials in a distillation column that boil at the highest temperature, found at the bottom of a distillation column.

Trays metal plates with openings (valve, bubble cap, or sieve design), placed within a distillation column, instead of packing, to increase contact between vapors and liquid in order to facilitate separation of components.

Weir flat or notched dams or barriers used to maintain a given depth of liquid on a tray.

Tray downcomer the component of a tower tray located below the outlet weir that directs flow to the next succeeding tray.

The **flash zone** is the section around the feed entry point. The material can be vapor, liquid, or a combination of both. Lighter ends flash, or rise upward, when exposed to the heat within the tower.

The **rectifying (fractionating) section** is the area of a fractionating column located between the feed tray and the top of the column where the overhead product is purified. **Light ends** are the product or stream that come off the top of a fractionation tower (also called *distillate*, *overhead takeoff*, or *overhead*). Light ends are the materials in a distillation column that boil at the lowest temperature.

The **stripping section** is the section of a distillation column where the heavier components with higher boiling points (bottoms products) are located. Bottoms products (also called heavy ends) are the materials collected in the stripping section of the distillation column. These materials move down the column as a liquid. **Heavy ends** are the substances that boil at the highest temperature.

Trays

Trays (Figure 24.3) are internal parts of the distillation column that improve contact between the liquid and vapor elements. A liquid level is maintained on each tray with the use of a weir. **Weirs** are flat or notched dams or barriers that are used to maintain a given depth of liquid on the trays. The liquid level is prevented from draining through the tray, due to the differential pressure across the tray (i.e., the pressure on the bottom of the tray is greater than the pressure on top of the tray).

The liquid flows across the tray, over the outlet weir to the downcomers. **Tray downcomers** are located below the outlet weir. They direct flow to the next succeeding tray. Meanwhile, vapor flows up through openings in the deck area of the tray.

Figure 24.3 Most common distillation tray designs. **A.** Sieve tray. **B.** Valve tray. **C.** Bubble cap tray.

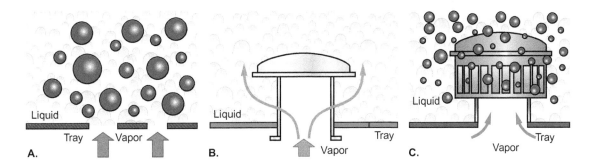

The three most common types of tray design are the sieve tray, valve tray, and bubble cap tray.

- A sieve tray is designed with perforated openings, typically circular, in the tray deck.
- A valve tray contains valve lifters, similar to those in a car engine. The valve is lifted, or opened, as the vapor rate increases.
- A bubble cap tray is designed with stationary risers in the deck area of each tray. Slotted caps top the risers, and vapor travels up through the riser and out of the slots in the cap. Figure 24.4 shows an example of downcomers in a bubble cap tray tower.

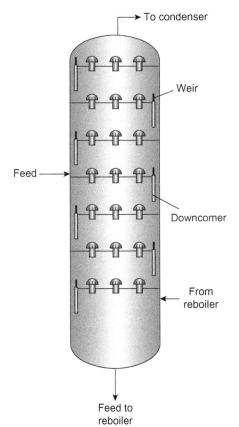

Figure 24.4 Distillation column showing tray downcomers and bubble cap trays.

Trayed columns (towers) tend to be larger than packed columns (discussed in the following section) and are used for larger scale operations, such as separating the different components in crude oil, or refining and purifying a chemical product.

Feed tray the tray located immediately below the feed line in a distillation tower; this is where liquid enters the column.

The **feed tray** is the tray located immediately below the feed line in a distillation column. Fluid enters the column at this point.

As mentioned previously, lighter vapors move up the column through openings in the tray such as bubble caps, perforations, or valves. Heavier liquid flows down the column to lower trays via downcomers.

24.3 How the Distillation Process Works

During the distillation process, a liquid mixture is heated in a distillation column until it reaches its boiling point and the various components of the mixture begin to vaporize into fractions. A **fraction** has one component of the distillate with a particular boiling point that is different from the boiling point of other fractions in the column.

Fraction one component of the distillate that has a particular boiling point that is different from the boiling point of the other fractions in the column.

A generalized overview of the distillation process follows:

1. The feed may be preheated to begin the vaporization of volatile components in the feed stream as it enters the process unit.
2. The vapor and liquid are fed into the distillation column.
3. The lighter component travels up to the higher part of the column.
4. The heavier component falls to the bottom of the column.
5. Contact between the vapor and the liquid occurs on each tray. The lower boiling compounds migrate to the vapor phase, and the higher boiling compounds migrate to the liquid phase.

Did You Know?

The distillation process has been in use for a very long time.
    The earliest recorded large-scale distillation of Cognac was in 1411.

CREDIT: Givaga/Fotolia.

Product purity and efficient system operation are dependent on the process technician's knowledge and understanding of distillation and the chemical properties of the substances being distilled. By understanding the concept of the distillation process, process technicians are better able to troubleshoot and to work safely and effectively to produce cleaner, purer products.

Distillation Methods

Several different methods of distillation may be used in process industries.

- *Binary distillation.* Some materials are easily separated because their boiling points are far enough apart that the lower boiler of the material is easily separated from the high boiler. This application method is generally called *binary distillation*.
- *Azeotropic distillation.* Materials with similar volatility points are more difficult to separate and require that the boiling point of one of the materials (generally the low boiler in the material) be changed. This is done by adding a third material to the lighter one. This mixture of two liquids is called an *azeotrope*. An azeotrope has a constant boiling point and composition throughout distillation, allowing easier separation. This distillation method is called *azeotropic distillation*.
- *Extractive distillation.* Another method of separating materials that have close boiling (or volatility) points is to add a solvent with a high boiling point to form a miscible mixture with the high boiler in the material. This allows the process of separation by conventional distillation. The low boiler leaves the top of the tower, and the high boiler mixture is separated by the binary distillation method.
- *Reactive distillation.* In this method, the reaction process and the distillation process are combined in the same unit. This specialty application reduces the need for energy (heat) and can be more cost effective in regard to the materials being separated.

24.4 Packed Columns

Another type of column used in the distillation process is a **packed tower**. Packed towers are filled with specialized material instead of trays. This material can be either structured in form or random (individual) packing placed in sections. In the process industries, packed columns tend to be smaller and are used less often than trayed towers.

Packed towers use the same principles as a trayed tower, but packing is used instead of trays to separate liquids. **Packing** is material used in a distillation column to maximize the contact area between the liquid and gas to effect separation. Packing comes in many different configurations and sizes. Figure 24.5 shows examples of some types of packing.

Packed tower a distillation column that is filled with specialized packing material instead of trays.

Packing material used in a distillation column to maximize the contact area between the liquid and gas to effect separating.

A.   B.   C.   D.

Figure 24.5 Examples of structured and random packing. **A.** Structured packing. **B.** Raschig rings (random packing). **C.** Berl/Intalox saddles (random packing). **D.** Pall ring (random packing).

Components in packed towers include packing materials, packing supports, liquid and gas distributors, and redistributors. The type of packing and the packing support depend on the material being processed in the tower. The packing support must be designed for the free flow of liquid and gas.

The supports must be able to support the packing and the weight of the liquid flowing through the packing.

Packed columns can be used for the following:

- Absorption of gases in liquid
- Stripping
- Evaporation of liquids
- Condensation of liquids
- Drying operations
- Scrubbing of gases to remove impurities

24.5 Operational Hazards

Working with the distillation process is an advanced job position. Process technicians will be responsible for monitoring and controlling all process variables and should always be aware of potential hazards such as overheating, leakage, and over pressurization. Table 24.1 lists some of these hazards and their impacts.

Table 24.1 Operational Hazards Associated with Distillation

Improper Operation	Possible Impacts			
	Individual	Equipment	Production	Environment
Introduction of steam to reboiler (too fast or too much) during startup	Exposure to chemicals Burn or injury Discipline Retraining	Upset conditions Damage to trays or packing	Extended start-up time and increased start-up cost Loss of production Off-spec product	Possible atmospheric release Spill into the environment
Tower not vented	Exposure to chemicals Burn or injury Discipline Retraining	Upset conditions Damage to trays or packing	Extended start-up time and increased start-up cost Loss of production Off-spec product	Possible atmospheric release Spill into the environment
No cooling liquid on condenser	Exposure to chemicals Burn or injury Discipline Retraining	Upset conditions Damage to trays or packing	Extended start-up time and increased start-up cost Loss of production Off-spec product	Possible atmospheric release Spill into the environment

Table 24.1 Operational Hazards Associated with Distillation (cont.)

Improper Operation	Possible Impacts			
	Individual	Equipment	Production	Environment
Rapid changes in column control and set points	Employee accountability discipline and possible retraining	Upset conditions Damage to trays or packing	Extended start-up time and increased start-up cost Loss of production Off-spec product	Possible atmospheric release Spill into the environment
Operation of column bottoms level too high or too low	Employee accountability discipline and possible retraining	Upset conditions Damage to trays or packing	Extended start-up time and increased start-up cost Loss of production Off-spec product	Possible atmospheric release Spill into the environment
Opening a bleed valve on a vacuum column	Employee accountability discipline and possible retraining	Upset conditions Damage to trays or packing Air in column—possible flammable mixture in system	Extended start-up time and increased start-up cost Loss of production Off-spec product	
Opening a bleed valve on a positive pressure column	Exposure to chemicals Burn or injury Discipline Retraining		Loss of production	Possible atmospheric release Spill into the environment

24.6 Monitoring and Maintenance Activities

When monitoring and maintaining distillation columns, process technicians must always remember to look, listen, and check for the items in Table 24.2.

Failure to perform proper maintenance and monitoring could negatively impact the process and result in equipment damage.

Table 24.2 Monitoring and Maintenance Activities for Distillation

Look	Listen	Check
■ Look for leaks anywhere in the system. ■ Monitor column pressure and temperature. ■ Monitor column/tower differential pressure. ■ Monitor reboiler shell pressure. ■ Monitor levels of bottoms products in drums. ■ Monitor condenser temperature drops. ■ Check reboiler condensate trap for malfunctions. ■ Conduct material balance check. ■ Monitor feed and product distribution flows. ■ Monitor overhead and bottom product specifications.	■ Listen for any unusual noises.	■ Using the proper equipment, check for temperature changes in equipment. ■ Using the proper equipment, check for unusual vibration in equipment.

Distillation Column Symbols

In order to accurately locate distillation columns on a piping and instrumentation diagram (P&ID), process technicians need to be familiar with the different types of columns and their symbols.

Figure 24.6 shows examples of distillation column symbols. However, these symbols may differ from facility to facility.

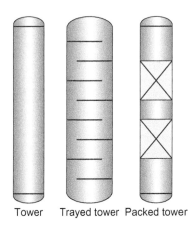

Figure 24.6 Distillation column symbols.

Tower   Trayed tower   Packed tower

Summary

Distillation is a commonly used technique in the process industries. Through distillation, process technicians are able to use heat, pressure, and boiling point to separate volatile liquids into their different components. These components, called fractions, have unique boiling points that are different from the boiling point of the other fractions in the column.

Distillation is used for many processes, such as water treatment, food and beverage manufacturing, oil refining, chemical and pharmaceutical manufacturing, and paper and pulp manufacturing.

Distillation columns (towers) are the equipment component used to facilitate distillation. Column shape and size are determined by the type of separation needed. Columns are often used in conjunction with preheaters, condensers, and reboilers. They may use trays or packing to effect separation of the various fractions.

Distillation columns are divided into three sections: the feed section, the rectifying section (top), and the stripping section (bottom). The rectifying section contains the substances with the lowest boiling point, which are called overhead products or light ends. The stripping section contains the substances with the highest boiling point, which are called heavy ends or bottoms.

There are many hazards associated with improper distillation column operation. These include overheating, leakage, and overpressurization.

When monitoring and maintaining distillation columns, process technicians must always remember to look, listen, and check to ensure that there are no leaks, the pressures and temperatures are within normal operating range, and the equipment is not producing abnormal noises or unusual vibrations.

Checking Your Knowledge

1. Define the following terms:
   a. flash zone
   b. fraction
   c. heavy ends
   d. light ends
   e. packing
   f. rectifying section
   g. stripping section
   h. tray

2. Which of the following is *not* a characteristic of bottoms products?
   a. They are referred to as heavy ends.
   b. They have the lowest boiling point of all the fractions in the column.
   c. They are located in the stripping section of the distillation column.
   d. They move down the column as a liquid.

3. List at least three things a process technician should look, listen, and check for during normal distillation column monitoring and maintenance.

4. *(True or False)* Reflux helps to increase the temperature in a distillation column.

5. Match the following illustrations with the correct type of packing.

1.	A. Berl/Intalox saddle
2.	B. Structured packing
3.	C. Raschig ring
4.	D. Pall ring

NOTE: Answers to Checking Your Knowledge questions are in Appendix I.

Student Activities

1. List and explain the steps in the distillation process.

2. Draw a simple diagram of a distillation column, preheater, condenser, and reboiler. Identify where each item is located in the process, and explain what the function of each component is.

3. List at least three hazards associated with improper distillation column operation, and explain the impacts these hazards might have on individuals, equipment, production, or the environment.

4. List five processes that utilize packed towers.

5. Given a piping and instrumentation diagram (P&ID), identify the distillation column and any associated equipment.

Chapter 25
Process Service Utilities

 Objectives

Upon completion of this chapter, you will be able to:

25.1 Discuss the different types of process utilities and their applications:

Water

Steam

Air

Gas

Electricity (NAPTA Process Utilities 1) p. 324

25.2 Describe the different types of equipment associated with each of the utility systems found in the process industries. (NAPTA Process Utilities 2, 3) p. 331

25.3 Identify the potential hazards and maintenance issues involved with process service utilities. p. 333

Key Terms

Boiler feed water—deionized water that is sent to a boiler to produce steam, **p. 325**.

Condensate—liquid resulting from cooled or condensed water vapor; frequently refers to condensed steam, **p. 329**.

Cooling water—water that is sent from a cooling tower to unit heat exchangers and other pieces of process equipment to cool process fluids, and that is then returned to the cooling tower for cooling and reuse, **p. 325**.

Fire water—pressurized water that is used to extinguish fires, **p. 325**.

Fuel gas—gas used as fuel in boilers and other types of furnaces, **p. 331**.

Instrument air—air that has been filtered and dried for use by instrument signal lines and valve actuators, **p. 330**.

Nitrogen—an inert (nonreactive) gas that is used to purge, or remove, explosive gases and air from process systems and equipment; also used as a backup to the instrument air supply, **p. 330**.

Plant air—See *utility air*, **p. 329**.

Potable water—water that is safe for human consumption, **p. 325**.

Process water—water that is used in any level of the manufacturing process and is not suitable for consumption, **p. 325**.

Service (raw) water—general-purpose water that may or may not have been treated, **p. 325.**

Sour water—wastewater produced from gas processing, sulfur recovery, and refining operations; it often contains hydrogen sulfide, ammonia, hydrogen cyanide, and phenol, **p. 325.**

Treated water—water that has been filtered, cleaned, and chemically treated, **p. 325.**

Utility air—compressed air header system that is used for all compressed air needs except instrumentation air (e.g., cyclone air movers, pneumatic power tools, and other air-operated equipment. It is not necessarily filtered and dried to instrumentation air standards. Also called *plant air* or *service air*, **p. 329.**

Wastewater—water that contains waste products in the form of dissolved or suspended solids and that must be treated prior to its return to the natural water source, **p. 325.**

Introduction

Process utilities are integral to the daily operations of a facility. These utilities provide essentials such as water, steam, compressed air, nitrogen and other gases, fuel gas, or electricity to a process unit.

Because these utilities are common throughout the process industries and play a vital role in almost every process, process technicians must understand their purpose, the equipment associated with each utility, potential hazards, and how to monitor and maintain each system. Many of these utilities are used daily by the process technician for cleaning and purging equipment, as part of a start-up or shut down procedure, and for other purposes in maintaining the safe operation of equipment and processes.

25.1 Types of Process Service Utilities

Service utilities are secondary processes that are part of the infrastructure of a plant facility. They are used to help maintain processes and equipment in correct working order. Some utilities are incorporated as part of the process flow. The most common service utilities include the following:

- Water
- Steam
- Compressed air/utility air
- Instrument air
- Nitrogen
- Fuel gas
- Electricity

The following sections explain the applications of each utility. Some facilities may group some of these utilities together for general use in centrally located stations that provide access to steam, utility air, nitrogen, and service water.

Water

Water is a critical utility in the process industries with many applications (Figure 25.1). Depending on the application, water may come directly from a lake, river or other waterway and be used with minimal or no treatment, or it may need to be purified before use. For example, water can be used for:

- Heating and cooling (e.g., process streams, cooling equipment for maintenance)
- Safety (e.g., fire protection, showers and eyewashes)
- Power generation (e.g., steam used to drive turbines, and moving water used in the hydroelectric process)

- Drinking and cooking (e.g., potable water)
- Washing and cleaning (e.g., using water as a solvent)
- Irrigation
- Transportation (e.g., slurry)
- Propulsion (e.g., hydraulic jets)

Figure 25.1 Some uses of water in the process industries. **A.** Safety shower. **B.** Process pump. **C.** Cooling towers.
CREDITS: A. Choksawatdikron/Shutterstock. **B.** MrPK/Shutterstock. **C.** Cpaulfell/Shutterstock.

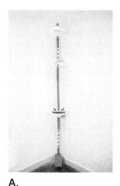
A.

B.

C.

The water used in a process facility comes from a source, such as a municipal supply or lake, and is then subjected to treatment, filtration, de-aeration, pressurization, or other changes as necessary. Different systems are used to distribute the water and make sure that it is in the proper condition.

- **Service (raw) water**—general-purpose water that may or may not have been treated.
- **Treated water**—water that has been filtered, cleaned, and chemically treated.
- **Fire water**—pressurized water that is used to extinguish fires.
- **Potable water**—water that is safe to drink and can be used for cooking.
- **Boiler feed water**—deionized water that is sent to a boiler in order to produce steam.
- **Cooling water**—water that is sent from a cooling tower into unit heat exchangers, and other pieces of equipment, to cool process fluids, and is then returned to the cooling tower for cooling and reuse.
- **Process water**—water that is used in any level of the manufacturing process; may contain contaminants and is not suitable for consumption. This water may become a stream that is considered to be wastewater.
- **Sour water**—water that contains contaminants as a result of process contact and becomes a stream considered wastewater.
- **Wastewater**—water that contains waste products in the form of dissolved or suspended solids.

Water quality is crucial for many uses in a facility. Water quality analysis and testing are imperative as much of the used water is recycled for use in the plant facility or returned to its natural source. Different process standards apply depending upon whether the water is returned to its natural source or used for service application in the plant. Some processes must meet government guidelines and specifications for process testing before the water can be returned to a natural waterway. Following are some of the water property requirements:

- Free of air or low in oxygen (to prevent corrosion, to avoid robbing a natural waterway of oxygen)
- Free of dissolved substances such as chlorine or salts

Service (raw) water general-purpose water that may or may not have been treated.

Treated water water that has been filtered, cleaned, and chemically treated.

Fire water pressurized water that is used to extinguish fires.

Potable water water that is safe for human consumption.

Boiler feed water deionized water that is sent to a boiler to produce steam.

Cooling water water that is sent from a cooling tower to unit heat exchangers and other pieces of process equipment to cool process fluids, and that is then returned to the cooling tower for cooling and reuse.

Process water water that is used in any level of the manufacturing process and is not suitable for consumption.

Sour water wastewater produced from gas processing, sulfur recovery, and refining operations; it often contains hydrogen sulfide, ammonia, hydrogen cyanide, and phenol.

Wastewater water that contains waste products in the form of dissolved or suspended solids and that must be treated prior to its return to the natural water source.

- Free of suspended solids
- Free of hydrocarbons and minerals
- pH controlled (pH level can vary based on the use of the water)
- Proper temperature and pressure range

SERVICE OR RAW WATER   Water comes into a process facility from a source (or sources) such as a municipal supply, lake, river, well, or the sea. This supply, called service or raw water, is the primary source of water for multiple uses around a facility.

Service water is usually treated and filtered either by a municipal source or internally in the facility. Typically, service water is stored in a reservoir and is then pumped to other water utility systems throughout the facility. Water pressure is maintained in accordance with facility usage.

Water from the service system can be treated using various techniques, such as chemical treatment and/or filtration, to achieve a high degree of purity. This treated water, sometimes called filtered water, is used for purposes such as:

- Boiler feed water
- Process water
- Seal water for rotating equipment
- Cooling tower water

FIRE WATER   Fire water is used during safety-related incidents such as fires, spills, and releases. The fire water system provides a high-pressure emergency water supply. Many facilities, depending on the location and process involved, place additives into the water to help extinguish flames. Fire water systems can be readily identified as they are color-coded red and are explicitly labeled. The fire water utility is restricted to the purpose of extinguishing fires and cooling equipment located in close proximity to a blaze.

Fire water pumping systems utilize a combination of electric motors, natural gas engines, steam turbines, and diesel engines to run the high-pressure supply pumps that can extinguish fires even in the event of loss of a major utility. These pumps are designed to start automatically if water pressure drops below a predetermined set point.

Fire water lines typically run underground and connect to hydrants. This location helps prevent them from freezing, mechanical damage, and fire damage. Systems are usually laid out in a gridlike pattern throughout the facility so that they can supply each area from two different lines, directions, or sources. Take-offs from the header supply hydrants and hose stations, sprinkler and deluge systems, water wall curtains, sprinkler risers, and other firefighting equipment. Some firefighting systems are initiated automatically based on fire, smoke, heat, or gas detectors. Systems can also be manually tripped.

POTABLE WATER   Potable water means water that is safe for human consumption. If a water supply is labeled non-potable, it cannot be used for drinking. Potable water is sanitary, so it can also come into contact with skin and not cause harm. Potable water is used for emergency eyewashes and safety showers (personal protective equipment).

A facility can either purchase potable water (from a municipal water source) or process it onsite from other sources (e.g., a well) using a water treatment system.

The potable water system delivers water to drinking fountains, restrooms, sinks, eyewash stations, emergency showers, and other similar fixtures.

BOILER FEED WATER   Boiler feed water is used as the supply for boilers to produce steam. Boiler feed water is deionized and filtered to remove chemicals, solids, dissolved gases, and impurities. This water is also treated for oxygen removal. The water quality must be consistent and fall within tight chemical parameters. Treatment prevents corrosion problems and increases efficient operation and safety of the unit.

Water testing is essential to proper boiler performance, so frequent tests must be done. Although it may vary depending on the facility's process requirements, typical testing performed on boiler feed water is for:

- Conductivity
- Alkalinity
- Hardness
- pH
- Phosphates

COOLING WATER Cooling water is circulated through equipment components to help remove heat from different stages of a process. Two widely used types of cooling water systems are open (evaporative) and closed (dry).

In an open cooling water system, water is taken from a source (river, lake, pond, or sea) and supplied to a cooling tower where the water is treated to help prevent the formation of algae and scale. It is pumped to the process unit and used for equipment cooling, and then returned to the cooling tower where evaporative cooling takes place. The water is then recycled for use again in the process unit.

In a closed system, water flows through heat exchangers and other pieces of process equipment and cooling towers in order to remove heat from the process. The heated water is then returned to a cooling tower and circulated through a heat coil, which is sprayed with water. Air flow removes heat from the spray through evaporation and indirectly cools the water in the heat coil. (The cooling water does not directly contact the air flow.) The water is then recirculated to the heat exchangers. See Chapter 21 for details.

In the past, a once-through system was used in power plants to cool the steam from turbines. Older power plant sites on the East Coast may still use this system, and cooling water is monitored closely for contaminants and stringent discharge standards. The once-through method was simple. Water was taken from a lake, river, or ocean, passed through the equipment, and returned to the natural waterway. This system has largely been discontinued due to its requirement for massive quantities of water for daily use.

Newer power plants use the evaporative or closed system. In a closed or evaporative system, cooling water is monitored for tower basin level, pump discharge pressure, water pH, and water impurities. Cooling water is filtered at the intake source to prevent fish and other wildlife from entering the system and, possibly, the unit equipment. Cooling water is treated with chemicals to reduce or prevent the buildup of algae, metal corrosion, and scale in the cooling tower and process unit equipment. Chemical addition must be strictly controlled to prevent foaming of the water. This foam can be carried away from the cooling tower and come in contact with humans or equipment.

PROCESS WATER Water can be used as part of the process. Process water is any water used to process materials into an end product. Depending on the nature of the process, this water may be called by different names. An example of process water is sour water, which is the water produced from gas processing, sulfur recovery, and refining operations often containing hydrogen sulfide, ammonia, hydrogen cyanide, and phenol.

As mentioned previously, cooling water and boiler feed water play an integral part in processes. But, water can also be used as feedstock. For example, water is integral to processes that make the following:

- Cement
- Medicines
- Foods
- Pulp and paper
- Hydrocarbons

The ability of water to evaporate makes it useful for many processes. Water is mixed with other feedstocks and then evaporated, forming an end product.

WASTEWATER Because many processes rely on water, much of the water used in a plant becomes contaminated in some way. Wastewater is water that has been used by processes, such as hydrocarbon cracking, distillation, reforming, and *alkylation* (a process that increases octane levels in gasoline and aviation fuel), that have affected the water quality.

Process water and sour water become, or are considered part of, wastewater after use. Federal and state agency regulations state that wastewater must be treated before the water can be discharged back to its source. Water discharge permits must be obtained to ensure that discharged water meets specific parameters, such as temperature, bacteria level, pH level, and so on.

The Environmental Protection Agency (EPA), U.S. Fisheries and Wildlife Service, and state water commissions set wastewater guidelines. Specific federal legislation addressing wastewater treatment includes the Clean Water Act and the Resource Conservation and Recovery Act (RCRA).

At a wastewater treatment facility, water is treated in several ways. For example, the water is filtered, the temperature is lowered, pH is adjusted, suspended and dissolved solids are removed, biological elements are adjusted, and clarification is performed.

Process Sewer In a plant, industrial sewer systems collect water and other process fluids from drains around the facility (Figure 25.2). The process sewer is separate from the storm sewer and should be marked or painted to ensure that personnel can tell the difference between it and other sewer systems. This sewer is located around all operating equipment, in area lab buildings, and in other areas where process fluids could reach the ground surface. Once collected, this water is sent to the facility's wastewater management system.

Figure 25.2 Wastewater flowing into a treatment plant.

CREDIT: Bogdan Wankowicz/Shutterstock.

Some process sewer water consists of oil and other hydrocarbons and may contain suspended solids. Typically, these fluids and suspended solids are separated from the water via the corrugated plate interceptor (CPI) system before being sent to the wastewater facility. They are recovered and reused if possible. Sanitary (septic) sewers can be handled through a municipal source or routed to the facility's wastewater management system, depending on plant site location.

Storm Sewer Storm sewers are open-graded sewer systems that collect and distribute surface water runoff through a series of lift stations and pumping stations into large holding ponds or tanks until treatment can occur. Storm water, especially water collected from within processing areas, must be analyzed and treated before being sent to a river, lake, or sea.

Steam and Condensate Systems

Steam has many applications in the process industries. For example, steam can be used as a driver or mover for process equipment, such as turbines or compressors, as a cleaning or purging medium, or as a source of heat for heat exchangers, reactors, and other types of equipment.

Steam is generated at a variety of temperatures and pressures, depending on the process requirements:

- Super high pressure: 1,000–1,600 psig
- High pressure: 400–800 psig
- Medium pressure: 150–200 psig
- Low pressure: 15–60 psig

As discussed in the Physics chapter (Chapter 11), pressure and temperature are related. Therefore, high-pressure steam has a higher temperature than low-pressure steam.

Facilities generate and distribute steam using a system composed of a boiler, steam header, piping, and other equipment. Before water is sent to the boiler, water must be treated to remove impurities and prevent scale and corrosion that can shorten the life of equipment. Steam condensate is also collected and reused.

Condensate is a liquid resulting from cooled or condensed vapor. It is a pure and valuable source of water in a process facility. Condensate is water that has condensed from steam equipment, unit processes, and steam distribution systems. It is collected, recovered, and reused because it requires minimal treatment.

Condensate liquid resulting from cooled or condensed water vapor; frequently refers to condensed steam.

Compressed Air

Compressed air can be used in a variety of ways at a process facility (Figure 25.3). For example, it can be used to power tools and equipment, operate control instrumentation, and process materials.

Figure 25.3 Compressor used in process industries.

CREDIT: Vereshchagin Dmitry/Shutterstock.

Compressed air is typically provided for the entire facility, using compressors placed in a central location. Air produced in this central location is called utility air. The air is typically filtered, and the water is removed. The air is then compressed to a pressure of around 125 psig (though this may vary from one facility to another). Compressed air is used in one of two ways: as utility air for power tools and general use, or as instrument air for control instrumentation.

Often, utility air compressors are configured so that loss of power does not affect the critical instrument air system.

UTILITY AIR **Utility air** (plant air) is compressed air that is used to power equipment, tools (e.g., cyclone air movers and pneumatic power tools), and dry equipment; to purge fireboxes prior to light-off; and to offload nonflammable delivery transports. Utility air is not pressure controlled or dried to the same degree as instrument air. It is also used to handle process needs and provide air to utility stations.

Utility air compressed air that is used to power equipment, tools (e.g., cyclone air movers and pneumatic power tools), and dry equipment; to purge fireboxes prior to light-off; and to off-load nonflammable delivery transports.

Instrument air air that has been filtered and dried for use by instrument signal lines and valve actuators.

INSTRUMENT AIR **Instrument air** is plant air that has been filtered and/or dried for use by instruments. Often, the instrument air system has a separate set of compressors to ensure purity and freedom from contaminants. It serves as the pneumatic power source for process control instrumentation signal lines within a control loop and for the actuators of control valves. Instrument air is cleaned and dried repeatedly, depending on facility usage.

The pressure is maintained at all times to ensure that instruments are provided a continuous supply for operation. It can also be used as clean, dry purge air. Some plants use a nitrogen system as a backup in the event that the instrument air system fails.

Breathing Air

A separate breathing air system provides an air supply to personnel working in oxygen-deficient or high-particulate environments. Hoses from the breathing system connect to face masks and tanks, called self-contained breathing apparatus (SCBA), which are worn by employees (Figure 25.4). A breathing air system can also consist of a full body suit with air lines connected to a central supply tank. The particular job assignment dictates the personal protective equipment (PPE) used as breathing air protection. Purity and oxygen concentration (i.e., percentage of oxygen) are important in maintaining a breathing air system.

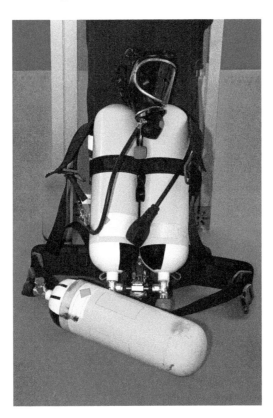

Figure 25.4 Self-contained breathing apparatus (SCBA).

CREDIT: Baloncici/Shutterstock.

Nitrogen System

Nitrogen an inert (nonreactive) gas that is used to purge, or remove, explosive gases and air from process systems and equipment; also used as a backup to the instrument air supply.

Nitrogen is an inert (nonreactive) gas that has various uses within a process facility. It can be used to purge (remove) explosive gases from process systems and equipment (e.g., furnaces, reactors). It can serve as a backup for instrument air in emergencies. It can be used as an inert blanket gas for storage tanks and lines (see the *Vessels* chapter). It also can be used for cooling and cryogenics.

Facilities can generate nitrogen or purchase it from a vendor. Nitrogen can be delivered by high-pressure pipeline, truck, or rail. Nitrogen is delivered at lower pressure to units for use at utility hose stations or in equipment. Nitrogen-rich environments can result in asphyxiation (suffocation). Care should be taken to monitor any environment, especially closed areas, where nitrogen is being used to ensure that the oxygen content is at least 21 percent. Meters are provided that check the oxygen content of a vessel or other closed area.

Fuel Gas

Fuel gas is used to operate fired equipment (e.g., furnaces, boilers). This gas can also be used as a blanket gas or as a supply for the pilot light on a flare system or other types of equipment containing burners. Some process facilities purchase natural gas from a supplier who delivers the gas by pipeline, truck, or rail. Other facilities produce their own fuel gas (e.g., as byproducts of their processes) and supplement it with natural gas.

Fuel gas gas used as fuel in boilers and other types of furnaces.

Electricity

The electrical power and distribution system provides operating power to systems and equipment throughout a facility (Figure 25.5).

Figure 25.5 High-voltage electrical transmission lines.
CREDIT: MEzairi/Shutterstock.

Many operating components, such as pumps and compressors, are driven by electric motors. Control rooms and lab facilities need electricity to function, and instrumentation components are also driven by electricity. Electricity can be provided by a variety of sources including the following:

- Power generation company
- Steam (driving turbines)
- Cogeneration (electricity production in which waste heat from the process is used to create steam)
- Diesel gas generators
- Emergency generators
- Portable generators
- Battery systems

See Chapter 19 for more information about electricity.

25.2 Types of Equipment Associated with Utility Systems

There are common types of equipment associated with utility systems. All systems need tanks, pumps, piping, and instrumentation to maintain the processes as directed by the specific task at hand. Some systems need other components, such as filtration, chemical treatment, and collection areas, as part of the process purification. Specialty systems need components that ensure they are in optimum operating order. Specialty systems are those that affect human safety and safe plant operation, such as electricity, breathing air, nitrogen, and instrument air.

The following lists identify equipment that is a part of different systems.

WATER EQUIPMENT

- Tanks and drums
- Pumps
- Instrumentation and control (pressure, level, flow, and temperature)
- Pipes and hoses
- Valves
- Filtration system
- Chemical addition and other treatment systems (chlorine, clarifiers, pH treatment, and deaerators)
- Cooling towers (cooling water systems)
- Heat exchangers (cooling water systems)
- Boilers (boiler feed water)
- Firefighting equipment and systems (fire water)

STEAM AND CONDENSATE EQUIPMENT

- Tanks
- Pumps
- Boilers
- Turbines (steam)
- Collection systems
- Pipes and hoses
- Valves
- Instrumentation for flow and pressure control

COMPRESSED AIR EQUIPMENT

- Compressors
- Tanks
- Dryers, moisture separators
- Filters
- Pipes
- Valves
- Instrumentation for flow and pressure control (instrument air only)
- Breathing equipment (e.g., hoses, masks, SCBAs, and suits) for breathing air only

NITROGEN EQUIPMENT

- Compressors
- Pipes
- Valves
- Instrumentation

GAS EQUIPMENT

- Compressors
- Pipes
- Valves
- Instrumentation

ELECTRICITY EQUIPMENT

- Turbines
- Boilers
- Generators
- Motors
- Motor control centers and circuit breakers
- Instrumentation

25.3 Operational Hazards

There are various operational hazards associated with process utility systems. Table 25.1 describes these hazards and their potential impact on individuals, equipment, production, and the environment. Specific hazards are also identified in the lists below and in the appropriate equipment chapters of this textbook.

Table 25.1 Operational Hazards Associated with Process Utilities

Improper Operation	Possible Impacts			
	Individual	Equipment	Production	Environment
Water/steam/condensate systems				
High or low pressure operation	Burns or other injury if exposed to hot temperatures Bodily injury	Damage to equipment Water hammer creates vibration and shock to equipment and associated lines	Loss of product due to downed equipment Low fire water pressure could result in loss of equipment and delayed production Equipment shutdown	Noise pollution from safety valve release during high pressure situation Pollution by smoke from fires Release of hot steam/condensate into the atmosphere
Leaks	Injury caused by slippery surfaces	Hazardous operating situations	Downtime for repair	
Cross-connecting systems				
Oxygen imbalance	Injury to personnel from possible exposure to low O_2 atmosphere	Corrosion due to loss of correct gas purge Possible rust, scale, and other equipment damage	Product contamination	Possible release of process to atmosphere
Cross-connection of potable water to a non-human consumption source	Injury to personnel, possible death	Possible damage to equipment	Product contamination	Possible release of process into environment
Gas systems				
Nitrogen-rich environment; cross-connection of utility air with instrument air, nitrogen, or fuel gas; contact with cooled nitrogen	Suffocation (e.g., nitrogen is an asphyxiant), injury, death Burns (liquid nitrogen burns skin on contact; gases can ignite) Injury from loose hoses whipping around under pressure Breathing difficulties from flammable gas Death (minimal breathing exposure to hydrogen sulfide gas can be lethal)	Plugged instrumentation lines from cross connection between plant air and instrument air Hoses whipping under pressure can damage equipment Equipment damage or loss of function from loss of gas purge or blanket	Possible loss of instrumentation and/or equipment control Loss of production Loss of gas purge or blanket can create loss of product specification	Possible release of process into the atmosphere With flammable gas, possible fire and explosion hazard Pollution from inefficient fuel burning
Electricity				
Contact with electrical lines and/or associated equipment Contact with water	Shock, burns, death of personnel Water and electricity do not mix; take care during wash down of equipment	Loss of electricity creates downtime of equipment Surges and arcs can damage motors and equipment	Loss of production due to loss of power	Possible release of process into atmosphere during outage conditions

Water

- Pressure loss can damage equipment and impact the process.
- High-pressure water can cause injuries or death, damage equipment, and impact the process.
- Hot water can cause burns, injury, or death.
- Low pressure in the fire water system can impede firefighting activities.
- Connecting a low-pressure system to a high-pressure system without a check valve can cause injuries or death, damage equipment, and impact the process.
- Scaling and corrosion can damage equipment and negatively affect the process.
- Leaks can cause slips and injuries, damage equipment, and impact the process.
- Collected water can result in drowning.
- Crossing a non-potable water system with a potable water system can cause illness or death.
- Chemicals used in water treatment can cause respiratory problems, illness, or death.

Steam and Condensate

- Water hammer (slugs of condensate in steam lines) can damage equipment and impact the process.
- High-pressure or high-temperature steam can cause injuries or death, damage equipment, and impact the process.
- Loss of steam can damage equipment and negatively affect the process.
- Closed or stuck valves can result in a rupture or explosion, causing injuries or death, damaging equipment, and affecting the process.

Compressed Air

- Low pressure can damage equipment and negatively affect the process.
- Moisture or condensate can affect control instrumentation, which can damage equipment and impact the process.
- Compressor failure can damage equipment and shut down the process.
- Unfiltered air can damage equipment and impact the process.
- Ruptured or loose hoses under air pressure can result in hose whipping, which can cause injuries or death.
- Using utility air lines improperly (e.g., horseplay) can cause injuries or death.
- Cross-connecting compressed air to breathing air can cause personnel injury or damage equipment.

Breathing Air

- Breathing air with less oxygen purity than necessary can result in injuries or death.
- Cracks and leaks to breathing air hoses and masks can result in injuries or death.

Nitrogen

- Nitrogen-rich environments can result in asphyxiation
- Contact with cooled nitrogen can result in freeze-related burns.
- Cross-connection of nitrogen line to a breathing air system can cause personnel injury or possible death.

Gas

- Cross-connections with other systems can result in fires or explosions, which can cause injuries or death, damage equipment, and negatively affect the process.
- Gas leaks can cause injuries or death if oxygen content is lowered or the gas mix explodes.
- Inefficient burning of fuel can reduce process efficiency and cause air pollution.
- Buildup of hydrogen sulfide (a gas known for its rotten egg odor) can occur, causing injuries or death, damaging equipment, and affecting the process.

Electricity

- Loss of power can damage equipment and have a negative impact on the process.
- Overvoltage or undervoltage can damage equipment such as circuits or motors.
- Short circuits or arcing can cause injuries or death, damage equipment, and affect the process.

Monitoring and Maintenance Activities

Scheduled maintenance activities are an essential part of keeping process utility systems in good condition and preventing failure. The most common monitoring and maintenance activities are described in the specific chapters related to the equipment listed under the section *Monitoring and Maintenance Activities*, as well as here in Table 25.2. Any other activities are advanced and beyond the scope of this text book.

Table 25.2 Monitoring and Maintenance Activities for Process Utilities

Look	Listen	Check
■ Visually check the appearance of lines and equipment components to make sure they are free of leaks, rust or other corrosion, bulging insulation, and other issues. ■ For flammable systems, inspect to see whether fog has appeared close to the ground. This could be a sign of a spill or process leak. ■ Ensure that systems are marked and labeled as appropriate. ■ Check to ensure system lines are not cross connected (e.g., plant air to breathing air).	■ Are there unusual sounds associated with the lines and equipment? ■ Is there extreme vibration associated with rotating equipment or the system piping? ■ Is there evidence of water hammer in steam system lines?	■ Using appropriate PPE and the correct monitoring equipment, check the area for correct oxygen content. ■ Depending on the job and the system, check the equipment to make sure that it is free of oxygen prior to start up. ■ Check the system for flammable content prior to entry and/or maintenance work.

Symbols Associated with Process Utilities

Process utilities generally do not have universal symbols. Symbols associated with process utility systems may vary by facility. Process utilities are called *secondary processes* on a P&ID.

Summary

Process utilities are integral to the daily operations of a facility. These utilities provide essentials to a process unit. The most common utilities include water, steam, compressed air, nitrogen, fuel gas, and electricity.

Water can be service or raw water, fire water, potable water, boiler feed water, cooling water, process water, and wastewater. Steam has many applications in the process industries. For example, steam can be used as a driver or mover for process equipment, such as for turbines or compressors; or can provide heat for heat exchangers and reactors.

Compressed air is used in a variety of ways at a process facility. For example, it may be used to power tools and equipment, operate control instrumentation, and process materials. Compressed air is typically provided for the entire facility, using compressors placed in a central location. Air produced in this central location is called plant air. Compressed (plant) air is used in one of two main ways: as utility air (for power tools and general use) or as instrument air (for control instrumentation). Instrument air may be compressed through a separate system of compressors.

A separate breathing air system provides an air supply to personnel who work in oxygen-deficient or high-particulate environments. Hoses from the breathing system connect to face masks and tanks that workers wear.

Nitrogen has various uses in a process facility. It can be used to remove explosive gases from process systems and equipment (e.g., furnaces and reactors); as a backup for instrument air (in emergencies); as an inert blanket gas for storage tanks and lines; and for cooling and cryogenics.

Fuel gas is used to operate fired equipment (e.g., furnaces and boilers). This gas can also be used as a blanket gas or as a supply for the pilot light on a flare system.

The electrical power and distribution system provides operating power to numerous systems and equipment throughout a facility.

There is a wide variety of equipment associated with utility systems. Process technicians should understand the hazards associated with utility systems and the associated equipment. Each type of equipment must be properly monitored and maintained.

Checking Your Knowledge

1. Define the following terms:
   a. condensate
   b. cooling water
   c. fire water
   d. fuel gas
   e. instrument air
   f. potable water
   g. process water
   h. service water
   i. treated water
   j. utility air
   k. wastewater

2. Which of the following statements best applies to potable water?
   a. safe to drink
   b. used for fire deluge systems
   c. used for cooling towers
   d. collected as condensate

3. (True or False) The pressure for utility air is regulated.

4. Which type of air is more critically controlled: utility air or instrument air?

5. Name three uses for nitrogen in a process facility.

6. Name five types of equipment and components associated with water utilities.

7. If a person comes into contact with a nitrogen-rich environment and dies, what is the most likely cause of death?
   a. chemical burns
   b. asphyxiation
   c. explosion
   d. skin irritation

8. Which of the following is NOT a hazard of compressed air?
   a. unfiltered air
   b. high pressure
   c. freeze burns
   d. moisture

9. Name three electrical hazards.

NOTE: Answers to Checking Your Knowledge questions are in Appendix I.

Student Activities

1. Pick one of the process utility systems and write a one-page paper describing its uses, equipment, and hazards.

2. What is the purpose of utility (plant) and instrument air systems at a process facility? Describe how they are used and why each is important.

3. Select a utility system, and make a list of five hazards-maintenance priorities on the equipment used with the system. You will need to refer to the specific equipment chapter, such as *Compressors*, for details.

Chapter 26
Process Auxiliaries

 Objectives

Upon completion of this chapter, you will be able to:

26.1 Describe the purpose or function of the different process auxiliary systems and their applications:

Flare system
Refrigeration system
Lubrication system
Hot oil system
Amine system
Fluidized bed system
Nitrogen header system (NAPTA Auxiliaries 1) p. 338

26.2 Discuss the equipment associated with flare systems found in the process industries. (NAPTA Auxiliaries 3) p. 339

26.3 Discuss the components associated with refrigeration systems found in the process industries. (NAPTA Auxiliaries 4) p. 340

26.4 Discuss the components associated with lubrication systems found in the process industries. (NAPTA Auxiliaries 5) p. 341

26.5 Discuss the components associated with hot oil systems found in the process industries. (NAPTA Auxiliaries 6) p. 342

26.6 Discuss the components associated with amine, fluidized bed, and nitrogen header systems found in the process industries. p. 343

26.7 Describe operational hazards, monitoring, and maintenance related to auxiliary processes. p. 344

Key Terms

Amine system—a system used to remove CO_2 or other acid gases from a gas stream, **p. 343**.
Evaporator—a heat exchanger in a refrigeration system that uses pressurized refrigerant to remove heat, **p. 341**.

Expansion valve—a valve that reduces the pressure of a liquid refrigerant, causing it to vaporize and produce a cooling effect, **p. 341.**

Flare header—a pipe system that connects several area/unit vents to the flare stack, **p. 339.**

Flare stack—the riser section or body of the flare system that supports the steam ring and pilot burner, **p. 339.**

Flare system—the connected system of individual lines, pressure control valves, and relief/safety valves used in both overpressure and purging situations to eliminate volatile process gases by combustion before they are released into the atmosphere, **p. 339.**

Fluidized bed system—a system that uses a motive gas to move particulate solids into motion as if they were a fluid, **p. 343.**

Hot oil system—a system that utilizes an organic or synthetic oil product as a heat transfer fluid; used in applications to prevent freezing of heavy oil products, **p. 342.**

Knockout drum—a vessel located between the flare header and the flare stack or between stages of a centrifugal compressor, where liquid drops out and is separated from the vapors being removed, **p. 339.**

Lubricant—a substance used to reduce friction between two contact surfaces, **p. 341.**

Lubrication system—a system that supplies oil or grease to rotating equipment parts, such as bearings, gearboxes, sleeve bearings, and seals on operating machinery, **p. 341.**

Nitrogen header system—a system used to purge hydrocarbons or oxygen so that they do not mix, **p. 344.**

Receiver—a tank that stores the liquid phase refrigerant once it leaves the condenser, **p. 341.**

Refrigerant—a fluid with a low boiling point that is circulated throughout a refrigeration system, **p. 341.**

Introduction

Auxiliary systems, secondary support systems in a process unit, are integral to the daily operation of a facility. They include, but are not limited to, flares, refrigeration, lubrication, amine systems, fluidized bed systems, nitrogen headers, and hot oil systems. Depending on the geographical location and season of the year, facilities may employ other systems to ensure safe and cost-effective operation of process units. Process technicians must understand these auxiliaries, their purpose, the equipment associated with each one, their potential hazards, and how to monitor and maintain them.

26.1 Types of Process Auxiliaries

This chapter describes the use, equipment, and components of the most common auxiliary systems:

- Flare system
- Refrigeration system
- Lubrication system
- Hot oil system

In addition to these process auxiliaries, the following process auxiliaries might also be used:

- Amine system
- Fluidized bed system
- Nitrogen header system

26.2 Flare Systems and Associated Equipment

A **flare system** burns off unwanted process gases before they are released into the atmosphere. They do not burn off flammable excess (or waste) substances. For that, an incinerator is used. Flare systems are used heavily during plant commissioning procedures, startup of equipment, shutdown of equipment, and purging of equipment for safety reasons.

Flare systems are generally tall, vertical pipes (although they can also be horizontal) called stacks, located a safe distance from process units. Flare systems consist of the associated process control elements, including both control and safety relief valves, piping and vents, knockout tanks, and a stack. Flares burn flammable and toxic substances from a process, converting them into environmentally acceptable substances, such as carbon dioxide (CO_2) and water (H_2O). Figure 26.1 shows an example of a flare system.

Flare system the connected system of individual lines, pressure control valves, and relief/safety valves used in both overpressure and purging situations to eliminate volatile process gases by combustion before they are released into the atmosphere.

A.

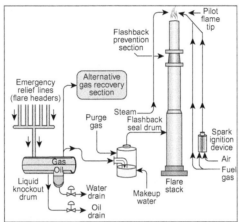
B.

Figure 26.1 Examples of a flare system. **A.** Flare stack. **B.** Drawing of a flare system with knockout drum.

SOURCE: A. photostock77/Shutterstock.

In a flare system, flammable substances are routed from a process unit to a **flare header**, a pipe system that connects process unit vents to the flare. A **knockout drum** (a vessel located between the flare header and the flare stack) separates the liquids from the vapors being sent into the flare stack. These liquids are either vaporized or pumped out into a storage tank.

In vertical flares, vapors travel up the **flare stack**, the riser section or body of the flare system that supports the steam ring and burner. Vapors are ignited by a pilot burner (an ignition source that is used for ignition at the flare tip as gases are released from the system). Figure 26.2 shows an example of a flare tip with a pilot burner.

As the substances are combusted, steam is distributed through a steam ring at the tip of the flare. The steam ring helps create a draft, reduce smoke, and protect the flare tip from the heat of the flame. A required safety procedure used with flare stacks is to have a

Flare header a pipe system that connects several area/unit vents to the flare stack.

Knockout drum a vessel located in the flare system, between the flare header and the flare stack, that is designed to separate liquid substances from the vapors being sent into the flare stack.

Flare stack the riser section or body of the flare system that supports the steam ring and pilot burner.

Figure 26.2 Flare tip with pilot burner.

CREDIT: Leonid Eremeychuk/Shutterstock.

continuous gas purge. This prevents oxygen from migrating into the flare stack piping and keeps the flame from flowing back down the flare stack. Figure 26.3 shows an example of a flare system with a steam ring.

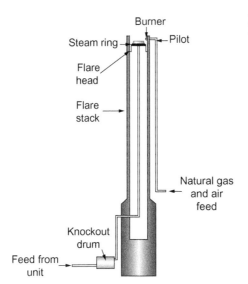

Figure 26.3 Flare system with steam ring.

Flare systems can be designated as either hot-service or cold-service, depending on the temperature of the substances being flared. Hot-service flare systems collect fluids above 32 degrees F (0 degrees C), while cold-service flare systems collect fluids below this temperature. This is important because construction materials are designed for specific temperature ranges.

Flare systems are essential during facility start-ups, normal shutdowns, and emergency shutdowns. They are also essential in the event of process upsets. If a process upset occurs that results in a dangerous pressure buildup, safety relief valves open and vent to the flare system.

Flare systems are regulated by clean air and safety laws set forth by various government agencies. Because of this, flare systems must be located a safe distance from process units, and the flame cannot endanger people, the surrounding area, or equipment. In addition, substances must be burned with a minimum amount of smoke and noise.

26.3 Refrigeration Systems and Associated Components

Cooling is a major element in many processes. Some processes require feed and/or product temperatures to be lower than temperatures that can be reached using cooling methods such as air or cooling water. In these instances, refrigeration systems are used to cool feed or products (Figure 26.4). The two most common types of refrigeration systems are mechanical and absorption.

Mechanical systems use a refrigerant (a fluid that boils at low temperatures) for cooling. In this type of system, refrigerant is cycled through a closed loop system that contains a low

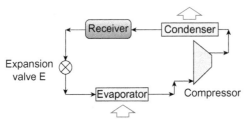

Figure 26.4 Drawing of a refrigeration system.

pressure cold side (evaporator) and a high pressure hot side (condenser). A compressor and expansion valve maintain the system pressure. Car and home air conditioners are examples of mechanical refrigeration systems.

Absorption systems cycle water and a heat absorbent material (typically lithium bromide, LiBr) through a closed loop system. The loop has a low-pressure side (evaporator) and a high-pressure side (condenser), but it adds a low-pressure absorber and high-pressure generator with a heater to complete the cycle. This type of system is used less often than mechanical systems.

> **Did You Know?**
>
> The first patent for a mechanical refrigeration system was issued in Great Britain in 1834 to the American inventor Jacob Perkins.
>
> Mechanical refrigeration systems are based on the principle that absorption of heat by a fluid (refrigerant) as it changes phase to a gas will lower the temperature of the objects around it.
>
> CREDIT: Zentilia/Fotolia.

Components of a Mechanical Refrigeration System

Mechanical refrigeration systems typically consist of a closed loop process that includes components for compression, condensation, and evaporation. The components associated with mechanical refrigeration systems are as follows:

- **Refrigerant**—a fluid with a low boiling point that is circulated throughout the refrigeration system.
- **Evaporator**—a heat exchanger that uses the refrigerant to remove heat and, in the process, causes the refrigerant to evaporate.
- **Compressor**—a device used to increase the pressure of a refrigerant in the gas phase.
- **Condenser**—a device used to cool the gas phase refrigerant from the compressor so that it condenses back into a liquid phase.
- **Receiver**—a tank that stores the liquid phase refrigerant once it leaves the condenser.
- **Expansion valve**—a valve that reduces the pressure of the liquid refrigerant, causing it to vaporize and produce a cooling effect.

Refrigerant a fluid with a low boiling point that is circulated throughout a refrigeration system.

Evaporator a heat exchanger in a refrigeration system that uses pressurized refrigerant to remove heat.

Receiver a tank that stores the liquid phase refrigerant once it leaves the condenser.

Expansion valve a valve that reduces the pressure of a liquid refrigerant, causing it to vaporize and produce a cooling effect.

26.4 Lubrication Systems and Associated Components

A **lubricant** is a substance used to reduce friction between two contact surfaces. Oil and grease are common lubricants. A **lubrication system** (Figure 26.5) supplies oil or grease to rotating equipment parts, such as bearings, gearboxes, sleeve bearings, and the seals of operating equipment.

Some process units have a centrally located lubrication system that atomizes (reduces to a fine spray) lubricants before they are distributed to equipment parts. The components associated with a typical lubrication system include the following:

- Lube oil reservoir tank—a vessel that contains lubricating oil for a system.
- Cooler—a device that removes heat from lubricant circulating through a system.
- Filter—a device that removes contaminants and particulates from the lubricant prior to circulation through a system.
- Pump—a device used to increase pressure in order to move the lubricant through a system to the process equipment, and back again.

Lubricant a substance used to reduce friction between two contact surfaces.

Lubrication system a system that supplies oil or grease to rotating equipment parts, such as bearings, gearboxes, sleeve bearings, and seals on operating machinery.

Other ways of supplying lubricants to equipment are as follows:

- Grease guns—handheld devices used to apply grease to equipment components (e.g., bearings).

> **Did You Know?**
>
> According to the American Petroleum Institute (API), motor oil doesn't wear out; it just gets dirty.
>
> Used motor oil can be cleaned and reprocessed for use in furnaces for heat or in power plants to generate electricity. It can also be recycled into lubricating oils that meet the same specifications as virgin motor oil.

Figure 26.5 Drawing of a lubrication system.

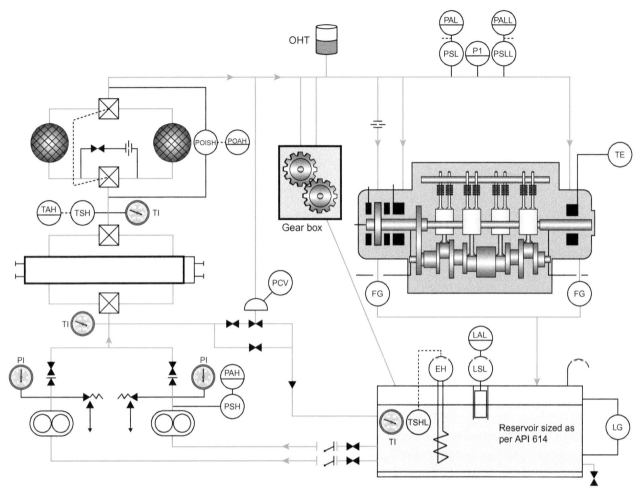

- Glass bottle oiler cup—an oil reservoir permanently connected to a specific part of a piece of equipment so that it remains lubricated.
- Automatic grease lubricator—(also called a grease cup) a device that slowly forces grease from a reservoir into a part so that it remains lubricated.

26.5 Hot Oil Systems and Associated Components

Hot oil system a system that utilizes an organic or synthetic oil product as a heat transfer fluid, used in applications to prevent freezing of heavy oil products.

A **hot oil system** utilizes an organic or synthetic oil product as a heat transfer fluid (Figure 26.6). Hot oil systems are used in applications such as heating jacketed reactors, heating tanks, heat exchangers, and pipeline tracing, to prevent freezing of heavy oil products.

The following components are associated with hot oil systems:

- Storage tank—a vessel that contains oil.
- Oil heating system—a heater or furnace that raises the temperature of the oil to the desired temperature. This furnace can burn a variety of fuels from natural gas (methane), oil, diesel, or biofuels, or it may be fueled by incinerating trash or waste.
- Oil circulation system—piping that transfers oil from the tank to the heater, to the process equipment, and back again.
- Oil pump—equipment that moves oil through the circulation and heating systems.

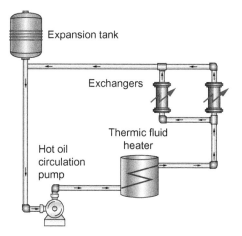

Figure 26.6 Drawing of a hot oil system.

26.6 Other Common Auxiliary Systems

Amine System

Amine system is a system used to remove acid gases from a gas stream (Figure 26.7). A common use of amine would be to remove CO_2 from a synthesized gas stream. This is done with a countercurrent flow of amine flowing down though a vessel (commonly called a *contactor* or *absorber*) while the synthesized gas flows in the bottom of the contactor, up through trays, and out the top. The result is considered "sweet" synthesized gas and "rich" amine.

Amine system a system used to remove CO_2 or other acid gases from a gas stream.

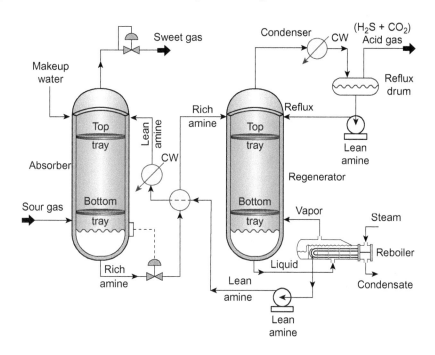

Figure 26.7 Process drawing of an amine system used to remove CO_2 and produce "sweet" gas and "rich" amine..

Rich amine is amine saturated with an acid gas like CO_2 or hydrogen sulfide (H_2S). The rich amine is then dropped lower in pressure and heated in a vessel (commonly called a *stripper* or *regenerator*) to boil out the entrained acid gases. Next, the "lean" amine (void of acid gases) is cooled and circulated back to the contactor. The recirculating amine is not part of the process synthesized gas flow, so the starting up of the amine circulation is considered an auxiliary process.

Fluidized Bed System

A **fluidized bed** is a system that uses motive gas to make a solid particulate substance flow like a liquid (Figure 26.8). Commonly, the size of the solid is small, like dust or sand. The

Fluidized bed system a system that uses a motive gas to move a solid like it was a fluid.

Figure 26.8 Process drawing of a fluidized bed.

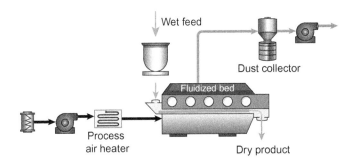

solid particles are on a bed that has a motive gas flowing up through small holes in the floor. As the gas flow is adjusted, the solid particles move like a liquid.

A few of the uses of fluidized beds are: cleaning, coating, granulation (growing particles), drying, filtration, blending, classification, transporting, reactions, separation, and cracking. The motive gas is commonly not part of the unit process, so starting up the motive gas in a fluidized bed is considered an auxiliary process.

Nitrogen Header System

Nitrogen header system
a system used to purge hydrocarbons or oxygen so that they do not mix.

A **nitrogen header** is a system used when an inert gas is needed (Figure 26.9). Nitrogen is often used to purge oxygen before a combustible gas or liquid is introduced into piping or vessels. This ensures that the combustible gas or liquid will not react with oxygen.

Purging is a commonly required step during startup of a system. Purging of combustible gas or liquid with nitrogen is also performed when shutting a system down. Purging on shutdown removes the entire combustible gas or liquid from piping or vessels in a system before opening them to the atmosphere. A nitrogen header is commonly maintained by allowing liquid nitrogen to warm up and flash into nitrogen gas. The nitrogen gas pressure is then regulated to the nitrogen header pressure. Because the system of flashing the nitrogen from liquid to gas is a secondary support process, it is considered an auxiliary process.

Figure 26.9 Drawing of a nitrogen header used to purge piping or vessels.

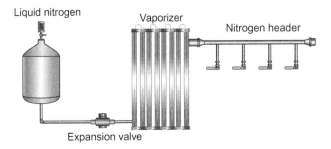

26.7 Operational Hazards

As an aspect of the safe operating culture mandated in process plants, process technicians must understand the hazards associated with the improper operation of process auxiliaries. Table 26.1 lists some of these hazards and their impacts.

Table 26.1 Operational Hazards Associated with Process Auxiliaries

Improper Operation	Possible Impacts			
	Individual	Equipment	Production	Environment
Amine systems				
Amine flow too low or process gas flow too high			Poor-quality process gas	
Regenerator temperature too low			Poor-quality process gas	
Amine temperature in contactor too high			Poor-quality process gas	

Flare				
Flare pilot or burner not lit	Possible exposure to toxic fumes	Fire or explosion	Process upset	Potential environmental release
Failure to line up purge gas to flare stack	Possible exposure to toxic fumes	Fire or explosion	Process upset	Potential environmental release
Flare knockout drum allowed to fill with liquid and spill out of the flare stack	Possible exposure to toxic fumes and/or liquids; potential for burns and injury	Back pressure on the flare can create dangerous system pressure	Process upset	Potential environmental release, spill, or fire
Refrigeration				
Refrigerant leak	Possible exposure to toxic fumes and/or liquids; potential for chemical burns and injury	Fire or explosion	Process upset	Potential environmental release
Failure to line up cooling water to condenser in refrigeration system		Equipment damage	Process upset	Potential environmental release
Lubrication				
Loss of lubrication on equipment		Bearing failure	Process upset	Potential environmental release
Hot oil				
Hot oil pump drains and vents left open	Potential for burns and injury	Equipment failure	Process upset	Potential environmental release
Hot oil header drains left open	Potential for burns and injury	Equipment failure	Process upset	Potential environmental release

Monitoring and Maintenance Activities

When monitoring and maintaining process auxiliary systems, process technicians are required to perform, but are not limited to, the tasks described in Table 26.2.

Failure to perform proper maintenance and monitoring could cause personal harm or injury, impact the process, and result in equipment damage.

Table 26.2 Monitoring and Maintenance Tasks for Process Auxiliaries

Flares	Refrigeration	Lubrication	Hot Oil
▪ Maintain level in the knockout drum. ▪ Check knockout drum pump operation. ▪ Check pilot gas regulator for proper flow. ▪ Check that the pilot is lit. ▪ Check purge gas for proper flow.	▪ Check the cooling water in the condenser. ▪ Check compressor operation. ▪ Check the expansion valve. ▪ Check the system temperature profile.	▪ Maintain proper oil levels in reservoirs. ▪ Check bearing lubrication. ▪ Check grease fittings. ▪ Check inlet and outlet temperature for excess heat. ▪ Check lubrication flow to equipment. ▪ Check pressure differential across filter. ▪ Check pump for proper operation.	▪ Check pump for proper operation. ▪ Check tank temperature. ▪ Check oil circulating temperature. ▪ Check oil tank level. ▪ Check circulating valves for proper alignment.

Symbols for Process Auxiliaries

Auxiliaries are not individual pieces of equipment but are rather systems, so there are no universal symbols that apply. These systems utilize many of the same types of equipment covered by symbols discussed in other chapters.

Summary

Process auxiliary systems include flare, refrigeration, lubrication, hot oil systems, amine systems, fluidized bed systems, and nitrogen header systems.

Flare systems burn unwanted process gases before they are released into the atmosphere. Equipment associated with flare systems includes the flare header, knockout drum, flare stack, pilot, steam ring, and flare tip.

Cooling is a major element in many processes. Refrigeration systems are used to cool feed or products to required temperatures during the process or for storage. Mechanical refrigeration systems use a refrigerant (a fluid that boils at low temperatures) for cooling. In this type of system, refrigerant is cycled through a closed loop system that contains a low-pressure cold side (evaporator) and a high-pressure hot side (condenser). A compressor and expansion valve maintain the system pressure.

Lubricants are substances used to reduce friction between two contact services. Common types of lubricants are oil and grease. Lubrication systems supply oil or grease to rotating equipment parts such as bearings, gearboxes, sleeve bearings, and seals. The components associated with lubrication systems are the lube reservoir tank, cooler, filter, and pump.

Hot oil systems utilize an organic or synthetic oil product as a heat transfer fluid, used in applications, such as heating jacketed reactors, heating tanks, in heat exchangers, and for pipeline tracing, to prevent freezing of heavy oil products. The components associated with hot oil systems are the storage tank, oil heating system, oil circulation system, and oil pump.

Other auxiliary systems that may be used for particular processes are amine systems, fluidized bed systems, and nitrogen header systems.

Technicians must understand the hazards associated with the improper operation of process auxiliaries. Technicians must also remember to perform monitoring and maintaining tasks specific to each process auxiliary system.

Checking Your Knowledge

1. Define the following terms:
   a. evaporator
   b. expansion valve
   c. flare system
   d. refrigerant
   e. fluidized bed
   f. amine system
   g. nitrogen header

2. Which of the following flare system components is a vessel designed to separate liquid hydrocarbons from vapors being sent to the flare for burning?
   a. reboiler
   b. knockout drum
   c. flare stack
   d. steam ring

3. What is the purpose of a refrigeration system expansion valve?
   a. acts as a heat exchanger that uses the refrigerant to remove heat
   b. cools the gas phase refrigerant from the compressor and causes it to condense back into a liquid phase
   c. stores the liquid phase refrigerant once it leaves the condenser
   d. reduces the pressure of the liquid refrigerant, thereby vaporizing it and causing it to cool

4. Which of the following refrigeration system components is a tank that stores liquid refrigerant after it leaves the condenser?
   a. evaporator
   b. receiver
   c. filter
   d. compressor

5. (True or False) Lubrication systems are used to heat oil for the hot oil system.

6. Which component of a lubrication system removes contaminants and particulates?
   a. filter
   b. lubricator
   c. scrubber
   d. cooler

7. Name two common times when nitrogen header systems are utilized in process industries.

8. Which of the following is NOT part of a hot oil system?
   a. storage tank
   b. oil pump
   c. heater
   d. refrigerant

9. *(True or False)* The motive gas is commonly part of the process when using fluidized bed systems.

10. Match the diagram label numbers with the components listed below.
    a. _____ flare tip
    b. _____ knockout drum or sump system
    c. _____ pilot
    d. _____ flare stack
    e. _____ steam ring

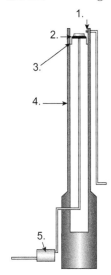

11. In which auxiliary system would a process technician check purge gas for proper flow?
    a. lubrication
    b. fluidized bed
    c. hot oil
    d. flare

NOTE: Answers to Checking Your Knowledge questions are in Appendix I.

Student Activities

1. Write five to six paragraphs on TWO of the following process auxiliary systems. Describe each system, how it is used, and why it is important to the process industries:
   - flare system
   - refrigeration system
   - lubrication system
   - hot oil system
   - amine system
   - fluidized bed system
   - nitrogen header

2. Describe the steps that occur before ignition at the flare tip.

3. Divide into small groups. In each group, list the monitoring and maintenance activities for one of the following: flare systems, refrigeration systems, lubrication systems, and hot oil systems.

Chapter 27
Instrumentation

 Objectives

Upon completion of this chapter, you will be able to:

27.1 Describe the purpose or function of process control instrumentation in the process industries. (NAPTA Instrumentation 1) p. 350

27.2 Identify the key process variables that are monitored and/or controlled by instrumentation:

Temperature

Pressure

Level

Flow

Analytical variables (NAPTA Instrumentation 5) p. 350

27.3 Identify typical process control instruments, their major types, applications, and functions:

Sensor (e.g., thermocouple)

Transmitter

Indicator

Recorder

Controller

Transducer

Alarm

Control valve

Actuator

Valve positioner (NAPTA Instrumentation 3, 6) p. 354

27.4 Describe types of process control instrumentation signals. (NAPTA Instrumentation 2) p. 356

27.5 Define a generic control loop and provide an example. (NAPTA Instrumentation 4) p. 357

27.6 Describe distributed control systems and how they are applied in the process industries. (NAPTA Instrumentation 7) p. 358

27.7 Discuss the hazards associated with process control instruments. (NAPTA Instrumentation 8) p. 359

27.8 Describe the monitoring and maintenance activities associated with process control instrumentation. (NAPTA Instrumentation 9, 10) p. 359

Key Terms

Actuator—an electrical, pneumatic, or hydraulic device that manipulates a control device, usually a valve, **p. 356.**

Alarm—a device, such as a horn, flashing light, whistle, or bell, that provides visual and/or audible cues of conditions outside the normal operating range, **p. 356.**

Analytical variable—a measurement of the chemical or physical properties of a substance, **p. 354.**

Control loop—a series of instruments that work together to monitor and control a process variable, **p. 357.**

Control valve—the final component of a control loop, operated either remotely or by hand, that receives its signal from the controller and manipulates the process variable in response to a system disturbance or change, **p. 355.**

Controller—the component of a control loop that receives information from the transmitter in the form of a pneumatic or electronic signal, compares that signal to a prescribed setting, and sends a correcting output signal to the final control element in order to maintain the process at that set point, **p. 355.**

Distributed control system (DCS)—a computer-based system consisting of a series of controllers used to monitor and control a process, **p. 358.**

Flow rate—a quantity of fluid that moves past a specific point within a given amount of time, usually expressed in volume units per unit of time, such as gallons per minute (GPM) or cubic feet per minute (CFM), **p. 352.**

Hydraulic instruments—devices that use pressurized fluid (hydraulic fluid) as the power source, **p. 355.**

Indicator—an instrument that shows the current condition of a process variable, usually through a visual representation such as a needle on a scale in a gauge or a digital readout, **p. 355.**

Instrumentation—the set of devices used to measure, control, or indicate flow, temperature, level, pressure, and analytical data, **p. 350.**

Level—a position of height or depth along a vertical axis; in the process industries, level usually means the height of the surface of a material compared to a zero reference point, **p. 352.**

Local—related to instrumentation in close physical proximity to the process, generally found outside the control room, **p. 355.**

Pneumatic instruments—devices that use air pressure or inert gas as their power source, **p. 355.**

Process variable— a measurement of the chemical or physical properties of a process stream, generally relating to operating parameters or product specifications, **p. 350.**

Recorder—an instrument, either mechanical or electronic, used to document variables such as pressure, speed, flow rate, temperature, level, and electrical units; it also provides a history of the process over a given period of time, which is called a trend, **p. 355.**

Remote— relating to instrumentation located away from the process (e.g., in a control room) and controlled by the DCS or other panel board mounted controller, **p. 355.**

Sensor—(also known as a *primary element*) the control loop component that is in direct contact with the process variable to be measured; it senses or detects change and provides a signal to the transmitter, **p. 355.**

Transducer—the control loop component that converts one instrumentation signal type to another type (e.g., electronic to pneumatic), **p. 356.**

Transmitter—the control loop component that receives a signal from the sensor, measures that signal, and sends it forward to the controller or indicator, **p. 355.**

Valve positioner—a device that ensures a valve is positioned properly in reference to the signal received from the controller, **p. 356.**

Introduction

The process technician is responsible for knowing and understanding instrumentation that he or she will use to monitor and control processes in a facility. Instrumentation is integral to the safe and efficient operation of a process. Efficient operation of a process unit allows companies to produce more cost-effective and higher-quality products in a safer environment. Without instrumentation, the process industries would need numerous workers to monitor and control even the simplest process.

Automation of process instrumentation and advances in technology have helped streamline process production and have increased safety in many operating plants. Processes that were once manually monitored and controlled are now done automatically. This change allows a process technician to oversee increasingly complex processes from a single source.

Process instrumentation can be found on local panelboards in the process area or in control rooms central to the unit location. Some companies utilize a "super control room" where entire plants are monitored and controlled from a single building.

27.1 Purpose of Process Control Instrumentation

Instrumentation the set of devices used to measure, control, or indicate flow, temperature, level, pressure, and analytical data.

Process variable a measurement of the chemical or physical properties of a process stream, generally relating to operating parameters or product specifications.

The process industries use **instrumentation**, which is configured in specific groups of devices used to measure, control, or indicate flow, temperature, level, pressure, and analytical data. These conditions are called process variables (PVs). A **process variable** is a varying operational condition associated with a processing operation, such as temperature, pressure, flow rate, level, and composition. It may measure chemical or physical properties of a process stream.

The process technician oversees instrumentation used to measure, compare, compute, and correct process variables. Instrumentation can be considered the eyes and ears of a process, constantly reporting the status of variables. It also plays a role in the decision making involved with keeping a process operating efficiently and effectively, oftentimes directly controlling it.

Even a simple process may contain multiple process variables that need to be monitored and controlled. In a complex process, the number of process variables can require management by more than one process technician.

Some processes are completed in a series of automated actions initiated by the process technician. The instrumentation is programmed to begin each step in the process upon completion of the previous one. Examples of this type of process are dryer regeneration, filter backwash, and regeneration of a catalyst bed. Some everyday processes are a car wash, clothes dryer, and automatic lawn sprinklers.

27.2 Key Process Variables Controlled by Instrumentation

The five most common process variables controlled by instrumentation are pressure, temperature, level, flow, and analytical values. In addition, speed, vibration, and a range of other variables are also monitored and controlled, depending on the process requirements.

Pressure

Pressure is the force exerted on a surface divided by its area. Pressure is measured on different scales in a process unit given a reference point. Many different pressure scales are used in industry depending on unit need.

Pounds per square inch (psi) is the most common measurement found in industry. Pounds per square inch gauge (psig) is a pressure measurement referenced to ambient air pressure. Pounds per square inch absolute (psia) is a pressure measurement referenced to a vacuum. (See also Chapter 11 for discussion of pressure.)

Differential pressure is the difference between measurements taken from two separate points (the difference between two related pressures) in a process unit. There are two ways to express differential pressure: the Greek letter delta (Δ) with a P following it (ΔP), or the letter combination D/P.

Local instruments designed to measure pressure include manometers, pressure gauges (containing a Bourdon tube, diaphragm, or bellows), and differential pressure (D/P) cells. D/P cells are most commonly used as the measuring component of the control loop instrumentation configuration. Figure 27.1 shows examples of some of these pressure-measuring devices.

Figure 27.1 A. Manometer. B. Pressure gauge. C. Differential pressure (D/P) cell.
CREDIT: B. Vlad Ivantcov/Fotolia; C. ekipaj/Shutterstock.

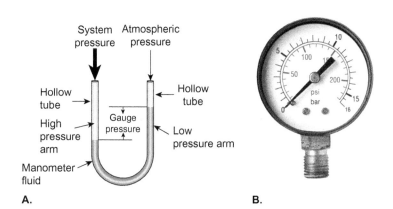
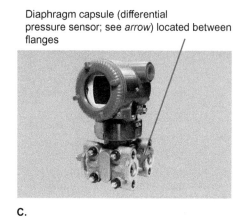

Temperature

Temperature is the specific degree of heat or cold as indicated on a reference scale. The two most common temperature measurement scales used in the process industries are Fahrenheit (degrees F) and Celsius (degrees C). A common term associated with temperature is differential (ΔT). *Differential temperature* is the difference between measurements taken at two separate points (the difference between two related temperatures). Differential temperature is measured in many process applications, such as during the measurement and comparison of cooling tower water inlet and outlet temperatures.

Temperature-sensing instruments include thermocouples, resistance temperature devices (RTDs), thermistors, and filled systems.

Figure 27.2 A. Portable infrared thermometer (pyrometer). B. Thermal imaging camera. C. Bimetallic temperature gauge.
CREDIT: A. AlexLMX/Shutterstock; B. Joyseulay/Shutterstock.

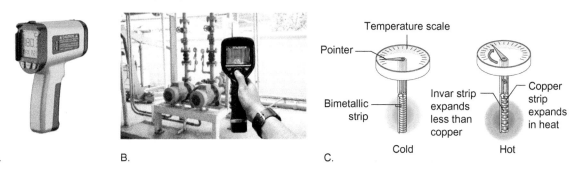

Local temperature-measuring instruments include liquid thermometers, infrared pyrometers, thermal imaging cameras, change-of-state sensors, and bimetallic temperature gauges. Figure 27.2, shows examples of some of these devices.

Level

Level a position of height or depth along a vertical axis; in the process industries, level usually means the height of the surface of a material compared to a zero reference point.

Level is a position of height or depth along a vertical axis. In the process industries, level usually means the height of the surface of a material compared to a zero reference point. Local level-measuring instruments include sight glasses, level gauges, float and tape devices, and radar gauges. Figure 27.3 shows examples of some of these devices. Displacers (buoyancy devices), delta P cells, and capacitance probes work as part of an indicating or control loop. These will be discussed more in the Equipment textbook.

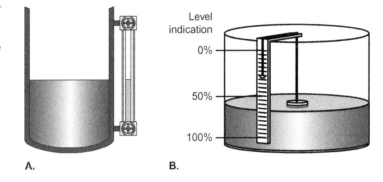

Figure 27.3 Two instruments for determining level. **A.** Tubular type sight glass. **B.** Float and tape gauge.

Flow

Flow rate a quantity of fluid that moves past a specific point within a given amount of time, usually expressed in volume units per unit of time, such as gallons per minute (GPM) or cubic feet per minute (CFM).

Flow (also **flow rate**) is a quantity of fluid that moves past a specific point within a given amount of time, usually expressed in volume units per unit of time, such as gallons per minute (GPM) or cubic feet per minute (CFM). Flow can be measured by using differential pressure, positive displacement, velocity, mass, or open channel, depending on the process need and function.

Flow can also be measured as the total amount of flow through a unit, such as production flows that are charged to a customer. This is commonly called *mass flow* and is generally calculated against a 24-hour time period.

Common flow-sensing elements that create a difference in pressure that can be measured by a D/P cell include pitot tubes (Figure 27.4), orifice plates, flow nozzles, flow tubes, and

Figure 27.4 Pitot tube.
CREDIT: Fouad Saad/Shutterstock.

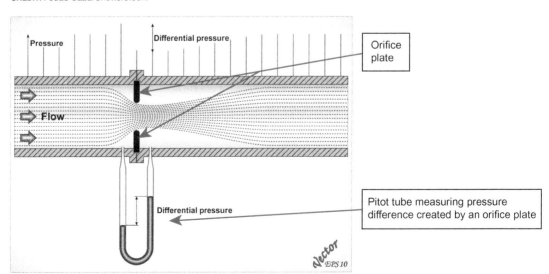

Venturi tubes. The D/P measurement corresponds to a specific flow rate. D/P cells are used with local indicators or, more commonly, as one of the four major components of a control loop. This type of measurement is commonly used in process units.

Some common instruments used to measure flow by positive displacement are the rotating piston, nutating disk, oval gear, and rotating vane meters.

The vortex shedding, electromagnetic, and turbine flow meters measure flow velocity. The Coriolis meter measures mass flow. Figure 27.5 shows examples of some of these devices.

Figure 27.5 Devices for assisting with flow measurement. **A.** Orifice plate. **B.** Venturi tube.

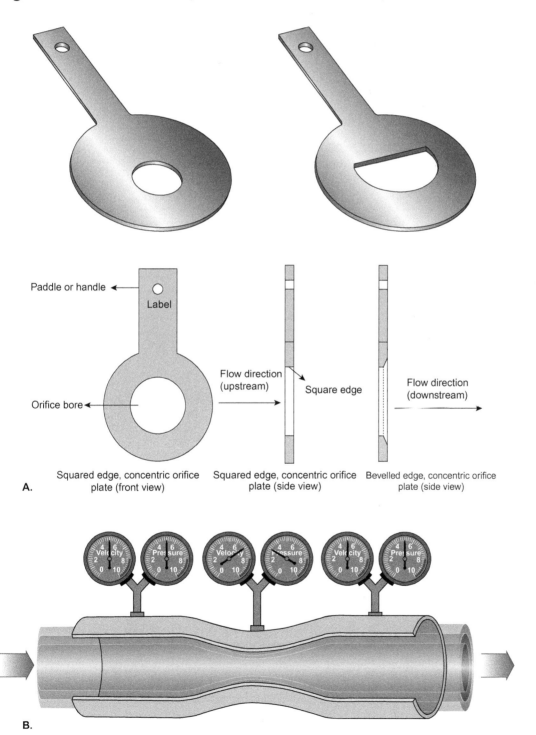

Analytical variable
a measurement of the chemical or physical properties of a substance.

Analytical Variables

An **analytical variable** is a measurement of the chemical or physical properties of a substance. Analytical variables provide process technicians with knowledge that is used to maintain quality control and consistency in the process. Some analytical-related instruments include pH meters, chromatographs, and viscosity meters. Figure 27.6 and Figure 27.7 show some examples of these devices.

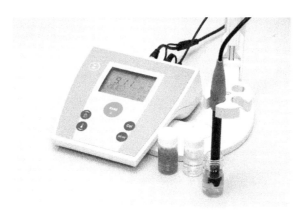

Figure 27.6 Device for analytical measurements: pH meter.
CREDIT: photongpix/Fotolia.

Figure 27.7 Device for analytical measurements: **A.** Gas chromatograph. **B.** Diagram of a gas chromatograph.
CREDIT: **A.** Kadmy/Fotolia.

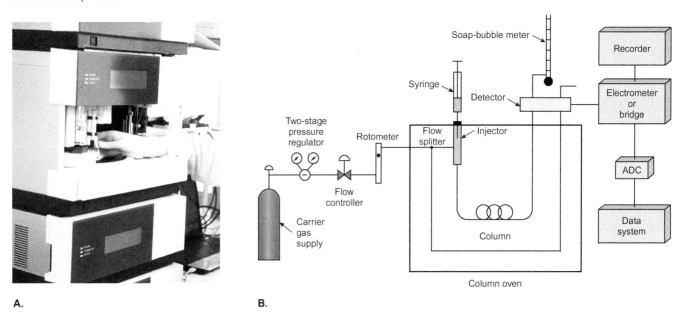

A.

B.

27.3 Typical Process Control Applications

Process instrumentation is used to monitor and control both critical and noncritical processes. Instrumentation can also be used for redundancy or as a backup to the main source as a safety practice. Critical processes are controlled in a control room where plant sections are generally divided into separate workstations. Those same processes may have local mounted instruments on the equipment in the field. Some processes have local panel boards that can be used to help with the equipment startup and monitoring process.

Early instrumentation was manually controlled. Adjustments were completed by the operator (process technician) who opened or closed a valve while watching a dial or gauge. An advance in technology in the late 1930s brought the advent of the pneumatic power source. Pneumatic power is now the main signal for process instrumentation lines and control loop instruments. The most common range of operation is 3 psi to 15 psi.

Further evolution saw electricity implemented as the power source for process instrumentation. With electricity, the common operating range is considered to be 4 mA to 20 mA. Electronic instrumentation is more cost effective and more precise in its operation and instrumentation-control application than pneumatic signals.

Instrumentation configuration depends on the nature of the process, whether it only needs to be monitored or needs to be both monitored and controlled. A simple process, for example, a water tank level, may not require more than a sight or gauge glass in the field. However, that same tank could have the level indication sent to the control room via a D/P cell transmitter and a signal line (pneumatic or electronic), ending with a panel board indication. This would be considered an *open instrumentation loop*.

If that same tank level is more critical and needs to be controlled at a specific amount, there might be a gauge glass in the field for redundancy. The tank would also have a control loop that includes a sensor, transmitter, controller, and a control valve at the tank in the field. This is called a *closed loop of instrumentation*.

Location

An instrument may be categorized as **local**, meaning in the field at the process equipment, or **remote**, meaning located in a control room either on a panel board or within the distributed control system (DCS). DCS will be discussed later in this chapter.

Function

An instrument can function as a sensor, transmitter, indicator, recorder, controller, or control valve.

A **sensor** (or *primary element*) is an element in direct contact with the process variable to be measured. It senses a change and provides a signal to the transmitter. The **transmitter** is a device that receives its signal from the sensing element, measures that signal, and sends the signal to the controller or indicator. Sensors and transmitters work together to measure and transmit data. Some devices combine sensors and transmitters. For example, a sensor/transmitter can measure pressure and then transmit a signal representing that measurement.

An **indicator** shows the current condition of a process variable, usually through a visual representation such as a needle on a scale in a gauge or a digital readout. Sometimes, transmitters, recorders, and controllers include an indicator.

A **recorder** is an instrument used to document variables such as pressure, speed, flow, temperature, level, and electrical units. It also provides a history of the process over a given period of time, called a *trend*.

The **controller** receives information from the transmitter in the form of a pneumatic or electronic signal, compares that input signal to the reference set point, and sends a corrective output signal to the final control element, typically a control valve.

The final control element, usually a **control valve**, responds to the signal from the controller and manipulates the process to help maintain the process setpoint.

Power Source

A power source is what provides the instrumentation with the power to work. Power sources can be pneumatic, hydraulic, electronic, or a combination of these sources.

Pneumatic instruments are devices that use air pressure, or a gas such as nitrogen, traveling through copper or stainless tubing lines, as the power source for control loop instrumentation. This type of power source is low cost and widely used in process industries.

Hydraulic instruments use pressurized liquid as the power source. This source is used when heavy equipment, such as large valves, must be controlled or manipulated open and closed.

Electronic instruments use electricity as the power source (grouped as analog, digital, or a hybrid of the two). The *analog signal* uses a continuously variable electrical current, which represents the change in variable and can be measured. The *digital signal* is a series of binary numbers; it uses microprocessors (mini computers) to measure and generate an

---

Local related to instrumentation in close physical proximity to the process, generally found outside the control room.

Remote related to instrumentation located away from the process (e.g., in a control room) and controlled by the DCS or other panel board mounted controller.

Sensor (also known as a *primary element*) the control loop component that is in direct contact with the process variable to be measured; it senses or detects change and provides a signal to the transmitter.

Transmitter the control loop component that receives a signal from the sensor, measures that signal, and sends it forward to the controller or indicator.

Indicator an instrument that shows the current condition of a process variable, usually through a visual representation such as a needle on a scale in a gauge or a digital readout.

Recorder an instrument, either mechanical or electronic, used to document variables such as pressure, speed, flow, temperature, level, and electrical units; it also provides a history of the process over a given period of time, which is called a trend.

Controller the component of a control loop that receives information from the transmitter in the form of a pneumatic or electronic signal, compares that signal to a prescribed setting, and sends a correcting output signal to the final control element in order to maintain the process at that set point.

Control valve the final component of a control loop, operated either remotely or by hand, that receives its signal from the controller and manipulates the process variable in response to a system disturbance or change.

Pneumatic instruments devices that use air pressure or inert gas as their power source.

Hydraulic instruments devices that use pressurized fluid (hydraulic fluid) as the power source.

output signal. Digital signals are used more frequently due to the increased use of automation in the process industries. DCSs (discussed later) utilize digital signals to and from the controllers. The common operating range for electronic instrumentation signals is 4 mA to 20 mA. Figure 27.8 shows examples of analog and digital instruments.

Figure 27.8 A. Analog instrument. **B.** Digital instrument.

CREDIT: **A.** Yuri Shebalius/Shutterstock; **B.** Dmitry Kalinovsky/Shutterstock.

A.   B.

Other Components

Other components typically found in a process control instrumentation system include, transducers, alarms, control valves, actuators, valve positioners, and signal control elements.

A **transducer** converts one instrumentation signal type into another type. This conversion can be electronic to pneumatic or pneumatic to electronic. A major use for this component at the control valve is to convert an electronic signal from the controller into a pneumatic signal to operate the actuator. The **actuator** (an electrical, pneumatic, or hydraulic device that manipulates a control device, usually a valve), moves the valve stem to open or close the valve. A signal of 3 psi would equal a 4 mA signal. A 15 psi pneumatic signal equates to the 20 mA signal, signifying the device (control valve) is either fully open or fully closed.

An **alarm** is an instrumentation that provides visual and/or audible cues that alert technicians of a condition outside the normal operating range, such as a tank reaching capacity. An alarm can be a computer display, horn, flashing light, whistle, bell, or other similar device. Depending on the criticality of the process, there may be multiple alarms, such as "soft" alarms for deviation from the set point, and hard alarms for processes that have moved beyond normal operating ranges.

A control valve is operated either remotely or by hand. Control valves consist of the actuator and valve body. The actuator and valve body are connected by a stem. The actuator may be operated by an electronic, pneumatic, or hydraulic signal. The signal causes the valve in the body to open or close in response to a signal change.

A **valve positioner** is a device that ensures a valve is positioned properly in reference to the signal received from the controller.

Transducer the control loop component that converts one instrumentation signal type to another type (e.g., electronic to pneumatic).

Actuator an electrical, pneumatic, or hydraulic device that manipulates a control device, usually a valve.

Alarm a device, such as a horn, flashing light, whistle, or bell, that provides visual and/or audible cues of conditions outside the normal operating range.

Valve positioner a device that ensures a valve is positioned properly in reference to the signal received from the controller.

27.4 Process Control Instrumentation Signals

Instrumentation can also be categorized by the type of signal it produces. The two main types of electronic signals are analog and digital. An analog signal is a continuously variable representation of a process variable. A digital signal uses the language of computers, called binary numbers (on/off switches that represent data), to represent continuous values or distinct (discrete) states.

Although digital signals are used more frequently, analog signals continue to be used in process industries for instrumentation applications. Some digital-based instruments (with a microprocessor inside) can generate either a digital or analog signal.

Fiber optics, a system that uses light frequency signals in either analog or digital mode, have gained popularity in industry use. Fiber optics do not pose an intrinsic danger in areas of hazardous chemical production. This application is safe, economical, and reliable for use in the process industries.

Pneumatic signals are also common in process facilities. These signals are generated by compressing and cleaning air to make it free of particulates and moisture. Compressed inert gas (like nitrogen) may be used as a backup for instrument air when needed.

Some plants may even have mechanical methods of transmitting signals, such as levers, cables, and pulleys. Hydraulic systems can also be used to transmit signals.

27.5 Instrumentation and Control Loops

A **control loop** is a group of instruments that work together in a system to monitor and control a process. At a fundamental level, a control loop consists of a process variable, a device to measure it, and a device to control it.

Control loop a series of instruments that work together to monitor and control a process variable.

The following steps describe how a simple control loop works:

1. A sensor in direct contact with the process sends out (outputs) a signal to the measuring device, which is the transmitter.
2. The process variable is measured and the measurement is converted into a proportional pneumatic or electronic signal in the transmitter that represents the process. This signal is then sent to the controller.
3. The controller has a set point, or optimum value, for the variable, which is compared to the current input signal from the transmitter. Note: The difference between the set point and the current process variable value is called the *error*.
4. The controller produces an output signal that is sent to the final control element (e.g., a valve or speed governor) to eliminate the error. Most control valves are actuated pneumatically, even if the loop signal is electronic (Figure 27.9). If the signal is electronic, a

Figure 27.9 Diagram highlighting simple control loop (see red rectangle).

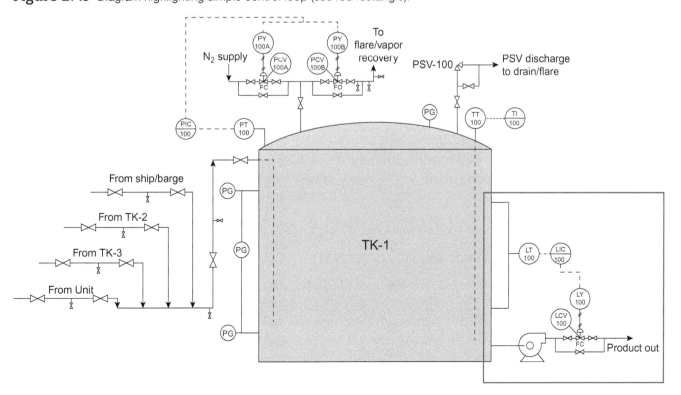

transducer converts the signal into a pneumatic signal. In newer applications, the positioner and transducer are a combined component.

5. As the error is corrected, the process variable returns to match the set point.

To illustrate a simple control loop, think about your home air conditioning system. The thermostat contains a sensor that detects the room temperature, compares it to your controller setting (set point), and turns the unit on or off depending on whether the room has reached the temperature setting. Figure 27.10 shows another example of a simple control loop.

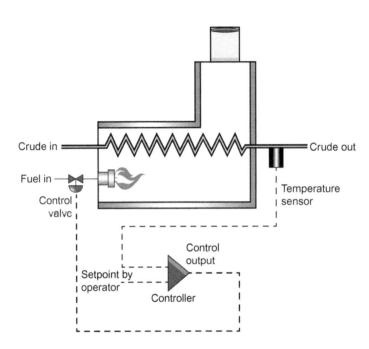

Figure 27.10 Simple temperature control loop example.

In this example, there is a furnace that heats crude oil to 400 degrees F (204.4 degrees C). A sensor located in the effluent crude oil flow sends that signal to a transmitter, which measures the crude oil temperature. The transmitter sends that signal to a controller. The controller compares the temperature reading to the set point value. If the temperature is below the set point, the controller opens the fuel valve to add more fuel. If the temperature is above the set point, the controller sends a signal to lower the valve output to reduce the fuel.

A control loop can either be a two-position control (also called *discrete*, i.e., can be on/off, open/shut, or start/stop), or a continuous (*modulated*) control. If the loop is on/off, the control element can only be in the on or off position (such as a heating system in a home). If the loop is continuous, control is variable. A continuous control element can be placed at any position between fully open or fully closed (like a valve).

27.6 Distributed Control Systems and Their Application

Distributed control system (DCS) a computer-based system consisting of a series of controllers used to monitor and control a process.

A **distributed control system (DCS)** is a computer-based system that is used to monitor and control processes. Because they are computer-based, DCSs can be connected to other DCSs and related systems, providing flexibility, accuracy, precision, and reliability.

The modern DCS can process and store data quickly and efficiently, sharing the workload among systems and offering redundancy (backup or fail-safe) systems.

Although DCS terminology varies based on the company using it and the manufacturer that provides it, there are some common components in all DCS systems:

- Input/output (I/O) devices—field instrumentation (sensors, transmitters, and final control elements such as valves and motors)
- Network—a data highway that connects the systems
- Workstations—computers that allow process technicians to monitor and control processes
- Supervisory computers—computers accessed by engineers and management at the programming level to monitor production targets
- Central processor—a computer used to configure the DCS
- Accessory devices—devices such as printers, data loggers/recorders, and storage devices.

27.7 Operational Hazards

Process technicians should understand the hazards associated with the improper operation of instrumentation. Table 27.1 lists some of these hazards and their impacts.

Table 27.1 Operational Hazards Associated with Instrumentation

Improper Operation	Possible Impacts			
	Individual	Equipment	Production	Environment
Opening bypass valve around control valve (only done with approval or supervision)		Possible damage to downstream equipment due to higher flows and pressures	Loss of process control	Release of a hazardous substance
Adjusting leaking control valve packing	Burns from product Appropriate training as needed for this task	Valve may not open, close, or operate properly Overtightening may cause valve to hang or stick	Unit could shut down with resulting loss of production Loss of process control	Release of a hazardous substance
Attempting to open a motor controlled valve manually	Hand and/or back injuries	Damage to valve and motor	Loss of process control	Release of a hazardous substance
Bypassing alarms (generally requires written permission of supervisor; mid-level management and safety department may be involved)	Injury due to malfunctioning equipment Disciplinary action	Shutdown of process unit	Loss of production	Release of a hazardous substance
Failure to respond to an alarm	Injury due to malfunctioning equipment Disciplinary action	Shutdown of process unit	Loss of production	Release of a hazardous substance
Turning off air supplies to control devices	Disciplinary action	Equipment shutdown	Could result in unit shutdown and loss of production Loss of process control	Release of a hazardous substance
Opening housings on electronic transmitters	Shock hazard Appropriate training as needed for this task	Equipment damage (e.g., blown fuses)	Ignition source for an explosion	

27.8 Monitoring and Maintenance Activities

When monitoring and maintaining instrumentation, process technicians should always remember to look, listen, and check for the items listed in Table 27.2.

Failure to perform proper maintenance and monitoring could impact the process and result in instrumentation and equipment damage.

360 Chapter 27

Table 27.2 Monitoring and Maintenance Activities Associated with Instrumentation

Look	Listen	Check
■ Monitor and control process variables as required. ■ Check for proper control mode. ■ Check set point readings. ■ Watch and listen for alarms. ■ View historical data to analyze trends. ■ Visually check control valve packing for leaks. ■ Inspect field instrumentation for broken parts. ■ Check for broken gauges. ■ Look at wiring and connections for potential problems. ■ Using the appropriate plant alarm management program, test alarms to ensure they are working. ■ Inspect auxiliary equipment.	■ Check for noisy valve seats and disks. ■ Check for control line leaks (pneumatic or hydraulic).	■ Check locally mounted gauges are reading correctly. ■ Check for valve and instrumentation vibration.

Symbols for Process Control Instruments

To locate instrumentation on a piping and instrumentation diagram (P&ID) accurately, process technicians must be familiar with the symbols that represent different types of instrumentation.

Figure 27.11 and Figure 27.12 show examples of instrumentation symbols and how to interpret them.

Figure 27.11 Standard instrumentation symbols.

Line symbols	Line type
—//—//—	Pneumatic signal
 — — — —	Electrical signal
—∼—∼—	Electromagnetic or sonic signal
—✕—✕—	Capillary tubing
—L—L—	Hydraulic signal
—o—o—	Software link
——————	Connection to process; secondary line; utility line
▬▬▬▬▬▬	Process line
 —●—●—	Mechanical link
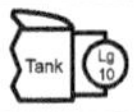 Tank [Lg 10]	Local level sight or gauge glass
○	Local mounted instrumentation symbol
⊖	Panel mounted instrumentation symbol
⊝	Remotely mounted instrumentation behind panel board
⊖	Auxiliary panel mounted instrumentation
⊟	DCS mounted instrumentation

Figure 27.12 Instrument symbol interpretation key.

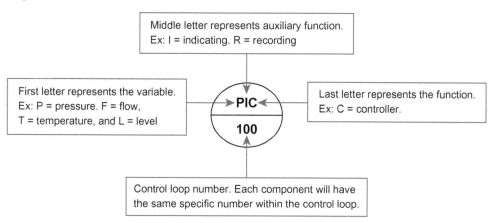

Summary

Instrumentation is a vital part of the process industries, allowing companies to produce higher quality products more efficiently and in a safer environment. The process industries use instrumentation to measure, control, indicate rate, or record factors that may change within a process. These factors are called process variables (PVs).

The five most common types of process variables are pressure, temperature, level, flow rate, and analytical measurements.

Pressure is the force exerted on a surface divided by its area. A common unit of pressure measurement is psi, or pounds per square inch. A differential is the difference between measurements taken from two separate points. Differential pressure devices are commonly used in the process industries, and differential temperatures are measured in many applications.

Temperature is the specific degree of heat or cold as indicated on a reference scale. The two most common temperature measurement scales are Fahrenheit (degrees F) and Celsius (degrees C).

Level is a position of height or depth along a vertical axis. In the process industries, level usually means the height of the surface of a material compared to a zero reference point.

Flow (also flow rate) is a quantity of fluid that moves past a specific point within a given amount of time, and is usually expressed in volume units per unit of time, such as gallons per minute (GPM) or cubic feet per minute (CFM).

Analytical measurements analyze the chemical or physical properties of a substance. Common analytical measuring instruments include pH meters, chromatographs, and viscosity meters.

The instrumentation configuration depends on the nature of the process, whether it only needs to be monitored or needs to be both monitored and controlled. An instrument can be categorized as local (field mounted in the process area), or remote (located in the control room). An instrument can function as a sensor, transmitter, indicator, recorder, controller, or control valve. The power source for the instrument signal in a control loop can be pneumatic, hydraulic, or electronic.

Other components typically found in an instrumentation system are transducers, alarms, control valves, actuators, valve positioners, and signal control elements.

A control loop is a group of instruments that work together in a system to monitor and control a process.

A distributed control system (DCS) is a series of controllers in a computer-based system used to monitor and control a process.

Process technicians need to understand operational hazards associated with instrumentation. When monitoring and maintaining instrumentation, technicians must always remember to look, listen, and check for abnormal conditions.

To locate instrumentation on a piping and instrumentation diagram (P&ID) accurately, process technicians must be familiar with the symbols that represent different types of instrumentation.

Checking Your Knowledge

1. Define the following terms:
   a. actuator
   b. alarm
   c. control valve
   d. controller
   e. indicator
   f. recorder
   g. sensor
   h. transducer
   i. transmitter
   j. valve positioner
2. In instrumentation, what do the initials PV stand for?
   a. pneumatic value
   b. process variation
   c. pressure value
   d. process variable
3. Which of the following is NOT a key process variable?
   a. pressure
   b. temperature
   c. flow
   d. surface area
4. A(n) _____ measures the chemical or physical properties of a substance.
5. A differential is:
   a. the difference between measurements taken from two separate points.
   b. the similarity between measurements taken from two separate points.
   c. the difference between measurements taken from the same point at different times.
   d. the similarity between measurements taken from the same point at different times.
6. An instrument is considered _____ if it is in close proximity to the process, but considered _____ if it is located away from the process, like in a control room.
7. Which of the following power sources for an instrument uses air pressure or a gas such as nitrogen?
   a. hydraulic
   b. electronic
   c. digital
   d. pneumatic
8. *True or False:* A transmitter is a device that shows the current condition of a process variable, usually through a visual representation.
9. A(n) _____ is a group of instruments that work together in a system to monitor and control a process.
10. What do the initials DCS stand for?
    a. dynamic collaboration system
    b. dual control sensors
    c. distributed control system
    d. disturbed control sensor
11. Which of the following are considered potential hazards if a process technician attempts to open a motor-controlled valve manually? Select all that apply.
    a. damage to the valve and motor
    b. personal injury
    c. loss of process control
    d. potential hazardous substance release
12. What do the following symbols represent?
    A. ─//──//─
    B. ─ ─ ─ ─ ─ ─
    C. ──◠──◠──

NOTE: Answers to Checking Your Knowledge questions are in Appendix I.

Student Activities

1. List four of the most common process control variables and describe each one. List one example of a sensor that is used in a control loop for each process.
2. Write a paragraph that describes the functions of each of the following: sensor, transmitter, indicator, recorder, controller, and control valve.
3. Describe the principles of operation of a simple control loop.

Appendix I

Answers for Checking Your Knowledge

Chapter 1	Answer
1.	See Key Terms list
2.	process
3.	Any of the following: oil and gas, mining, chemical, pharmaceutical, food and beverage, power generation, water treatment, pulp and paper
4.	d. an integrated group of process equipment used to produce a specific product or products for distribution and sale
5.	Any of the following: priority on safety, appropriate education, including a foundation in science and math, technical knowledge and skills, interpersonal skills, communication skills, ability to work on and contribute to a team, computer skills, physical capabilities, ability to deal with change, technology skills, ability to stay current in work skills, ability to troubleshoot
6.	Any of the following: strong work ethic, ability to deal with change, flexibility and self-direction as team player, ability to appreciate diversity, ability to follow SSHE procedures and policies, sense of responsibility, positive attitude, respect for others, ability to accept criticism and feedback
7.	True
8.	c. 2 a.m. to 6 a.m. and 2 p.m. to 6 p.m.
9.	b. take naps when possible c. avoid caffeine and alcohol d. eat lightly from 2-6 a.m./p.m.
10.	a. sample process streams, c. analyze data, d. make process adjustments
11.	b. lubrication, d. monitoring and analyzing equipment performance
12.	a. apply processes such as root cause analysis, b. use statistical tools, d. participate in corrective action teams
13.	b. search for ways to increase efficiency, e. plan for computerized instrumentation

Chapter 2	
1.	See Key Terms list
2.	Petroleum
3.	b. first refinery built in the United States, d. first successful commercial oil well drilled in Pennsylvania, c., a. automobile invented, c. improvements in refining during World War II
4.	Upstream/midstream/downstream or exploration & production/transportation/refining & marketing
5.	a. kerosene
6.	b. pipelines
7.	d. refining
8.	Minerals and gems
9.	d. metallic and nonmetallic
10.	c. surface and underground
11.	b. the Industrial Revolution
12.	a. 50%; one-half
13.	c. pipes and chutes

Chapter 3	
1.	See Key Terms list
2.	Petrochemical
3.	See page 47
4.	d. mustard gas
5.	c. nylon
6.	c. safety
7.	c. was the first book of drug standards to gain national acceptance
8.	b. World War II

Chapter 4	
1.	See Key Terms list
2.	Electricity
3.	d. Heron
4.	d. voltaic pile battery created, a. first generator created, c. water power used to turn wheat into flour, b. atomic bomb tested
5.	b. Faraday
6.	False
7.	c. distillation
8.	b. fossil fuels
9.	Benefit: cleaner; Drawback: radioactive waste
10.	process technicians may need to obtain specialized licensing from a government agency
11.	d. EPA

Chapter 5	
1.	See Key Terms list
2.	a. 1906
3.	contaminated environments; animals before slaughter
4.	d. destroying disease-causing organisms
5.	b. reverse osmosis
6.	Any of the following: FDA, CDC, USDA, EPA, NOAA, ATF
7.	c. NOAA
8.	b. USDA
9.	b. technicians can set up all products the same way for production
10.	True

Chapter 6

1. See Key Terms list
2. c. nineteenth century
3. e. all of the above
4. c. the Safe Drinking Water Act of 1974
5. c. removes settleable solids
6. c. secondary treatment

Chapter 7

1. See Key Terms list
2. b. the Chinese
3. b. papermaking process invented, a. printing machine invented, c. paper mill built in America, d. papermaking machine invented
4. a. Fourdrinier
5. a. 1690
6. cellulose
7. Any of the following: recycled paper, linen, cotton, synthetics
8. True
9. Any of the following: cutting, folding, coating, gluing, screening, embossing

Chapter 8

1. See Key Terms list
2. a. members are answerable to each other
3. c. members responded constructively to others' views
4. b. operations, c. safety, e. maintenance
5. Synergy
6. b. how the team functions interpersonally
7. c. reforming
8. Any of the following: tackling tasks that are not appropriate, lacking a clear purpose, having the wrong members, having poor team dynamics
9.
    - Describe
    - Express
    - Specify
    - Contract
    - Consequences
10. a. unique differences
11. salad bowl
12. I. d.
    II. a.
    III. b.
    IV. c.

Chapter 9

1. See Key Terms list
2. physical hazard
3. air pollution
4. OSHA 1910.120 or 29 CFR 1910.120 or HAZWOPER
5. process safety management
6. See Key Terms list
7. d. fuel, oxygen, and heat
8. cover goggles
9. b. recognize and promote effective safety and health management
10. environmental management system (EMS) document
11. c. protecting remote-control cameras from damage

Chapter 10

1. See Key Terms list
2. Any of the following: satisfying and retaining customers, maintaining a competitive advantage and responding to rapidly changing markets, capturing a leading position in the global marketplace, improving profitability, managing change, maintaining or bolstering the organization's reputation, offering standardized products, improving efficiency of operations and maintenance, reducing waste of resources, decreasing downtime, ensuring certifications necessary to trade internationally, tapping new technologies and methods
3. Deming
4. 80% of the problems come from 20% of the causes
5. c. Philip B. Crosby
6. total quality management
7. d. customer focus
8. provides global standards of product and service quality
9. See Key Terms list (SPC)
10. a. control charts, b. cause and effect diagrams, d. histograms
11. a. to maximize equipment effectiveness (overall effectiveness)
12. d. preventing potential equipment failures

Chapter 11

1. See Key Terms list
2. b. sublimation
3. a. conduction
4. False
5. a. 273 K
   b. 0° C
   c. 32° F
   d. 492° R
6. a. 352 K
   b. 154.4° F
   c. 660° R
   d. 83.3° C
   e. 311.8 K
7. absolute zero
8. b. British thermal unit (BTU)
9. c. vacuum pressure
10. a. increases
11. solid, liquid, gas, plasma
12. transition of a substance from one physical state to another
13. See Key Terms list
14. See Key Terms list
15. inversely proportional
16. direct relationship
17. the sum of the individual partial pressures
18. general or combined gas law
19. pressure is least where speed is greatest; inverse relationship
20. point A

Chapter 12

1. See Key Terms list
2. Organic chemistry studies substances that contain carbon; inorganic chemistry studies substances that do not contain carbon.
3. Chemical properties are characteristics associated with chemical reactions; physical properties are observable and are not associated with chemical reactions.
4. a. physical
   b. chemical
   c. chemical
   d. physical
   e. chemical
   f. physical
   g. physical
   h. physical
5. a. homogeneous
6. a. atom
7. 28; surrounding the nucleus
8. To separate the components of a liquid.
9.

	Calcium	Lead	Chlorine
Column number (e.g., 1, 2, 3)	2	14	17
Group number (e.g., I, II)	IIIB	IVA	VIIA
Atomic number	20	82	17
Symbol	Ca	Pb	Cl
Atomic weight	40.078	207.2	35.453

10. image B
11. NH_3 and HCl are reactants; NH_4Cl is the product
12. water

Chapter 13

1. See Key Terms list
2. process flow diagrams (PFDs) and piping and instrumentation diagrams (P&IDs)
3. b. BFD
4. Any of the following: symbols, legend, title block, application block
5. d. process flow diagram
6. edges of drawing; where a line comes from and where it goes on other drawings in a series
7. b. pressure and temperature variable listings
8. c. P&ID
9. a. 30-degree angle perspective

Chapter 14

1. See Key Terms list
2. They need to understand the process so they can choose materials that will be safe to use and will hold up well.
3. b. flanged pipe connection
4. A welded connection, because it is not prone to leaking and can stand up to high pressure.
5. butt weld
6. a. Wye (Y) fitting, b. cap, c. bushing, d. flange, e. plug, f. 45° elbow, g. cross, h. coupling, i. bell reducer, j. 90° elbow, k. union, l. tee, m. allthread, n. nipple
7. False
8. True
9. It could be damaged and may leak.
10. Excessive force can warp the valve or damage the seat, preventing a good seal.
11. ball, plug, butterfly
12. I. ball f, II. butterfly b, III. check i, IV. diaphragm d, V. gate e, VI. globe h, VII. plug c, VIII relief g, IX. safety a

Chapter 15

1. See Key Terms list
2. A (spherical)
3. False
4. contain chemical spills, protect environment from contaminants, protect humans from potential hazards, hold contaminated water until it can be drained or cleaned, protect against spread of fire
5. Any of the following: putting wrong material into tank, misaligning blanket, misaligning pump, pulling a vacuum on a tank while emptying, overpressurizing, lining valves up wrong, loss of nitrogen flow, cross-contamination, failed vent system, leaks, chemical reactions
6. Any of the following: abnormal noise, abnormal heat on vessels or piping, excessive vibration, abnormal odors

Chapter 16

1. To move liquid from one location to another.
2. positive displacement pump
3. c. positive displacement pump
4. a. C
   b. G
   c. F
   d. B
   e. D
   f. A
   g. E
5. a. C
   b. G
   c. B
   d. F
   e. D
   f. A
   g. E
6. Any of the following: overheating, overpressurization, cavitation, leaks, vibratioin
7. Any of the following: oil levels, no water collecting under oil, no leaks, suction and discharge pressure, abnormal sounds, excessive vibration, excessive heat

Chapter 17

1. To increase the pressure of gases by using a series of rotor blades with stator blades between them, creating gas flow that is parallel to the compressor shaft.
2. It uses a series of rotor blades with a set of stator blades between them to move gases parallel to the shaft.
3. c. positive-displacement
4. a. H casing
   b. A discharge
   c. B impeller
   d. E shaft
   e. C suction
   f. G packing
   g. D diffuser plates
   h. F packing gland

5.
   a. A discharge port
   b. F casing
   c. E diffuser
   d. C suction port
   e. B impeller
   f. D shaft
6. Any of the following: overpressurization, surging and operating at critical speeds, rotating equipment, overheating, leaks
7. Any of the following: check oil levels, check for leaks, check liquid level in suction drum, listen for abnormal noises, check for excessive vibrations, check for excessive heat
8. a.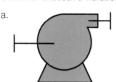
   b.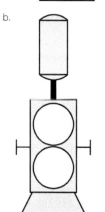
   c.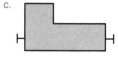
   d.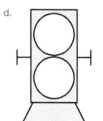

Chapter 18

1. See Key Terms list
2. False
3. a. gas turbine
4. b. hydraulic turbine
5. a. low steam inlet pressure
   d. running at set point
   e. appropriate lubrication
6. Any of the following: check oil levels, check condition of insulation, check bearing temperatures at gauges, check seals and flanges for leaks, check for water in the oil reservoir, check steam pressure and temperature, check operation of governor, check RPMs in operating range, check valves in appropriate position, check for abnormal noises, excessive heat, excessive vibration

Chapter 19

1. See Key Terms list
2. False
3. generators
4. motors
5. c. shroud
6. A stationary part of the motor where the alternating current flows in and creates a large magnetic field using magnets and coiled wire.
7. c. pump
8. d. all of the above
9. a. are caused by loss of attention
10. body
11. 1000 ohms, wet skin; 100000 ohms, dry skin
12. Any of the following: wires insulated and not cracked or worn, no loose covers and shrouds, bearings not showing wear, equipment preparation for maintenance check (bearings and bushings), no signs of corrosion, motor shaft not worn or bent, oil passages not plugged, oil wells or holes are clean
13. Circuit breaker; interrupts the flow of electricity

Chapter 20

1. See Key Terms list
2. b. baffle
3. False
4. a. convection
5. reboilers, preheaters, aftercoolers, condensers, chillers, interchangers
6. Any of the following: external leaks, internal tube leaks, abnormal pressure changes, temperature changes, hot spots or uneven temperatures, leaky gaskets, insulation, inlet and outlet temperature gauges and pressure gauges, inlet and outlet flow rate, inlet and outlet samples, abnormal noises, excessive vibration, excessive heat
7. c. shell and tube
8. b. U-tube

Chapter 21

1. See Key Terms list
2. Heat exchangers lower the temperature of a water stream by evaporating part of the stream and having a small amount of the heat transfer take place via mixing of fluids (convection).
3. natural draft, induced draft, forced draft
4. ambient air temperature, relative humidity of ambient air, wind velocity and direction, water contamination, cooling tower design
5. Any of the following: look for leaks, adequate basin water levels, chemical balance (pH and hardness), filter screens not plugged, temperature differentials, broken fill materials to fix at next turnaround, ice buildup or ice fog, foaming from improper chemical treatment or incorrect blowdown stream, proper water distribution on top of the tower, abnormal noises, excessive heat or vibration
6. b. grinding

Chapter 22

1. See Key Terms list
2. a. separating components of crude oil, b. superheating steam, c. removing metal from ore, e. raising the temperature of a process stream
3. b. balanced, c. forced, e. natural
4. a. convection
5. a. octane
6. b. box

Appendix I   367

7. Any of the following: opening inspection ports when the firebox has a positive pressure, failure to follow flame safety procedures, poor control of excess air and draft, bypassing safety interlocks, failure to wear PPE
8. Any of the following: secondary combustion, flame impingement, leaks, hot spots, pressure balance, temperature, firing efficiency, burner balance, burners lit, controlling instruments, flame pattern, flame color, fireballs, abnormal noise, excessive vibration

Chapter 23

1. See Key Terms list
2. d. the heat source is located on the tube side
3. risers and downcomers
4. Any of the following: heat and cool process fluids, drive equipment, fight fires, and purge equipment, provides energy to drive equipment
5. Any of the following: fossil fuel (e.g., crushed coal), mix of fuels such as oil sprayed in a fine mist, or combination fuels such as methane, off-gas, or natural gas
6. Any of the following: opening pressured header drains and vents, failing to purge firebox, poor control of excess air and draft, loss of boiler feed water, loss of fuel gas or oil
7. Any of the following: flame impingement, flame color, wall hot spots, draft balance (pressure), temperature gradient, firing efficiency, burner balance, controlling instruments, abnormal noise, excessive vibration

Chapter 24

1. See Key Terms list
2. b. lowest boiling point of all the fractions in the column
3. Any of the following: leaks, temperature, tower differential pressure, reboiler shell pressure, level of bottoms products, condenser pressure, reboiler condensate trap, material balance, feed and product distribution, product specifications, unusual noises, temperature changes, unusual vibrations
4. False
5. 1. B structured packing, 2. C Raschig ring, 3. D pall ring, 4. A Berl/Intalox saddle

Chapter 25

1. See Key Terms list
2. a. safe to drink
3. False
4. instrument air

5. Any of the following: removal of explosive gases from process systems and equipment, backup for instrument air, inert blanket gas for storage tanks and lines, cooling and cryogenics
6. Any of the following: tanks and drums, pumps, instrumentation and control, pipes and hoses, valves, filtration system, chemical addition and other treatment systems, cooling towers, heat exchangers, boilers, firefighting equipment
7. b. asphyxiation
8. c. freeze burns
9. Any of the following: shocks, burns, death, loss of power, over- or undervoltage, short circuits, arcing

Chapter 26

1. See Key Terms list
2. b. knockout drum
3. d. reduces the pressure of the liquid refrigerant, causing it to cool
4. b. receiver
5. False
6. a. filter
7. startup and shutdown
8. d. refrigerant
9. True
10. a. 3 flare head, b. 5 knockout drum, c. 1 burner, d. 4 flare stack, e. 2 steam ring
11. d. flare

Chapter 27

1. See Key Terms list
2. d. process variable
3. d. surface area
4. analytical value
5. a. the difference between measurements taken from two separate points.
6. local; remote
7. d. pneumatic
8. False (indicator is correct)
9. control loop
10. c. distributed control system
11. a. damage to the valve and motor, b. personal injury, c. loss of process control, d. potential release
12. a. pneumatic signal, b. electrical signal, c. electromagnetic or sonic signal

Glossary

Absolute pressure (psia) gauge pressure plus atmospheric pressure; pressure referenced to a total vacuum (zero psia).

Absorb to draw inward.

Acid a substance with a pH less than 7 that releases hydrogen (H+) ions when mixed with water.

Actuate to put into action.

Actuator an electrical, pneumatic, or hydraulic device that manipulates a control device, usually a valve.

Administrative controls the implementation programs (e.g., policies and procedures) and activities to address a hazard.

Adsorb to stick together.

Aftercooler a heat exchanger, located on the discharge side of equipment, with the function of removing excess heat from the system generated by the process.

Agglomerating gathering materials into a mass (instant coffee crystals are made using this process).

Air pollution the contamination of the atmosphere, especially by industrial waste gases, fuel exhausts, smoke, or particulate matter (finely divided solids).

Air register also called the *burner damper*, a device located on a burner that is used to adjust the primary and secondary airflow to the burner; air registers provide the main source of air entry to the furnace.

Air-rumble the process of introducing bursts of compressed air in the cooling water to remove contaminants.

Alarm a device, such as a horn, flashing light, whistle, or bell, that provides visual and/or audible cues of conditions outside the normal operating range.

Alchemy a medieval practice that combined occult mysticism and chemistry.

Alkaline having to do with a base (i.e., a substance with a pH greater than 7).

Alloy a compound mixture composed of two or more metals or a mixture of a metal and another element that are mixed together when molten to form a solution; not a chemical compound.

Alternating current (AC) an electrical current that reverses direction periodically. This is the primary type of electrical current used in the process industries and in residential homes.

American Society of Mechanical Engineers (ASME) organization that provides laws of regulation for boilers and pressure vessels.

Amine system a system used to remove CO_2 or other acid gases from a gas stream.

Ammeter a device used to measure current in amperes.

Amperes (amps) a unit of measure of the electrical current that flows in a wire; similar to "gallons of water" that flow in a pipe.

Analytical variable a measurement of the chemical or physical properties of a substance.

Antibiotics substances derived from mold or bacteria that inhibit the growth of other microorganisms (e.g., bacteria or fungi).

API gravity the American Petroleum Institute (API) standard used to measure the density of hydrocarbons.

Apothecary a person who studies the art and science of preparing medicines; in modern times we call these individuals pharmacists.

Application block the main part of a drawing that contains symbols and defines elements such as relative position, types of materials, descriptions, and functions.

Arc-rated having protective characteristics of the fabric as determined by a hazard/risk assessment.

Articulated drain a hinged drain, attached to the roof of an external floating roof tank, that moves up and down as the roof and the fluid levels rise and fall.

Assignable variation statement that, when a product's variation goes beyond the limits of natural variation, it is the result of a worker-related cause.

Atmospheric pressure the pressure at the surface of the earth (14.7 psia at sea level).

Atmospheric tank an enclosed vessel that operates at atmospheric pressure; usually cylindrical in shape and equipped with either a fixed or floating roof, and containing nontoxic or very high pressure vapor liquids.

Atomic number the number of protons found in the nucleus of an atom.

Atomic weight   the sum of protons and neutrons in the nucleus of an atom.

Atom   the smallest particle of an element that still retains the properties and characteristics of that element.

Attitude   a state of mind or feeling with regard to some issue or event.

Attributes   also called discrete data; data that can be counted and plotted as distinct or unconnected events (such as percentage of late shipments or number of mistakes made during a process).

Axial compressor   a dynamic-type compressor that uses a series of rotor blades with a set of stator blades between each rotating wheel and creates gas flow that is axial (parallel to) the compressor shaft.

Axial pump   a type of pump classified in the dynamic category that uses a propeller or row of blades to propel liquids axially along the shaft.

Back-flush   the process of removing contaminants by reversing the normal flow; typically associated with the use of shell and tube exchangers or plate and frame exchangers that employ cooling water.

Baffle   a metal plate, placed inside a tank or other vessel, that is used to alter the flow of chemicals or facilitate mixing.

Baffles   partitions located inside a shell and tube heat exchanger that direct flow and increase turbulent flow to help reduce hot spots.

Balanced draft furnace   a furnace that uses fans to facilitate airflow, fan(s) inducing flow out of the firebox, through the stack (induced draft); and fan(s) providing positive pressure to the burners (forced draft).

Ball valve   a flow-regulating device that uses a flow-control element shaped like a hollowed-out ball, attached to an external handle, to increase or decrease flow; a ball valve requires only a quarter turn to go from fully open to fully closed.

Base   a substance with a pH greater than 7 that releases hydroxyl (OH−) anions when dissolved in water.

Basin   a compartment at the base of a cooling tower that is used to store cooled water until it is pumped back into the process.

Batch reaction   a carefully measured and controlled process in which raw materials (reactants) are added together to create a reaction that makes a single quantity (batch) of the final product.

Baume gravity   the industrial manufacturing measurement standard used to measure the gravity of nonhydrocarbon materials.

Behavior   an observable action or reaction of a person under certain circumstances.

Bernoulli's law   a physics principle stating that as the speed of a fluid in a constricted space increases, the pressure inside the fluid, or exerted by it, decreases.

Bin/hopper   a vessel that typically holds dry solids.

Biological hazard   any danger that comes from a living, or once living, organism such as viruses, mosquitoes, or snakes that can cause a health problem.

Biologicals   products (e.g., vaccines) derived from living organisms that detect, stimulate, or enhance immunity to infection.

Blanketing   the process of putting an inert gas, usually nitrogen, into the vapor space above the liquid in a tank to prevent air leakage into the tank.

Block flow diagram (BFD)   a very simple drawing that shows a general overview of a process, indicating the parts of a process and their relationships.

Blowdown   the process of taking basin water out of a cooling tower to reduce the concentration level of impurities.

Blower   a mechanical device, either centrifugal or positive displacement in design, that has a lower ratio of pressure change from suction to discharge and is not considered to be a compressor. The resistance to flow is downstream of the blower; also can be termed a *fan*.

Boiler   a closed pressure vessel in which water is boiled and converted into steam under controlled conditions.

Boiler feed water   deionized water that is sent to a boiler to produce steam.

Boiling point   the temperature at which liquid physically changes to a gas at a given pressure.

Boot   a section in the lowest area of a process drum where water or other liquid is collected and removed.

Bottoms   liquid materials collected at the bottom of a distillation column.

Boyle's law   a physics principle stating that at a constant temperature, as the pressure of a gas increases, the volume of the gas decreases.

Braze   to join two pieces of metal together by melting an alloy (filler) that bonds the metals together; a brazed joint can be stronger than the metals but is not stronger than a welded joint.

Breech   area below the stack where flue gases are collected past the last convection tubes.

British thermal unit (BTU)   the amount of heat energy required to raise the temperature of 1 pound of water 1 degree Fahrenheit.

Buoyancy   the principle that a solid object will float if its density is less than the fluid in which it is suspended; the upward force exerted by the fluid on the submerged or floating solid is equal to the weight of the fluid displaced by the solid object.

Burner   a device with open flame used to provide heat to the firebox; there may be one or many depending on the furnace's size and capacity.

Butt weld   a type of weld used to connect two pipes of the same diameter that are butted against each other.

Butterfly valve   a flow-regulating device that uses a disc-shaped flow-control element to increase or decrease flow; requires a quarter turn to go from fully open to fully closed.

CAD (or CADD)   computer aided design (and drafting), a software technology that replaces manual drafting with an automated process for design and design documentation.

Calorie   the amount of heat energy required to raise the temperature of 1 gram of water by 1 degree Celsius (centigrade).

Casing   the housing that contains the internal components of a turbine. The casing component of a steam turbine holds all moving parts, including the rotor, bearings, and seals.

Catalyst   a substance used to facilitate the rate of a chemical reaction without being consumed in the reaction.

Catalytic cracking   the process of adding heat plus a catalyst to facilitate a chemical reaction.

Cause and effect diagrams   graphics that show the relationship between an event or incident (effect) and the potential causes that created it; sometimes called a fishbone diagram.

Caustic   capable of destroying or eating away human tissue or other materials by chemical action; also a process industries term that refers to a strong base.

Cavitation   a condition inside a pump in which vapor bubbles develop in the liquid being pumped and then collapse, creating vibration and loss of flow.

Cellulose   the principal component of the cell walls in plants.

Centrifugal compressor   a dynamic compressor in which the spinning impeller creates outward (centrifugal) force, increasing the velocity of the gas flowing from the inlet to the outer tip of the impeller blade; it converts velocity to pressure.

Centrifugal force   an apparent force exerted on the fluid by the impeller rotating in a circular pattern to move it outward from the center.

Centrifugal pump   a type of pump classified in the dynamic category that uses an impeller on a rotating shaft to generate centrifugal force that is converted to pressure in the volute at the discharge to move liquids outward.

Chain reaction   a series of occurrences or responses in which each reaction is initiated by the energy produced in the preceding one.

Channel head   (also called the *exchanger head*); a device located at the end of a heat exchanger that directs the flow of the fluids into and out of the tubes.

Charles' law   a physics principle stating that, at constant pressure, the volume of a gas increases as the temperature of the gas increases.

Check valve   a type of valve that allows flow in only one direction and is used to prevent reversal of flow in a pipe.

Chemical change   a reaction in which the molecular bonds between atoms of a substance are altered and a new substance is produced.

Chemical formula   a shorthand symbolic expression that represents the elements in a substance and the number of atoms present in each molecule (e.g., water, H_2O, is two hydrogen atoms and one oxygen atom bonded together).

Chemical hazard   any danger or risk that comes from a solid, liquid, or gas element, compound, or mixture that could cause health problems or pollution.

Chemical properties   characteristics of elements or compounds that are associated with chemical reactions.

Chemical reaction   a chemical change or rearrangement of chemical bonds to form a new product.

Chemical symbol   one- or two-letter abbreviations for elements in the periodic table.

Chemical treatment   chemicals added to cooling water that are used to control algae, sludge, scale buildup, and fouling of exchangers and cooling equipment.

Chemical   a substance with a distinct composition that is used in, or produced by, a chemical process.

Chemistry   the science that describes matter, its chemical and physical properties, the chemical and physical changes it undergoes, and the energy changes that accompany those processes.

Chiller   a device used to cool a fluid to a temperature below ambient temperature; chillers generally use a refrigerant as a coolant.

Circuit   a system of one or many electrical components connected together to accomplish a specified purpose.

Closed circuit cooling tower   a cooling tower that uses a closed coil over which water is sprayed and a fan provides air flow, and in which there is no direct contact between the water being cooled and the water and air used to cool it.

Cogeneration station   a utility plant that produces both electricity and steam that can be used for heating and cooling.

Column   also called *tower*, the distillation component where separation of materials takes place.

Commodity chemicals   basic chemicals that are typically produced in large quantities and in large facilities. Most of these chemicals are inexpensive and are used as intermediates.

Compound   a pure and homogeneous substance that contains atoms of different elements in definite proportions and that usually has properties unlike those of its individual elements.

Compounding   mixing two or more substances or ingredients to achieve a desired physical form.

Compression ratio the ratio of discharge pressure (psia) to inlet pressure (psia).

Compressor mechanical device used to increase pressure of a gas or vapor.

Condensate liquid resulting from cooled or condensed water vapor; frequently refers to condensed steam.

Condenser a heat exchanger that is used to change the phase of a material by condensing vapor to liquid.

Condensing turbine a turbine that exhausts its discharge into a vacuum (pressure less than atmospheric) in order to gain maximum energy transfer.

Conduction the transfer of heat through matter via vibrational motion; exchange media must be touching.

Conductor a material that has electrons that can break free and flow more easily than other materials.

Containment wall an earthen berm or constructed wall used to protect the environment and people against tank failures, fires, runoff, and spills; also called *bund wall*, *bunding*, *dike*, or *firewall*.

Continuous reaction a chemical process in which raw materials (reactants) are continuously being fed in and products are continuously being formed and removed from the reactor vessel.

Control charts documents that establish operating limits for the amount of variation in a process.

Control loop the series of instruments that work together to monitor and control a process variable.

Control valve the final component of a control loop, operated either remotely or by hand, that receives its signal from the controller and manipulates the process variable in response to a system disturbance or change.

Controller the component of a control loop that receives information from the transmitter in the form of a pneumatic or electronic signal, compares that signal to a prescribed setting, and sends a correcting output signal to the final control element in order to maintain the process at that set point; device that monitors the speed in RPMs of the turbine; also called a *governor*.

Convection the transfer of heat through the circulation or movement of a liquid or a gas.

Convection section the upper and cooler section of a furnace where heat is transferred and where convection tubes absorb heat, primarily through the method of convection; the part of the furnace where process feed and/or steam enters.

Convection tubes furnace tubes located above the shock bank tubes that receive heat through convection.

Conversion a process that changes the size and/or structure of the petroleum components by breaking them down, combining them, or rearranging them.

Cooking and frying using heat (typically hot oil) to process foods such as chicken strips and potato chips.

Cooler a heat exchanger that uses a cooling medium to lower the temperature of a process material.

Cooling tower a structure designed to lower the temperature of a cooling water stream by evaporating part of the stream (latent heat of evaporation); these towers are usually made of fiber-reinforced plastic (FRP), galvanized metal, or stainless steel and are designed to promote maximum contact between water and air.

Cooling water water that is sent from a cooling tower to unit heat exchangers and other pieces of process equipment to cool process fluids, and that is then returned to the cooling tower for cooling and reuse.

Criticism a serious examination and judgment of something; criticism can be positive (constructive) or negative (destructive).

Cutaway drawing that shows internal elements and structures.

Cybersecurity measures intended to protect information and information technology from unauthorized access or use.

Cylinder a vessel that can hold extremely volatile or high-pressure materials.

Dalton's law a physics principle stating that the total pressure of a mixture of gases is equal to the sum of the individual partial pressures.

Damper a valve, movable plate, or adjustable louver used to regulate the flow of air or draft in a furnace.

Density the ratio of an object's mass to its volume.

Deposit a natural accumulation of ore.

Desalination the removal of salts and minerals from a target substance such as saltwater.

DESCC conflict resolution model a model for resolving conflict, comprising the following steps: describe, express, specify, contract, and consequences.

Desiccant a specialized substance contained in a dryer that removes hydrates (moisture) from the process stream.

Desuperheater a system that controls the temperature of steam leaving a boiler by using water injection through a control valve.

Diaphragm valve a flow-regulating device that uses a flexible, chemical resistant, rubber-type diaphragm attached to a stem that closes onto a weir located in the valve.

Direct current (DC) an electrical current that travels in the same direction through a conductor.

Discharge in process technology, a term that refers to the outlet side of a pump, compressor, fan, or jet.

Disinfecting destroying disease-causing organisms (e.g., through washing, irradiation, or ultraviolet exposure).

Disinfection a process of killing pathogenic organisms.

Dissolved solids   solids that are held in suspension indefinitely.

Distillation   the process (also called *fractionation*) of refining a compound such as petroleum into separate components. This process uses the different boiling points of liquids to separate components (called *fractions*); the separation of the components of a liquid mixture by their boiling points, completed by partial vaporization of the mixture and separate recovery of vapor and residue; a method for separating a liquid mixture using the different boiling points of each component.

Distributed control system (DCS)   a computer-based system consisting of a series of controllers used to monitor and control a process.

Diversity   the presence of a wide range of variation in qualities or attributes; in the workplace, it also can refer to antidiscrimination training.

DOT   U.S. Department of Transportation; a U.S. government agency with a mission of developing and coordinating policies to provide an efficient and economical national transportation system, taking into account need, the environment, and national defense.

Downcomers   tubes exposed to the firebox that transfer water from the steam drum to the mud drum.

Draft gauges   devices calibrated in inches of water that measure the firebox pressure, airflow, and differential pressure between the outside of the furnace and the flue gas inside.

Drift eliminators   devices that prevent water from being blown out of the cooling tower; the main purpose of a drift eliminator is to minimize water loss.

Drugs   substances used as medicines or narcotics.

Drum   a specialized type of storage tank or intermediary process vessel.

Dryer   a vessel containing desiccant and screens across which process streams flow to have moisture (hydrates) removed.

Drying   removing moisture from items such as rice (using a conveyor belt moving through a hot air tunnel) or sugar (dried in a turning drum).

Dynamic compressor   a category of compressor that adds kinetic energy/velocity (speed) to the gas and converts that speed to pressure at the discharge.

Dynamic pump   a category of pump that uses velocity to increase speed and converts kinetic energy to pressure at the discharge.

Economizer   the section of a boiler used to preheat feed water before it enters the main boiler system.

Elasticity   an object's tendency to return to its original shape after it has been stretched or compressed.

Electrical diagram   a drawing that shows electrical components and their relationships.

Electricity   a flow of electrons from one point to another along a pathway, called a conductor.

Electromotive force (EMF)   the force that causes the movement of electrons through an electrical circuit, measured in volts.

Electrons   negatively charged particles that orbit the nucleus of an atom.

Elements   substances composed of like atoms that cannot be broken down further without changing their properties.

Elevation diagram   a drawing that represents the relationship of equipment to ground level and other structures.

Endothermic   having to do with a chemical reaction that requires the addition or absorption of energy.

Engineering controls   equipment and/or standards that use technology and engineering practices to isolate, diminish, or remove a hazard from the workplace.

Engineering flow diagram   a high-level drawing that represents the overall process, its flow, and unit equipment, and their relationships to each other; similar to a PFD.

EPA   Environmental Protection Agency; a federal agency charged with authority to make and enforce the national environmental policy.

Equilibrium   a point in a chemical reaction at which the rate of the products forming from reactants is equal to the rate of reactants forming from the products.

Equipment location diagram   a drawing that shows the relationship of units and equipment to a facility's boundaries.

Ergonomic hazard   any danger or risk that can create physical and psychological stresses because of forceful or repetitive work, improper work techniques, or poorly designed tools and work spaces.

Ethnocentrism   belief in the superiority of one's own ethnic group; belief that others should believe and interpret things exactly the way you do.

Evaporating   removing moisture from items such as milk and coffee (to create powdered milk or coffee).

Evaporation   a process in which a liquid is changed into a gas through the latent heat process.

Evaporator   a heat exchanger in a refrigeration system that uses pressurized refrigerant to remove heat.

Exothermic   relating to a chemical reaction that releases energy in the form of heat.

Expansion valve   a valve that reduces the pressure of a liquid refrigerant, causing it to vaporize and produce a cooling effect.

Exploration   the process of locating oil and gas reservoirs by conducting surveys and studies, and drilling wells.

Facility   also called a plant; a place that is built or installed to serve a specific purpose.

Fan a device used to force or draw air; see *blower*.

Feed tray the tray located immediately below the feed line in a distillation tower; this is where liquid enters the column.

Feedback evaluative or corrective information provided to the originating source about a task or a process.

Feedstock a raw material (such as propane or ethane) or an intermediate component (such as plastic) that is used to create a product.

Fiber a long, thin filament, either plant-based or manmade, resembling a thread.

Fill the material that breaks water into smaller droplets as it falls inside the cooling tower.

Filter a device used to remove liquid, gas, or solid particulates from the process stream.

Filtration the process of removing particles from water, or some other fluid, by passing it through porous media.

Fin fan an air-cooled heat exchanger in which air moves across wide fins containing fluid to release heat.

Fire triangle/fire tetrahedron the elements of fuel, oxygen, and heat that are required for a fire to start and sustain itself; a fire tetrahedron adds a fourth element: a chemical chain reaction.

Fire water pressurized water that is used to extinguish fires.

Firebox the combustion area and hottest section of the furnace where burners are located and radiant heat transfer occurs.

Firewall an earthen bank or concrete wall built around oil storage tanks to contain the oil in case of a spill or rupture. Also called a *bund*.

Fishbone diagrams cause and effect diagrams (sometimes called Ishikawa diagrams, after Kaoru Ishikawa); used to help identify possible causes of an event or incident.

Fission the process of splitting the nucleus, the positively charged central part of an atom, which results in the release of large amounts of energy.

Fitting a piping system component used to connect two or more pieces of pipe together.

Fixed bed reactor a reactor vessel in which the catalyst bed is stationary and the reactants are passed over it; in this type of reactor, the catalyst occupies a fixed position and is not designed to leave the reactor.

Fixed blades blades inside a steam turbine that remain stationary when steam is applied.

Flame impingement a condition in which the flames from a burner touch the tubes in the furnace.

Flange a type of pipe connection, glued, welded, or threaded, consisting of two matching circular plates that are joined together with bolts.

Flare header a pipe system that connects several area/unit vents to the flare stack.

Flare stack the riser section or body of the flare system that supports the steam ring and pilot burner.

Flare system the connected system of individual lines, pressure control valves, and relief/safety valves used in both overpressure and purging situations to eliminate volatile process gases by combustion before they are released into the atmosphere.

Flash zone the section located between the rectifying and stripping sections in a distillation column.

Floating roof a type of vessel covering (steel or plastic), used on storage tanks, that floats upon the surface of the stored liquid and is used to decrease vapor space and reduce potential for evaporation.

Flow the movement of fluids.

Flow rate a quantity of fluid that moves past a specific point within a given amount of time, usually expressed in volume units per unit of time, such as gallons per minute (GPM) or cubic feet per minute (CFM).

Flowcharts documents that represent a sequence of operations schematically (or visually).

Fluidized bed reactor a reactor that uses high-velocity fluid to suspend or fluidize solid catalyst particles.

Fluidized bed system a system that uses a motive gas to move a solid like it was a fluid.

Fluid substances, usually liquids or vapors, that can be made to flow.

Foam chamber a reservoir and piping installed on liquid storage vessels and containing fire-extinguishing chemical foam.

Forced draft furnace a furnace that uses fans or blowers located at the base of the furnace to force the air required for combustion into the burner's air registers.

Forced draft tower a cooling tower that has fans or blowers at the bottom of the tower; the fans or blowers force air through the tower components.

Force energy that causes a change in the motion of an object, involving strength or direction of push or pull.

Formulating and blending a process that combines or mixes components and additives to produce finished products with specific performance requirements; also called *blending*.

Fourdrinier machine a papermaking machine, developed by Henry and Sealy Fourdrinier, which produces a continuous web of paper.

Fracking a process using pressurized liquid to create cracks in rock formations to release oil and natural gas (called *hydraulic fracturing*).

Fraction one component of the distillate that has a particular boiling point that is different from the boiling point of the other fractions in the column.

Friction   the resistance encountered when one material slides against another.

Fuel gas valve   a valve that controls the flow of fuel gas entering the furnace burners.

Fuel gas   gas used as fuel in boilers and other types of furnaces.

Fuel   any material that burns; can be a solid, liquid, or gas.

Furnace   a type of process equipment in which heat is liberated and transferred directly or indirectly to a fluid mass for the purpose of increasing the temperature of a process fluid.

Furnace purge   a method of using steam or air to remove combustibles from a furnace firebox prior to lighting the burner and startup.

Gas turbine   a device that consists of an air compressor, combustion chamber, and turbine. Hot gases produced in the combustion chamber are directed toward the turbine blades, causing the rotor to move. The rotation of the connecting shaft can be used to operate other equipment.

Gases   substances with a definite mass but no definite shape, whose molecules move freely in any direction and completely fill any container they occupy, and which can be compressed to fit into a smaller container (vapor).

Gasket   a flexible material placed between the two surfaces of a flange to seal against leaks.

Gate valve   a positive shutoff valve using a gate or guillotine that, when moved between two seats, causes tight shutoff.

Gauge hatch   an opening on the roof of a tank that is used to check tank levels and obtain samples of the tank contents.

Gauge pressure (psig)   pressure measured with respect to Earth's surface at sea level (zero psig).

General (or Combined) Gas Law   relationships among pressure, volume, and temperature in a closed container; pressure and temperature must be in absolute scale.

Generator   a device that converts mechanical energy into electrical energy.

Geology   the study of Earth and its history as recorded in rocks.

Geothermal   a power generation source that uses steam produced by the Earth to generate electricity.

Globe valve   a type of valve that uses a plug and seat to regulate the flow of fluid through the valve body, which is shaped like a sphere or globe.

Governor   a device used to control the speed of a piece of equipment such as a turbine.

Ground fault circuit interrupter (GFCI)   a safety device that detects the flow of current to ground and opens the circuit to interrupt the flow.

Grounding   the process of using the earth as a return conductor in a circuit.

Hazardous agent   the substance, method, or action by which damage or destruction can happen to personnel, equipment, or the environment.

HDPE   high-density polyethylene; a plastic material used to create water pipes and drains.

Heat exchanger   a device used to move heat from one fluid to another without direct physical contact of the fluids.

Heat tracing   a coil of heated wire or tubing that adheres to or is wrapped around a pipe and piping components in order to increase the temperature of the process fluid, reduce fluid viscosity, and facilitate flow.

Heat tracing   a coil of heated wire or tubing that is adhered to or wrapped around a pipe in order to increase the temperature of the process fluid, reduce fluid viscosity, and facilitate flow.

Heat   added energy that causes an increase in the temperature of a material (sensible heat) or a phase change (latent heat); the energy required by the fuel source to generate enough vapors for the fuel to ignite.

Heat   the transfer of energy from one object to another as a result of a temperature difference between the two objects.

Heavy ends   materials in a distillation column that boil at the highest temperature, found at the bottom of a distillation column.

Heterogeneous   having matter with properties that are not the same throughout.

Histograms   bar graphs of a frequency distribution in which the widths of the bars are proportional to the classes into which the variable has been divided, and the heights of the bars are proportional to the class frequencies.

Homogeneous   relating to matter that is evenly distributed or consisting of similar parts or elements.

Hot oil system   a system that utilizes an organic or synthetic oil product as a heat transfer fluid, used in applications to prevent freezing of heavy oil products.

Hot spots   areas within a radiant section furnace tube, or areas within a furnace firebox, that have been overheated due to flame impingement or some other cause.

Hunting   a term used to describe the condition when a turbine's speed fluctuates while the governor or controller is searching for the correct operating speed.

Hydraulic instruments   devices that use pressurized fluid (hydraulic fluid) as the power source.

Hydraulic turbine   a turbine that is moved, operated, or affected by a fluid usually water.

Hydrocarbon   organic compounds that contain only carbon and hydrogen; most often found occurring in petroleum, natural gas, and coal.

Hydrocarbons   compounds that contain only carbon and hydrogen.

Hydroelectric   a power generation source that uses flowing water to generate electricity.

Hydrometer   an instrument designed to measure the specific gravity of a liquid.

Immiscible   having to do with liquids that do not form a homogeneous mixture when put together.

Indicator   an instrument that shows the current condition of a process variable, usually through a visual representation such as a needle on a scale in a gauge or a digital readout.

Induced draft furnace   a furnace that uses fans, located in the stack, to pull air up through the furnace and induce airflow by creating a lower pressure in the firebox.

Induced draft tower   a cooling tower that has fans at the top of the tower; these fans pull air through the tower components.

Information technology   the equipment, tools, processes, and methodologies (coding/programming, data storage and retrieval, systems analysis, systems design, and so on) that are used to collect, process, and present information.

Inhibitors   substances that slow or stop a chemical reaction.

Inlet   the point of entry for the fluid.

Inorganic chemistry   the study of substances that do not contain carbon.

Insoluble   describing a substance that does not dissolve in a solvent.

Instrument air   air that has been filtered and dried for use by instrument signal lines and valve actuators.

Instrumentation   the set of devices used to measure, control, or indicate flow, temperature, level, pressure, and analytical data.

Insulator   a material that does not give up its electrons as easily; an insulator is a poor conductor of electricity.

Interchanger   (also called a cross exchanger or intercooler) one of the process-to-process heat exchangers.

Intermediates   substances that are not made to be used directly, but are used to produce other useful compounds.

Intrinsically safe   condition in which devices such as motors are designed so as not to ignite a spark that could create a fire; a requirement for all equipment operating in areas where flammable gases and liquids are present.

ISO 14000   an international standard that addresses how to incorporate environmental aspects into operations and product standards.

ISO 9000   an international standard that provides a framework for quality management by addressing the processes of producing and delivering products and services.

Isometric diagrams   drawings that show objects as they would be seen by the viewer (like a 3-D drawing, the object has depth and height). Isometrics are drawn on graph paper with equipment shown in relation to compass points and line relationships to the equipment.

ISO   taken from the Greek word *isos*, which means equal, ISO is the International Standardization Organization, which consists of a network of national standards institutes from more than 140 countries.

Kinetic energy   energy associated with mass in motion.

Knockout drum   a vessel located between the flare header and the flare stack or between stages of a centrifugal compressor, where liquid drops out and is separated from the vapors being removed.

Laminar flow   a condition in which fluid flow is smooth and unbroken; viewed as a series of laminations or thin cylinders of fluid slipping past one another inside a tube.

Latent heat of condensation   the amount of heat energy given off when a vapor is converted to a liquid without a change in temperature.

Latent heat of fusion   the amount of heat energy required to change a solid to a liquid without a change in temperature.

Latent heat of vaporization   the amount of heat energy required to change a liquid to a vapor without a change in temperature.

Latent heat   heat that does not result in a temperature change but causes a phase change.

Legend   a section of a drawing that explains or defines the information or symbols contained within the drawing (like a legend on a map).

Level   a position of height or depth along a vertical axis; in the process industries, level usually means the height of the surface of a material compared to a zero reference point.

Leverage   an assisted advantage, usually gained through the use of a tool (such as lever and fulcrum).

Light ends   also called *distillate, overhead,* or *overhead takeoff,* materials in a distillation column that boil at the lowest temperature and that come off the top of a fractionation tower.

Liquids   substances with a definite volume but no fixed shape that demonstrate a readiness to flow with little or no tendency to disperse, and are limited in the amount in which they can be compressed.

LNG   liquefied natural gas; a clear, colorless, nontoxic liquid formed when natural gas is purified and cooled to −162°C (−260°F).

Local   related to instrumentation in close physical proximity to the process, generally found outside the control room.

Loop diagram   a drawing that shows all components and connections between instrumentation and a control room.

Lubricant a substance used to reduce friction between two contact surfaces.

Lubrication system a system that supplies oil or grease to rotating equipment parts, such as bearings, gearboxes, sleeve bearings, and seals on operating machinery.

Lubrication a friction-reducing film placed between moving surfaces in order to reduce drag and wear.

Makeup water water brought in from an external source such as a lake or river that is used to replace the water lost during the blowdown and/or evaporation process.

Manufacturing making a product from raw materials by hand or with machinery (e.g., cooking, decorating, grinding, milling, and mixing).

Manway an opening in a vessel that permits entry for inspection and repair.

Mass flow rate amount of mass passing through a plane per unit of time.

Mass the amount of matter in a body or object measured by its resistance to a change in motion.

Matter anything that takes up space and has inertia and mass.

Mechanical energy energy of motion that is used to perform work.

Metal chemical elements that have luster (ability to reflect light) and can conduct heat and electricity (e.g., copper, bauxite, iron, lead, gold, silver, zinc, nickel, and uranium).

Mill a facility where a raw substance is processed and refined to another form.

Mine a pit or excavation from which minerals are extracted.

Minerals and gems naturally occurring inorganic substances that have a definite chemical composition and a characteristic crystalline structure.

Mining the extraction of valuable minerals or other geological materials from the earth.

Miscible having to do with liquids that form a homogeneous mixture when put together.

Mist eliminator a device in the top of a tank, composed of mesh, vanes, or fibers, that collects droplets of mist (moisture) from gas to prevent it from leaving the tank and moving forward with the process flow.

Mixer a device used to mechanically combine chemicals or other substances; also known as an agitator.

Mixing stirring materials to blend them (such as instant drink mixes).

Mixture two or more substances that are combined together but do not react chemically.

Molecule a set of two or more atoms held together by chemical bonds.

Motor a mechanical driver with rotational output; usually electrically operated.

Moving blades rotor blades inside a steam turbine that move or rotate when steam is applied.

Mud drum (1) the lower drum of a boiler that is used as a junction area for boiler tubes. (2) A low place in a boiler where heavy particles in the water will settle out so that they can be blown down.

Natural draft furnace a furnace that has no mechanical draft or fans; instead, the heat in the furnace causes draft.

Natural draft tower a cooling tower that uses temperature differences inside and outside the stack to facilitate air movement.

Net positive suction head (NPSH) pressure the amount of pressure needed at the suction of a pump to prevent cavitation.

Neutrons particles without electrical charge, found in the nucleus of an atom.

Nitrogen header system a system used to purge hydrocarbons or oxygen so that they do not mix.

Nitrogen an inert (nonreactive) gas that is used to purge, or remove, explosive gases and air from process systems and equipment; also used as a backup to the instrument air supply.

Noncondensing turbine a turbine that exhausts its discharge into a lower steam pressure that can be used elsewhere in the facility.

Nonmetals substances that conduct heat and electricity poorly; are brittle, waxy, or gaseous; cannot be hammered into sheets or drawn into wire (e.g., gems and precious stones, coal, gravel, sand, lime, stone, soda ash, phosphate rock, and clay).

Nozzle the device in the turbine entry piping that converts pressure to velocity and directs the flow of steam to the turbine blades. There can be several nozzles in the turbine, depending on design and desired speed.

NRC Nuclear Regulatory Commission; a U.S. government agency that protects public health and safety through regulation of nuclear power and the civilian use of nuclear materials.

Nuclear a power generation source that uses the heat from splitting atoms to generate electricity.

Ohm a measurement of resistance in electrical circuits; symbolized by the Greek letter omega: Ω.

Ohm term used to describe electrical resistance.

Open circuit cooling tower a cooling tower design that cools cooling water by evaporation and has the cooling water exposed to air.

Ore a metal-bearing mineral that is valuable enough to warrant mining (e.g., iron or gold).

Organic chemistry the study of carbon-containing compounds.

OSHA Occupational Safety and Health Administration; a U.S. government agency created to establish and enforce workplace safety and health standards, conduct workplace inspections, propose penalties for noncompliance, and investigate serious workplace incidents.

Outlet   the point of exit for the fluid.

Overall equipment effectiveness (OEE)   a concept that assesses availability (time), performance (speed), and quality (yield) to evaluate manufacturing productivity.

Overhead   the product or stream that comes off the top of a fractionation tower.

Packed tower   a distillation column that is filled with specialized packing material instead of trays.

Packing   material used in a distillation column to maximize the contact area between the liquid and gas to effect separating.

Pareto charts   graphics that rank causes from most significant to least significant; they represent the 80-20 rule described by Juran, stating that most undesired effects come from relatively few causes.

Pareto principle   a quality principle, also called the 80-20 rule, that states 80% of problems come from 20% of the causes.

Pasteurization   heating foods to kill organisms, such as bacteria, or make them less likely to cause disease (milk and orange juice are treated this way).

Pathogen   a disease-causing microorganism.

Periodic table   a chart of all known elements listed in order of increasing atomic number and grouped by similar characteristics.

Personal protective equipment (PPE)   specialized gear that provides a barrier between hazards and the body and its extremities.

Petrochemical   a chemical derived from fossil fuels or petroleum products.

Petroleum   a substance found in the earth, such as oil or gas, composed of chemical compounds consisting primarily of hydrogen and carbon.

pH   a measure of the quantity of hydrogen ions in a solution that can react and indicate whether a substance is an acid or a base.

Pharmaceuticals   manmade or naturally derived chemical substances with medicinal properties that can be used to treat diseases, disorders, and illnesses.

Phase change   transition of a substance from one physical state to another, such as when ice melts to form water.

Physical change   an event in which the physical properties of a substance (e.g., how it looks, smells, or feels) can be altered, but the change is reversible and a new substance is not produced.

Physical hazard   any danger or risk that comes from environmental factors such as excessive levels of noise, temperature, pressure, vibration, radiation, electricity, or mechanical hazards (Note: OSHA has its own definition of physical hazard that relates specifically to chemicals).

Physical properties   aspects of an element or compound that are observable and do not pertain to chemical reactions.

Physical security   measures intended to protect specific assets such as production facilities, pipelines, control centers, tank farms, and other vital areas.

Pilot   an initiating device used to ignite the burner fuel.

Piping and instrumentation diagram (P&ID)   also called a *process and instrumentation drawing*. A drawing that shows the equipment, piping, and instrumentation of a process in the facility, along with more complex details than a process flow diagram.

Plant air   also called *service air* or *utility air*, the supplied air system used for pneumatic tools, air hoists, and other air-operated equipment.

Plasma   a gas that contains positive and negative ions.

Plate tectonics   a theory that explains how the planet moves, as semi-rigid plates in the Earth's crust, drift or flow, causing geological changes such as volcanoes, mountains, and earthquakes.

Plot plan diagram   illustration drawn to scale, showing the layout and dimensions of equipment, units, and buildings; also called an *equipment location drawing*.

Plug valve   a flow-regulating device that uses a flow-control element shaped like a hollowed-out plug, attached to an external handle, to increase or decrease flow; requires a quarter turn to go from fully open to fully closed.

Pneumatic instruments   devices that use air pressure or inert gas as their power source.

Positive displacement pump   a category of pump that uses pistons, vanes, screws, or lobes to move a specific amount of liquid through a system at a given pump speed.

Positive-displacement compressor   a compressor that uses screws, pistons, sliding vanes, lobes, gears, or diaphragms to trap a specific volume of gas, and reduce its volume, thereby increasing pressure at the discharge.

Potable water   water designated as ingestible for human consumption or food preparation.

Potable water   water that is safe for human consumption.

Potential energy   the energy of a body as a result of its position or condition.

PPM   predictive/preventive maintenance, a program to identify potential issues with equipment and use preventive maintenance before the equipment fails.

Preheater   a heat exchanger used to warm liquids before they enter a distillation tower or other part of the process.

Prejudice   attitude toward a group or its individual members based on stereotyped beliefs.

Preserving preparing foods for long-term storage (e.g., canning, drying, freezing, salting).

Pressure the amount of force a substance or object exerts over a particular area.

Pressurized tank an enclosed vessel in which a greater-than-atmospheric pressure is maintained.

Priming the process of filling the suction line and pump casing with liquid to remove vapors, and to heat or chill the case metal to its service temperature.

Process a system of people, methods, equipment, and structures that creates products from other materials; method for doing something, generally involving tasks, steps, or operations that are ordered and/or interdependent.

Process flow diagram (PFD) a basic drawing that shows the primary flow of product through a process, using equipment, piping, and flow direction arrows.

Process industries a broad term for industries that convert raw materials, using a series of actions or operations, into products for consumers.

Process technician a worker in a process facility who monitors and controls mechanical, physical, and/or chemical changes, throughout many processes, to produce either a final product or an intermediate product, made from raw materials.

Process technology processes that take quantities of raw materials and transform them into other products.

Process variable a measurement of the chemical or physical properties of a process stream, generally relating to operating parameters or product specifications.

Process water water used in industrial processes that may contain inorganic and organic compounds not suitable for release to the environment; water that is used in any level of the manufacturing process and is not suitable for consumption.

Process water sewer water collection system of drains surrounding and under process equipment, which is directed to a wastewater treatment plant.

Product also called output; the desired end components from a particular process.

Production output, such as material made in a plant, oil from a well, or chemicals from a processing plant.

Products substances that are produced during a chemical reaction.

Protons positively charged particles found in the nucleus of an atom.

Pulp a cellulose fiber material, created by mechanical and/or chemical means from various materials (e.g., wood, cotton, recycled paper), from which paper and paperboard products are manufactured.

PVC polyvinyl chloride; a plastic-type material that can be used to create cold water pipes and drains and other low-pressure applications.

Quarry an open excavation from which stones are extracted.

Radiant section the lower portion of a furnace (firebox) where heat transfer occurs, primarily through radiation.

Radiant tubes tubes located in the firebox that receive heat primarily through radiant heat transfer.

Radiation the transfer of heat energy through electromagnetic waves.

Raw materials also called feedstock or input. The material sent to a processing unit to be converted into a different material or materials.

Reactants the starting substances in a chemical reaction.

Reaction furnace a reactor that combines a firebox with tubing to provide heat for a reaction that occurs inside the tubes.

Reactor a vessel in which chemical reactions are initiated and sustained.

Reboiler a tubular heat exchanger, placed at the bottom of a distillation column or stripper, used to supply the necessary column heat.

Receiver a tank that stores the liquid phase refrigerant once it leaves the condenser.

Reciprocating compressor a positive-displacement compressor that uses the inward stroke of a piston or diaphragm to draw gas into a chamber (intake) and then positively displace the gas using an outward stroke (discharge).

Reciprocating pump a type of positive displacement pump that uses the inward stroke of a piston or diaphragm to draw a specific amount of liquid into a chamber (intake) and then positively displace that volume of liquid using an outward stroke (discharge).

Recorder an instrument, either mechanical or electronic, used to document variables such as pressure, speed, flow, temperature, level, and electrical units; it also provides a history of the process over a given period of time, which is called a trend.

Rectifying (fractionating) section also called the enriching section, the section of a fractionating tower between the feed tray and the top of the tower where overhead product is purified.

Refining the process of purifying a substance being separated into other products, such as petroleum into gasoline, kerosene, gas, and oils.

Reflux cooled and condensed overhead vapor that is returned as a liquid to the tower to provide cooling and to help with efficient column operation.

Refractory referring to a type of insulation used inside reactors, heaters, reactors, high temperature boilers, and furnaces, commonly firebrick.

Refractory lining a form of insulation used inside high temperature boilers, incinerators, heaters, reactors, and furnaces; a common type is called fire brick.

Refrigerant fluid with a low boiling point that is circulated throughout a refrigeration system.

Relief valve a safety device designed to open slowly as the pressure of a fluid in a closed vessel exceeds a preset level; can be used in services where liquid expands to create a gas.

Remote relating to instrumentation located away from the process (e.g., in a control room) and is controlled by the DCS or other panel board mounted controller.

Resistance the force that opposes the push of electrons, measured in ohms.

Reverse osmosis method for processing water by forcing it through a membrane through which salts and impurities cannot pass (purified bottled water is produced this way).

Risers tubes that go into the side or top of the steam drum, allowing heated water and steam to flow up from the mud drum.

Roasting and toasting using heat (from an oven) to process types of food such as cereal flakes and coffee.

Rock cycle the process by which the different types of rocks are formed, exposed to weather and other forces that erode or break them down, and then are reformed by geological forces (such as heat and pressure).

Rock formations geology that is arranged or formed in a certain way; also, scenic outcroppings formed by weathering and erosion.

Rocks minerals or gems, or a combination of minerals or gems mixed with other chemical compounds.

Rotary compressor a positive-displacement compressor that uses a rotating motion and internal elements of either screw, sliding vane, or lobe to pressurize and move the gas through the device.

Rotary pump a type of positive displacement pump that moves fluids by trapping liquid in a specific area of a rotating screw or a set of lobes, gears, or vanes, then displacing the volume of liquid at the discharge.

Rotor the rotating member of a motor that is connected to the shaft.

Safety Data Sheet or MSDS (SDS) (formerly Material Safety Data Sheet) a document that provides key safety, health, and environmental information about a material.

Safety valve a device designed to open quickly as the pressure of a fluid in a closed vessel exceeds a preset level; typically used in gas service.

Scatter plots graphs using dots or a similar symbol to represent data.

Schematic a drawing that shows the direction of current flow in a circuit, beginning at the power source.

Self-managed team also called *self-directed team*; a small group of employees whose members determine, organize, plan, and manage their day-to-day activities and duties under reduced or no supervision.

Semiconductor a material that is neither a conductor nor an insulator.

Sensible heat heat transfer that results in a temperature change.

Sensor (also known as a *primary element*) the control loop component that is in direct contact with the process variable to be measured that senses or detects change and provides a signal to the transmitter.

Service (raw) water general-purpose water that may or may not have been treated.

Set point the desired RPM indication where a controller is set for optimum operation.

Settleable solids solids in wastewater that can be removed by slowing the flow in a large basin or tank.

Shaft a metal rod on which the rotor resides; suspended on each end by bearings; connects the driver to the shaft of the driven end or mechanical device.

Shell the outer housing of a heat exchanger that covers the tube bundle.

Shell inlet and outlet openings that allow process fluids to flow into and out of the shell side of a shell and tube heat exchanger.

Shock bank (shock tubes) also called the *shield section*, a row of tubes located directly above the firebox in a furnace that receives both radiant and convective heat; it protects the convection section from exposure to the radiant heat of the firebox.

Simulation realistic three-dimensional representation.

Six Sigma an advanced quality management method that increases output by minimizing variability and defects in manufacturing processes.

Socket weld a type of weld used to connect pipes and fittings when one pipe is small enough to fit snugly inside the other.

Soil pollution the accidental or intentional discharge of any harmful substance into the soil.

Solar (CSP) Concentrated Solar Power; a power generation source that uses the power of the sun to heat water and generate electricity.

Solar (PV) Photovoltaics solar power; generation of power when light excites semiconducting materials (solar panels) to generate electricity.

Solder a metallic compound that is melted and applied in order to join together and seal the joints and fittings in tubing systems and electrical components; soldered joints are not as strong as brazed joints.

Solids substances with a definite volume and a fixed shape that are neither liquid nor gas, and that maintain their shape independent of the shape of the container.

Soluble having to do with a substance that will dissolve in a solvent.

Solute the substance that is dissolved in a solvent.

Solution a homogeneous mixture of two or more substances.

Solvent the substance that is present in a solution in the largest amount.

Sour water wastewater produced from gas processing, sulfur recovery, and refining operations; it often contains hydrogen sulfide, ammonia, hydrogen cyanide, and phenol.

Spacer rods the rods that hold the tubes in a tube bundle apart so they do not touch one another.

SPC statistical process control; uses mathematical laws dealing with probability. Companies utilize SPC to gather data (numbers) and study the characteristics of processes, then use the data to make the processes behave the way they should.

Specialty chemicals chemicals that are produced in smaller quantities, are more expensive, and are used less frequently than commodity chemicals.

Specific gravity the ratio of the density of a liquid or solid to the density of pure water, or the density of a gas to the density of air at standard temperature and pressure (STP).

Specific heat the amount of heat required to raise a unit of mass by one degree (e.g., the temperature of 1 gram of a substance 1 degree Celsius).

Spherical tank a type of pressurized storage tank that is used to store volatile or highly pressurized material; also referred to as "round" tanks.

Stack a cylindrical outlet, located at the top of a furnace, through which flue (combustion) gas is removed from the furnace.

Stator a stationary part of the motor where the alternating current flows in, and a large magnetic field is created using magnets and coiled wire.

Steam drum the top drum of a boiler where all of the generated steam gathers before entering the separating equipment.

Steam turbine a rotating device driven by the pressure of steam (potential energy), discharged at high velocity (kinetic energy) against the turbine blades; used as the motive force (mechanical energy) for equipment.

Stereotyping maintaining beliefs about individuals or groups based on opinions, habits of thinking, or rumors, which lead to generalizations about all members of a group.

Stirred tank reactor a reactor vessel that contains a mixer or agitator to improve mixing of reactants.

Stripping section the section of a distillation tower below the feed tray where the heavier components with higher boiling points are located.

Suction in process technology, a term that refers to the inlet side of a pump, compressor, fan, or jet.

Suction screens devices located on the inlet side of equipment used to filter out debris; they are most often associated with pump suction.

Sump an area of temporary storage located at the bottom of a tank from which undesirable material is removed.

Superheated steam steam that has been heated to a very high temperature so that a majority of the moisture content has been removed (also called dry steam).

Superheater a set of tubes, located near the boiler outlet, that increases (superheats) the temperature of the steam.

Suspended solids solids that cannot be removed by slowing the flow.

Symbol figures used to designate types of equipment and instrumentation.

Synergy the total effect in which a whole is greater than the sum of its individual parts.

Synthetic a substance resulting from combining components, instead of being naturally produced.

Tank a large container or vessel for holding liquids and/or gases.

Task a set of actions that accomplish a job.

Team a small group of people, with complementary skills, committed to a common set of goals and tasks.

Team dynamics interpersonal relationships; ways in which workers get along with each other and function together.

Temperature the measure of the thermal energy of a substance.

Threaded (screwed) pipe piping that is connected using male and female threads.

Throttle valve a device, or series of devices depending on turbine power, that can be opened or closed slowly (throttled) to control steam flow to the turbine.

Throttling a condition in which a valve is partially opened or partially closed in order to restrict or regulate the amount of flow.

Title block a section of a drawing (typically located in the bottom right corner) that contains information such as drawing title, drawing number, revision number, sheet number, and approval signatures.

Tower see Column.

TPM total productive maintenance; an equipment maintenance program that emphasizes a company-wide effort to involve all levels of staff in various aspects of equipment maintenance.

TQM total quality management; a collection of philosophies, concepts, methods, and tools used to manage quality; TQM consists of four parts: customer focus, continuous improvement, managing by data and facts, and employee empowerment.

Transducer the control loop component that converts one instrumentation signal type to another type (e.g., electronic to pneumatic).

Transformer a device that will raise or lower the voltage of alternating current of the original source.

Transmitter the control loop component that receives a signal from the sensor, measures that signal, and sends it forward to the controller or indicator.

Transportation the oil and gas industry segment responsible for moving petroleum from wells to processing facilities and finished products to consumers. Transportation methods include pipelines, watercraft, railways, and trucks.

Tray downcomer the component of a tower tray located below the outlet weir that directs flow to the next succeeding tray.

Trays metal plates with openings (valve, bubble cap, or sieve design), placed within a distillation column, instead of packing, to increase contact between vapors and liquid in order to facilitate separation of components.

Treated water water that has been filtered, cleaned, and chemically treated.

Treatment a process that prepares the components for additional processing and creates some final products. Treatments can remove or separate various components or contaminants at this point.

Trip valve a device that can be closed quickly (tripped) to stop the flow of steam to a turbine.

Tube bundle a group of fixed or parallel tubes, such as are used in a heat exchanger, through which process fluids are circulated; the tube bundle includes the tube sheets with the tubes, the baffles, and the spacer rods.

Tube inlet and outlet openings that allow process fluids to flow into and out of the tube bundle in a shell and tube heat exchanger.

Tube sheet a flat plate to which the tubes in a heat exchanger are fixed.

Tubular reactor a continuously flowed vessel in which reactants are converted in relation to their position within the reactor tubes, not influenced by residence time in the reactor.

Turbidity cloudiness caused by particles suspended in water or some other liquid.

Turbine a machine for producing power; activated by the expansion of a fluid (e.g., steam, gas, air, or water) on a series of curved vanes on an impeller attached to a central shaft, which is used to create rotational mechanical energy to rotate shaft-driven equipment.

Turbulent flow a condition in which the fluid flow pattern is disturbed so there is considerable mixing.

Unit an integrated group of process equipment used to produce a specific product or products.

Utility air *see* Plant air; compressed air that is used to power equipment, tools (e.g., cyclone air movers and pneumatic power tools), and dry equipment; to purge fireboxes prior to light-off; and to off-load nonflammable delivery transports.

Utility flow diagrams (UFD) drawings that show the piping and instrumentation for the utilities in a process.

Vacuum pressure (psiv) any pressure below atmospheric pressure.

Valve positioner a device that ensures a valve is positioned properly in reference to the signal received from the controller.

Valve a piping system component used to control the flow of fluids through a pipe.

Vane separator a device, composed of metal vanes, used to separate liquids from gases or solids from liquids.

Vapor pressure a measure of a substance's volatility and its tendency to form a vapor.

Vapor recovery system the process of recapturing vapors by methods such as chilling or scrubbing; vapors are then purified, and the vapors or products are sent back to the process, sent to storage, or recovered.

Variables also called continuous data; pieces of information that can be measured and plotted on a constant scale (such as flow through a pipeline or liquid in a tank).

Velocity the distance traveled over time or change in position over time.

Vessel a container in which materials are processed, treated, or stored.

Viscosity the measure of a fluid's resistance to flow.

Volt the derived unit for electrical potential, electrical potential difference (voltage), and electromotive force; the electromotive force that will establish a current of one amp through a resistance of one ohm.

Voltmeter a device used to measure voltage.

Voluntary Protection Program (VPP) an OSHA program designed to recognize and promote effective safety and health management.

Volute a widening chamber in the casing of a centrifugal pump at the discharge side, designed so that velocity is converted to pressure without shock.

Vortex breaker a metal plate, or similar device, placed inside a cylindrical, cone-shaped, or other type operating unit, which prevents a vortex from being created as liquid is drawn out of the vessel.

Vortex the cone formed by a swirling liquid or gas.

Wastewater water that contains waste products in the form of dissolved or suspended solids and that must be treated prior to its return to the natural water source.

Water distribution header a device that distributes water evenly across the top of the cooling tower through the use of distribution nozzles; the header helps create more effective cooling and allows the water to flow evenly into the tower fill.

Water pollution the introduction, into a body of water or the water table, of any EPA-listed potential pollutant that affects the chemical, physical, or biological integrity of that water.

Weight a measure of the force of gravity on an object.

Watt a unit of measure of electric power; the power consumed by a current of one ampere using an electromotive force of one volt.

Weir a flat or notched dam or barrier to liquid flow that is normally used either for the measurement of fluid flows or to maintain a given depth of fluid as on a tray of a distillation column, in a separator, or other vessel.

Weir flat or notched dam or barrier used to maintain a given depth of liquid on a tray.

Wind power a power generation source that uses flowing air currents to push against blades of giant wind turbines.

Wind turbine a mechanical device that converts wind energy into mechanical energy.

Wiring diagram a drawing that shows electrical components in their relative position in the circuit and all connections in between.

Work group a group of people organized by logical structures within a company, having a designated leader, and performing routine tasks.

Zero defects the goal of a quality practice with the objective of reducing defects, thus increasing profits.

Glossary 383

Index

Page numbers with a *t* indicate tables.

A

29 CFR (Code of Federal Regulations), 114, 115, 116, 267
Absolute pressure (psia), 160
  definition, 148
Absorb, 76, 80
Absorption systems, 341
AC (alternating current), 56, 264–65
  definition, 53
Acids, 175–76
  definition, 168
Actuate, 248, 252
Actuators, 349, 356
Adhesives, 44
Administrative controls, 107, 121
Adsorb, 76, 80
Aeolipile, 254
Aftercoolers, 273, 280
Agglomerating, 66, 69
Agricultural chemicals, 44
Airplanes, 250
Air pollution, 107, 110
Air registers, 295, 299
Air-rumble, 273, 282
Alarms, 349, 356
Alchemy, 40–42
Alkaline, 168, 176
Alloys, 194, 196
Alternating current (AC), 56, 264–65
  definition, 53
American National Standards Institute (ANSI), 122
American Petroleum Institute (API), 148, 154, 341
American Society of Mechanical Engineers (ASME), 211, 214
Amine system, 337, 343
Ammeter, 258, 261
Ampere, André-Marie, 56
Amperes (amps), 53, 260–61
  definition, 258
origin of name, 56
Analog instrument, 356
Analytical variables, 349, 354
ANSI (American National Standards Institute), 122
Antibiotics, 40, 47
API (American Petroleum Institute), 148, 154, 341
API gravity, 148, 154
Apothecary, 40, 46
Application blocks, 179, 183
Arc-rated, 258, 269
Arsenic, 33
Articulated drains, 211, 215
ASME (American Society of Mechanical Engineers), 211, 214
Assignable variation, 130, 133
ATF (Bureau of Alcohol, Tobacco and Firearms), 16, 72–74
Atmospheric pressure, 148, 159
Atmospheric tanks, 211, 214
Atomic number, 168, 171
Atomic weight, 168, 171
Atoms, 57, 171–72
  definition, 53
Attitudes, 107, 117
Attributes, 130, 138
Automobile, invention of, 24
Axial compressors, 237, 241–42
Axial pumps, 226, 231–32
Azeotropic distillation, 318

B

Babcock, George Herman, 308
Back-flush, 273, 282
Baffles, 218, 277
  definition, 211, 273
Balanced draft furnaces, 295, 301
Baldrige, Malcolm, 141–42
Ball valve, 194, 200
Base, 176
  definition, 168
Basic chemistry. *See* Chemistry
Basic physics. *See* Physics
Basins, 286, 290
Batch reaction, 211, 220
Baume gravity, 148, 154
Beer production, 69
Behavior, 107, 117
Benz, Karl, 24
Bernoulli, Daniel, 163
Bernoulli's law, 148, 163–64
Bimetallic temperature gauges, 351, 352
Binary distillation, 318
Bins, 211, 214
Biological hazards, 107, 109
Biologicals, 40, 47
Blanketing, 211, 215

Block flow diagrams (BFDs), 183, 184
  definition, 179
Blowdown, 286, 291
Blowers, 237, 242–43
Body protection, 124
Boiler feed water, 325–27
  definition, 323
Boilers, 306
  definition, 305
  fuels used in, 309
  how they work, 308–9
  key terms, 305
  monitoring and maintenance activities, 311, 311t
  operational hazards, 310–11, 311t
  overview, 306, 312
  parts, 306–8
  purpose, 306
  symbols, 311
  types
    fire tube boiler, 309–10
    water tube boiler, 310
Boiling point, 148, 160
Boots, 211, 219
Bottoms (heavy ends), 313, 316
Bourdon tube, 351
Boyle, Robert, 161
Boyle's law (pressure-volume law), 148, 161
Braze, 194, 198
Breathing air, 330, 334
Breech, 295, 299
British thermal units (BTUs), 148, 157
Bubble cap trays, 317
Buoyancy, 148, 154
Bureau of Alcohol, Tobacco and Firearms (ATF), 72–74
Burners, 295, 298
Butterfly valve, 201
  definition, 194
Butt weld, 198
  definition, 194

C

Cabin furnace, 298
CAD (computer-aided design) 179, 180 or CADD
    (computer-aided design and drafting), 179, 180
Calories, 157
  definition, 148
Casings, 248, 253
Catalysts, 168, 174
Catalytic cracking, 168, 174
Cause and effect diagrams, 130, 139
Caustic, 168, 176
Cavitation, 226, 233
Cellulose, 83, 85–86
Celsius, 156
Centers for Disease Control and Prevention (CDC), 72–73
Centrifugal compressors, 237, 241
Centrifugal force, 226, 230
Centrifugal pumps, 230–31

components, 231
definition, 226
symbols, 234
CFR (code of federal regulation), 114, 116
Chain reactions, 107, 119
Channel heads, 273, 274
Charles, Jacques Alexander Cesar, 161
Charles' law (temperature-volume law), 148, 161
Check valve, 201–2
  definition, 195
Chemical change, 173
  definition, 168
Chemical, definition of, 40
Chemical formulas, 172, 174
  definition, 168
Chemical hazards, 107, 109
Chemical industry, 4
  chemicals, 41
  growth and development
    in the beginning, 41–42
    1000s–1600s, 42
    1600s–1800s, 42
    World War I, 42
    World War II, 43
    1950–2000, 43
    2000–present, 43–44
  key terms, 40–41
  origins of word, 42
  overview, 44–45
  process technicians
    duties, 45
    expectations of, 45
    responsibilities, 45
  products produced by (examples), 41
  sectors, 44
  summary of, 50–51
Chemical properties, 168, 173
Chemical reactions, 168
Chemical symbols, 168, 171
Chemical treatment, 286, 291
Chemical versus physical properties, 173
Chemistry, 41
  acids and bases, 175–76
  applications in the process industries, 169
  chemical reactions, 174
    catalysts, 174
    chemical formulas, 174
    definition, 173–74
    endothermic versus exothermic reactions, 174
    oxidation/reduction reactions, 174
    reactants, 174
  chemical versus physical properties, 173
  definition, 168
  elements and compounds
    atoms, 171–72
    compounds, 172–73
    elements, 170, 173
    periodic table, 170–71

Index **387**

key terms, 168–69
mixtures and solutions, 175
origins of word, 42
overview, 169, 176–77
Chernobyl, 57, 111
Chillers, 273, 280
Chlorine (as water disinfection agent), 77
Circuits, 261–62
  definition, 259
Cleaning materials, 44
Clean Water Act, 328
Closed circuit cooling tower, 286, 288
Coal, 58–59
Coating, 44
Cogeneration station, 53, 58
Column section, 313–14, 316
Combined gas law, 163
  definition, 149
Commodity chemicals, definition of, 40, 41
Communication model with feedback, 101
Compounding, definition of, 40, 46
Compounds, 168, 172–73
Compressed air, 329–30, 332
  instrument air, 330
  utility air, 329
Compression ratio, 237, 242
Compressors, 237, 341
  blowers, 242–43
  definition, 238
  differences between compressors and pumps, 238
  dynamic compressors, 240
    axial compressors, 241–42
    centrifugal compressors, 241
  flowchart, 239
  key terms, 237–38
  monitoring and maintenance activities, 245, 245t
  multistage compressors, 242
  operational hazards, 243t
    leakage, 245–46
    overheating, 244
    overpressurization, 244
    surging, 244
  overview, 238, 246
  positive displacement compressors
    definition, 239
    reciprocating compressors, 240
    rotary compressors, 239–40
  symbols, 245
  types, 239
Computer aided design (and drafting) CAD (or CADD), 179, 180
Concentrated solar power (CSP), 54, 61
Condensate, 323, 329
Condensers, 273, 280
Conduction, 158, 277–78
  definition, 148
Conductors, 259, 260
Conflict, resolving
  consequences phase, 100
  contract phase, 100
  describe phase, 100
  express phase, 100
  specify phase, 100
Connecting methods, piping and valves, 196–98
  bonds, 198
  flanges, 197
  threaded (screwed), 197
  welds, 198
Containment walls, 211, 220
Continuous reaction, 211, 220
Control charts, 130, 138
Controller, 248, 255, 349, 355
Control loops, 357–58
  definition, 349
Control valves, 349, 356
Convection, 158, 277–78
  definition, 148
Convection section, 295, 299
Convection tubes, 295, 299
Conversion, 19, 29
Cookie production, 70
Cooking, 66, 69
Coolers, 286, 292
Cooling towers, 287
  applications, 292
  component parts of, 289–90
  definition, 286
  factors that impact performance, 291–92
  key terms, 286–87
  monitoring and maintenance activities, 293, 293t
  operational hazards, 292–93, 292t
  overview, 287, 294
  principles of operation, 290–92
  purpose, 287
  symbols, 293
  types, 287–88
Cooling water, 325, 327
  definition, 323
Cracking furnace, 297
Criticism, 91, 101
Crosby, Philip B., 135, 141
CSP (Concentrated solar power), 54, 61
Cutaways, 179, 190
Cybersecurity, 107, 110, 126–27
Cylinders, 211, 215

D

Daimler, Gottlieb, 24
Dalton, John, 162
Dalton's law (partial pressures), 148, 162
Dampers, 295, 299
DC (direct current), 56, 264–65
  definition, 53
DCS (Distributed control system), 349, 358–59
Deadheading, 150

Delaney proviso, 67
Deming, Dr. W. Edwards, 133–34
Density, 148, 152
Department of Transportation (DOT). *See* DOT (Department of Transportation)
Deposit, 19, 31
Desalination, 76, 78
DESCC conflict resolution model, 99
    definition, 91
Desiccant, 211, 213
Desuperheaters, 305, 308
DHS (Department of Homeland Security), 16
Diaphragm valve, 202–3
    definition, 195
Differential pressure (D/P) cell, 351
Digital instrument, 356
Dikes, 211, 220
Direct current (DC), 56, 264–65
    definition, 53
Discharge, 238
Disinfection, 66, 69, 76–77
Dissolved solids, 76, 80
Distillation, 19, 22, 70, 280–81
    definition, 66, 274, 313
    food processing systems, 71
    how the process works, 318
    key terms, 313–14
    methods, 318
    monitoring and maintenance activities, 320, 320*t*
    operational hazards, 319–20, 319*t*–20*t*
    overview, 314, 321
    packed columns, 319
    purpose, 314–15
    symbols for distillation columns, 320
    system components, 315–18
        column sections, 316
        trays, 316–18
Distributed control system (DCS), 349, 358–59
Diversity, 102–3
    definition, 91
    respecting, 103
    terms associated with, 102–3
Dollar bills, composition of, 86
DOT (Department of Transportation), 16, 113
    definition, 107
DOT CFR 49.173.1—Hazardous Materials—General Requirements for Shipments and Packaging, 116
Downcomers, 305, 308
Draft gauges, 296, 299
Drake, Colonel Edwin, 22
Drift eliminators, 286, 290
Drugs, 40, 46
Drums, 211, 214
Dryer, 211, 214
Drying, 66, 69
Dynamic compressors
    axial compressors, 241–42
    centrifugal compressors, 241
    definition, 238, 240

Dynamic pumps
    centrifugal pumps, 230–31
    components, 231
    definition, 226, 229
    vs. positive displacement pumps, 231

E

Ear protection, 123
Economizers, 305, 307
Edison, Thomas, 56, 264
EFDs (Engineering flow diagrams), 179, 186
Einstein, Albert, 57
Elasticity, 148, 153
Electrical diagrams, 183, 187–90
    definition, 179
Electrical transmission, 262–63
Electricity, 53, 158, 331, 333, 335
    definition, 54
Electricity and motors
    alternating (AC) and direct (DC) current, 264–65
    electricity
        amps, 260–61
        circuits, 261–62
        electrical transmission, 262–63
        grounding, 262
        introduction, 259–60
        Ohm's law, 261
        resistance (ohms), 260
        volts, 260
        watts, 261
    electric motors in process industries, 265
        tips for reducing risk of, 268
    key terms, 258–59
    monitoring and maintenance activities, 270, 270*t*
    operational hazards, 267–69
    overview, 259, 271
    primary components of typical AC induction motor, 265–66
    principles of operation of AC electric motors, 266–67
    symbols, 270
Electromotive force, 259–60
Electrons, 168, 171
Elements, 170, 173
    definition, 168
Elevation diagrams, 179, 191
Endothermic reactions, 168, 174
Engineering controls, 107, 121
Engineering flow diagrams (EFDs), 179, 186
Environment. *See* Safety, health, environment, and security
EOWEO (every other weekend off), 12
EPA (Environmental Protection Agency), 16, 72–73, 107, 112–13, 328
EPA 40 CFR Parts 239–282—Resource Conservation and Recovery Act (RCRA), 116
EPA Clean Air and Clean Water Acts, 116–17
Equal Employment Opportunity Act, 114. *See* 29 CFR
Equilibrium, 168, 174
Equipment location diagrams, 180, 191
Ergonomic hazards, 107, 109

Index

E

Ethnocentrism, 91, 103
Evaporation, 66, 69, 148, 160, 290
  definition, 286
Evaporators, 337, 341
Exothermic reactions, 168, 174
Expansion valves, 338, 341
Exploration and production, oil and gas, 27–28
Exploration, definition of, 19, 27
Extractive distillation, 318
Eye protection, 123

F

Face protection, 123
Facility, 1, 6
Fahrenheit, 156
Fans, 286, 290
Faraday, Michael, 56, 264
FDA (Food and Drug Administration), 67, 72–73
Feedback, 101
  definition, 91
Feedstock, 1, 3
Feed trays, 313, 318
Fibers, 83, 85
Fill, 287, 290
Filter, 211, 214
Filtration, 76, 77
Fin fan, 274
Fireboxes, 296, 298
Fire classes, 119–20
Fire triangle and fire tetrahedron, 107, 118–20
Fire tube boiler, 309–10
Firewalls, 211, 220
Fire water, 325, 326
  definition, 323
Fishbone diagrams, 139
  definition, 131
Fission, 53, 57
Fittings, 195, 198
Fitting types (piping), 198–99
Five-step model, for conflict resolution, 100
Fixed bed reactors, 221
  definition, 212
Fixed blades, 249, 253
Flame impingement, 296, 302
Flange, 197–98
  definition, 195
Flare headers, 338, 339
Flare stacks, 338, 339
Flare systems, 338, 339–40
Flash zone, 313, 316
Floating roof, 212, 215
Flow, 154–58
  definition, 148
Flowcharts, 131, 138
Flow instrumentation, 352–53
Flow rate, 349, 352–53
Fluid, 148, 153
Fluidized bed reactors, 222
  definition, 212
Fluidized bed system, 338, 343–44
Foam chambers, 216
  definition, 212
Food Additives Amendment, 67
Food and beverage industry, 5
Food and Drug Act, 67
Food and Drug Administration (FDA), 67, 72–73
Food nutrition label, 68
Foot protection, 124
Force, 149, 158
Forced draft furnaces, 296, 301
Forced draft towers, 287, 288
Ford, Henry, 24
Formulating and blending, 19, 29
Fort Knox, 31
Fossil fuel industries
  mining industry
    economics, 32
    growth and development, 34–36
    how minerals are formed, 30–32
    overview, 38
    process technicians, 37
    sectors, 36–37
    types, 32–34
  oil and gas industry
    growth and development, 21–26
    overview, 20, 37–38
    process technicians, 30
    sectors, 26–29

389 Index

Fourdrinier, Henry and Sealy, 83
Fourdrinier machine, 83, 85
Fracking, 19, 26
Fraction, 313, 318
Franklin, Benjamin, 260
Friction, 149, 156
Frying, 66, 69
Fuel, 107, 118
Fuel gas, 323, 331
Fuel gas valves, 296, 298
Furnace purging, 296, 299
Furnaces
    components
        convection section, 299
        radiant section, 298–99
    definition, 296
    designs, 299–300
    draft types
        balanced draft, 301
        forced draft, 301
        induced draft, 301
        natural draft, 292, 299–300
    key terms, 295–96
    monitoring and maintenance activities, 302t
    operational hazards, 302, 302t
    overview, 296, 303
    principles of operation, 299
    purpose, 296–97
    symbols, 303

G

Galen, 46, 47
Gas, 58–59, 149, 150, 332, 335. *See also* Oil and gas industry
Gas chromatograph, 354
Gaskets, 195, 197
Gas law
    Bernoulli's law, 163–64
    Boyle's law (pressure-volume law), 161
    Charles' law (temperature-volume law), 161
    Dalton's law (partial pressures), 162
    General (or Combined) gas law, 162–63
Gas turbines, 249, 250–51
Gate valves, 195, 203
Gauge hatch, 212, 217
Gauge pressure (psig), 160
    definition, 149
Gems, 20, 31
General (or Combined) gas law, 162–63
    definition, 149
Generators, 53, 55
Geology, 19, 30
Geothermal energy, 53, 62–63
Globe valves, 195, 204
Good Manufacturing Processes (GMPs), 48–49
Governors, 249, 253
Gravity, 152
Ground fault circuit interrupter (GFCI), 259, 269
Grounding, 259, 262

H

Hand protection, 124
Hart, William Aaron, 21
Hazardous agents, 108–9
Hazards found in the process industries, 109–10
HDPE (high-density polyethylen), 195, 198
Head protection, 122–23
Health. *See* Safety, health, environment, and security
Heat, 118, 157
    definition, 108, 149
    transfer, 158–61
    types of, 157
Heat exchangers
    applications, 280–81
    components of typical shell and tube, 277
    definition, 274
    key terms, 273–74
    monitoring and maintenance activities, 281–82, 282t
    operational hazards, 281, 281t
    overview, 274, 284
    principles of operation, 277–80
    purpose, 274
    relationship between different types of heat exchangers, 280–81
    symbols, 282–83
    types, 274–77
Heat tracing, 149, 155, 212, 218
Heavy ends (bottoms), 313, 316
Hero, 254
Heterogeneous, 168, 175
Histograms, 131, 140
Homogeneous, 168, 172
Hoppers, 211, 214
Horton, Horace Ebenezer, 215
Hortonsphere, 215
Hot oil systems, 338, 342–43
Hot spots, 296, 302
Human heart, 203
Hunting, 249, 255
Hydraulic fracturing, 19, 26
Hydraulic instruments, 349, 355
Hydraulic turbines, 249, 251
Hydrocarbons, 19, 21, 168
Hydroelectric power, 53, 59
Hydrometer, 149, 153–54

I

Immiscible, 168, 175
Indicators, 349, 355
Induced draft furnace, 296, 301
Induced draft towers, 287, 288
Industrial wastewater treatment process, 79–80
Information technology, 108, 126
Infrared pyrometers, 351, 352

Index 391

Inlets, 249, 253
Inorganic chemistry, 168
Insoluble, 168, 175
Instrument air, 323, 330
Instrumentation, 350
  control loops, 357–58
  definition, 349
  distributed control systems and their application, 358–59
  key process variables controlled by instrumentation
    analytical variables, 354
    flow, 352–53
    level, 352
    pressure, 350–51
    temperature, 351–52
  key terms, 349–50
  monitoring and maintenance activities, 359–60, 360t
  operational hazards, 359, 359t
  overview, 350, 361
  process control instrumentation signals, 356–57
  purpose of process control instrumentation, 350
  symbols for process control instruments, 360–61
  typical process control applications
    actuators, 356
    alarms, 356
    control valves, 356
    function, 355
    introduction, 354–55
    location, 355
    power source, 355–56
    transducer, 356
    valve positioner, 356
Insulators, 259, 260
Interchangers, 274, 280
Intermediates, definition of, 40, 41
Intrinsically safe, 259
Ishikawa, Kaoru, 135
ISO (International Organization for Standardization), 131, 136–37
Isometric diagrams, 180, 183, 190–91
ISO 9000 Standard, 108, 125
ISO 14000 Standard, 108, 125–26

J
Japanese influence on quality control, 135
Juran, Joseph M., 134–35

K
Katzenbach, Jon, 94
Kelvin, 156
Kier, Samuel, 21
Kinetic energy, 249, 250
Knockout drums, 338, 339

L
Laminar flow, 274, 279
Latent heat, 157
  of condensation, 149, 157
  definition, 149
  of fusion, 149, 157
  of vaporization, 149, 157
Leakage
  compressors, 244–45
  pumps, 233
Legends, 181, 183
  definition, 180
Level, 349
Level instrumentation, 352
Leverage, 149, 158–59
Libau, Andreas, 42
Light ends, 313, 316
Liquids, 149–50
LNG (liquefied natural gas), 20
Local, 349, 355
Lockout/tagout device, 269
Loop diagrams, 180, 191
Lubricants, 338, 341
Lubrication, 149, 156
Lubrication systems, 338, 341–42
Lun, Ts'ai, 84

M
Maintenance programs, 142–43
  predictive/preventive maintenance (PPM), 143–44
  total productive maintenance (TPM), 143
Makeup water, 287, 290
Malcolm Baldrige National Quality Award, 141–42
Manometer, 351
Manufacturing, 66, 69
Manways, 217
Mass, 149, 151–52
  definition, 212
Mass flow rate, 149, 154
Material safety data sheet (MSDS). See Safety Data Sheet (SDS)
Matter, 150–51
  definition, 149
Meat Inspection Act, 67
Mechanical energy, 249, 250
Mechanical refrigeration system, 341
Metals, 20, 31
Mills, 83, 85
Mine Improvement and New Emergency Response (MINER) Act, 110
Minerals, 20, 31
Mines, 20, 31–32
Mining industry, 4, 20, 31
  economics, 32
  growth and development
    in the beginning, 34
    industrial revolution, 34–35
    World Wars I and II, 35
    1960s to 2000, 35
    2000 to present, 35–36
  how minerals are formed, 30–32

Mining industry (*Continued*)
  key terms, 19–20
  overview, 38
  processing operations, 36–37
  process technicians, 37
  sectors, 36–37
  types of mining
    surface mining, 32–33
    underground mining, 33–34
Miscible, 168, 175
Mist eliminators, 212, 220
Mixers, 212, 217
Mixing, 66, 69
Mixtures, 169, 175
Molecules, 169, 171
Motors, 259. *See also* Electricity and motors
Moving blades, 249, 253
Mud drums, 305, 307
Multistage compressors, 242
Municipal water treatment process, 78–79
Muspratt, James, 42

N
Natural draft furnace, 296, 299–300
Natural draft towers, 287, 288
Net positive suction head (NPSH) pressure, 227, 233
Neutrons, 169, 171
Nitrogen, 323, 330, 332, 334
Nitrogen header system, 338, 344
NOAA (National Oceanic and Atmospheric Administration), 72–74
Noise-induced hearing loss (NIHL), 122
Non-compliance with regulations, consequences of
  economic, 121
    legal, 120
    moral and ethical, 120
    safety, health, and environmental, 120
Nonmetals, 20, 31
Nozzles, 249, 253
NRC (Nuclear Regulatory Commission), 16, 108, 114
Nuclear, definition of, 54
Nuclear energy, 60–61
Nung, Shen, 46
Nutrition Labeling and Education Act, 67
Nylon, 42–43

O
Occupational Safety and Health Administration (OSHA). *See* OSHA (Occupational Safety and Health Administration)
Ocean thermal energy conversion (OTEC), 62
Offshore rigs, 27
Ohm, Georg Simon, 56
Ohms, 260
  definition, 54, 259
  origin of name, 56
Ohm's law, 261

Oil and gas industry, 3
  42-gallon barrel standard, 23
  growth and development
    early days of refining, 21–24
    1950s to 2000, 25–26
    2000 to present, 26
    World Wars I and II, 24–25
  key terms, 19–20
  overview, 20, 37–38
  petroleum, 20, 21–22, 27, 58–59
  process technicians, 30
  sectors
    exploration and production, 27–28
    refining, 29
    transportation, 28–29
OPEC (Organization of Petroleum Exporting Countries), 57
Open circuit cooling tower, 287, 288
Ore, 20, 31
Organic chemistry, 169
Orifice plate, 352, 353
OSHA (Occupational Safety and Health Administration), 16, 108, 113
  electrical hazards, 268
OSHA 1910.1000 Air Contaminants, 116
OSHA 1910.1200—Hazard Communication (HAZCOM), 115–16
OSHA 1910.120—Hazardous Waste Operations and Emergency Response (HAZWOPER), 116
OSHA 1910.132—Personal Protective Equipment (PPE), 115
OSHA 1910.119—Process Safety Management (PSM), 114–15
OSHA Voluntary Protection Program (VPP), 125
Osmosis, reverse, 67, 70
OTEC (Ocean thermal energy conversion), 62
Outlets, 249, 253
Overall equipment effectiveness (OEE), 131, 143
Overhead, 313, 316
Overheating
  compressors, 244
  pumps, 232
Overpressurization
  compressors, 244
  positive displacement pumps, 232–33

P
Packed towers, 314, 319
Packing, 314, 319
Paints, 44
Paper. *See* Pulp and paper industry
Papin, Frenchman Denis, 205
Papin's Bone Digester, 205
Papyrus Ebers, 46
Parallel circuits, 262
Pareto charts, 131, 139
Pareto principle, 131, 134
Pareto, Vilfredo, 134
Pasteurization, 66, 70
Pathogens, 76, 77

Index

Periodic table, 169–70
Perkins, Jacob, 341
Personal protective equipment (PPE), 121–24
  body protection, 124
  definition, 108
  ear protection, 123
  examples, 122
  eye protection, 123
  face protection, 123
  foot protection, 124
  hand protection, 124
  head protection, 122–23
  respiratory protection, 123–24
Petrochemicals, 40, 45
Petroleum
  definition, 20
  origins of word, 21
pH, 169, 176
PFDs (process flow diagrams), 180, 183, 184–85
Pharmaceutical industry, 4
  developing new medicines, 48
  drug manufacturing process, 48
  generic flow chart, 49
  growth and development
    in the beginning, 46
    1000 B.C., 46
    2000 B.C., 46
    A.D. 200, 46
    1600s–1700s, 47
    1800s, 47
    1900s–2000, 47
    2000–present, 47
  key terms, 40–41
  overview of, 45–46, 48
  pharmaceuticals, 40, 45
  process technicians
    equipment, 50
    job duties, 49
    responsibilities, 49
    workplace conditions and expectations, 50
  quality, 48–49
  summary of, 51
Pharmaceuticals, 40, 45, 47
Pharmacopoeia, 47
Phase changes, 149, 151
Photovoltaics (PV) solar power, 54, 61
Physical change, 169, 173
Physical chemistry, 169
Physical hazards, 108–9
Physical properties, 169, 173
Physical security, 108, 110, 126–27
Physics
  application in the process industries, 150
  electricity, 158

boiler feed water, 327
meter, 354, 361
scale, 176, 177
Pharmaceutical industry, 4
  developing new medicines, 48
  drug manufacturing process, 48
  generic flow chart, 49
  growth and development
    in the beginning, 46
  flow, 154–58
  force and leverage, 158–59
  friction, 156
  gas law
    Bernoulli's law, 163–64
    Boyle's law (pressure-volume law), 161
    Charles' law (temperature-volume law), 161
    Dalton's law (partial pressures), 162
    General (or Combined) gas law, 162–63
  heat transfer, 158–61
  key concepts for process industries, 151–54
    buoyancy, 154
    density, 152
    elasticity, 153
    mass, 151–52
    specific gravity, 153–54
    viscosity, 153
  key terms, 148–50
  matter, 150–51
  overview, 150, 164
  pressure
    atmospheric pressure, 159
    concept of, 159
    pressure gauge measurements, 159–160
    pressure impact on boiling, 161
    vapor pressure and boiling point, 160
    temperature, 156–58
  velocity, 155
Pilot, 296, 298
Pipelines, oil and gas, 28
Piping and instrumentation diagram (P&ID), 180, 183, 184–85
Piping and valves
  connecting methods, 196–98
    bonds, 198
    flanges, 197
    threaded (screwed), 197
    welds, 198
  construction materials, 196
  fitting types, 198–99
  function, 195–96
  key terms, 194–95
  monitoring and maintenance activities, 206–7, 207t
  operational hazards, 206, 206t
  overview, 195, 208
  purpose, 195–96
  symbols, 207
  valve types, 199–206
    ball valve, 200
    butterfly valve, 201
    check valve, 201–2
    diaphragm valve, 202–3
    gate valve, 203
    globe valve, 204
    plug valve, 201
    relief valve, 204–5
    safety valve, 204, 205
    valve actuators, 205–6

Piston-type reciprocating compressors, 240
Piston-type reciprocating pumps, 229
Plant air, 323, 329
Plasma, 149–50
Plate tectonics, 20, 30
Plot plan diagrams (PPDs), 180, 186
Plug valves, 201
   definition, 195
Pneumatic instruments, 349, 355
Polyvinyl chloride (PVC), 195, 198
Pop valves, 205
Positive-displacement compressors
   definition, 238, 239
   reciprocating compressors, 240
   rotary compressors, 239–40
Positive displacement pump, 227
Possum bellies, 216
Potable water, 76, 77, 323, 325–26
Power generation industry, 4–5
   coal, oil, and gas, 58–59
   electricity, 54
   fusion reaction, 61
   geothermal, 62–63
   growth and development
     early civilizations, 55–56
     1700s, 56
     1800s, 56–57
     1900s, 57–58
     2000–present, 58
   hydroelectric, 59
   key terms, 53–54
   major sectors of, 63
     distribution, 54
     generation, 54
     transmission, 54
   nuclear, 60–61
   overview, 58–63
   process technicians
     expectation of, 63
     future trends, 63
     job duties, 64
     skills, 64
   solar, 61–62
   summary of, 63–64
   wind, 62
PPE (personal protective equipment). See Personal protective equipment (PPE)
Predictive/preventive maintenance (PPM), 131, 143–44
Preheaters, 274, 280
Prejudice, 91, 103
Preserving, 66, 69
Pressure
   atmospheric pressure, 159
   concept of, 159
   definition, 149
   pressure gauge measurements, 159–60
   pressure impact on boiling, 161
   vapor pressure and boiling point, 160

Pressure gauge (Bourdon tube), 159–60, 351
Pressure impact on boiling, 161
Pressure instrumentation, 350–51
Pressurized tanks, 212, 214–15
Priming, 227, 233
Process, 1, 3, 92, 95
Process auxiliaries
   amine system, 343
   flare systems and associated equipment, 339–40
   fluidized bed system, 343–44
   hot oil systems and associated components, 342–43
   key terms, 337–38
   lubrication systems and associated components, 341–42
   monitoring and maintenance activities, 345, 345t
   nitrogen header system, 344
   operational hazards, 344–45, 344t–45t
   overview, 338, 345–46
   refrigeration systems and associated components, 340–41
   symbols, 345
   types, 338
Process control instruments, 356–57
   purpose of, 350
   symbols, 360–61
Process drawings
   application blocks, 183
   common components and information, 181–83
   example, 182
   key terms, 179–80
   legends, 181, 183
   overview, 180, 192
   purpose, 180–81
   symbols, 181
   title blocks, 182–83
   types and their uses
     block flow diagrams (BFDs), 183, 184
     electrical diagrams, 183, 187–90
     elevation diagrams, 191
     engineering flow diagrams (EFDs), 186
     equipment location diagrams, 191
     isometric drawings, 183, 190–91
     loop diagrams, 191
     piping and instrument diagrams (P&IDs), 183, 184–85
     plot plan diagrams (PPDs), 186
     process flow diagrams (PFDs), 183, 184–85
     utility flow diagrams (UFDs), 183, 186–87
Process flow diagrams (PFDs), 180, 183, 184–85
Process heaters, 296
Process industries, 1, 3
   definition, 3
   distillation methods, 318
   how process industries operate, 6
   impact on
     community, 13–14
     economy, 14
     environment, 14
     other industries, 14
   regulations. See Regulations affecting the process industries

Index

food and beverage industry
  equipment, 10
  duties, 9
  definition of, 3
  responsibilities, 45
  expectations of, 45
  duties, 45
chemical industry
Process technicians, 1
Process sewers, 328
wastewater, 328
utility in process industries, 324–25
treated water, 326
service or raw water, 326
process water, 327–28
potable water, 326
introduction, 324–25
fire water, 326
cooling water, 327
boiler feed, 326–27
water
steam and condensate, 332
nitrogen, 332
introduction, 331
gas, 332
electricity, 333
compressed air, 332
types of equipment associated with utility systems
types, 324
symbols, 335
steam and condensate systems, 328–29
overview, 324, 335–36
water, 334
steam and condensate, 334
nitrogen, 334
gas, 335
electricity, 335
compressed air, 334
breathing air, 334
utilities, 333t
operational hazards associated with process
nitrogen system, 330
monitoring and maintenance activities, 335, 335t
key terms, 323–24
fuel gas, 331
equipments associated with, 331–33
electricity, 331
utility air, 329
instrument air, 330
compressed air, 329–30, 332
breathing air, 330
Process service utilities
what process industries do, 3–6
safety and environmental regulations, 14
global competition, 14
for future changes in PT role, 14–15
response to current issues and trends

equipment, 72
job roles, 68
responsibilities, 71
workplace conditions and expectations, 72
mining industry, 37
oil and gas industry, 30
pharmaceutical industry
equipment, 50
job duties, 49
responsibilities, 49
workplace conditions and expectations, 50
power generation industry
expectation of, 63
future trends, 63
job duties, 64
skills, 64
pulp and paper industry
equipment, 89
workplace conditions and expectations, 89
quality and, 11, 144–45
role
future trends in, 15
in safety, health, environment, and security, 117
safety, 11
shift work, 12–13
skills, 8
teams, 11–12, 93–94
union versus non-union work environment, 13
water and wastewater treatment industry, 80–81
workplace conditions and expectations, 10
Process technology, overview of, 1, 3
individual expectations, 2
industry involvement, 2
key terms, 1–2
process industries. See Process technicians
process technicians. See Process technicians
program purpose, 2
program value, 2
summary, 16
Process variables, 349
Process water, 76, 78, 323, 325, 327–28
Process water sewer, 76, 79
Product, 2, 6, 169, 174
Production, 2, 3
Protons, 169, 171
Pulp, 83, 85
Pulp and paper industry, 5
growth and development
  in the beginning, 84
  1400s, 84
  1600s, 85
  1700s, 85
  1800s, 85
  1900s, 85
  2000–present, 85
key terms, 83
overview, 83–84, 89
paper and papermaking, 86

Pulp and paper industry (*Continued*)
  paper mills, 88
  paper processing/converting, 88
  process flow diagram, 87
  process technicians
    equipment, 89
    workplace conditions and expectations, 89
  pulp mills, 86–87
  segments, 88
  wood pulp in pulp processing plant, 87
Pumps
  axial pumps, 231–32
  differences between compressors and pumps, 238
  dynamic pumps, 229–31
    centrifugal pumps, 230–31
    components, 231
  "family tree," 227
  key terms, 226–27
  monitoring and maintenance activities, 234, 234*t*
  operational hazards, 232*t*
    cavitation, 233
    leakage, 233
    overheating, 232
    overpressurization, 232–33
    vibration, 234
  overview, 227, 234
  positive displacement pumps, 228–29
    definition, 228
    reciprocating pumps, 229
    rotary pumps, 228–29
  symbols, 234
  types, 227
PVC (polyvinyl chloride), 195, 198

Q
Quality, 11, 132
  industry response to quality issues and trends, 132
  ISO, 136–37
  key terms, 130–31
  maintenance programs, 142–43
    predictive/preventive maintenance (PPM), 143–44
    total productive maintenance (TPM), 143
  Malcolm Baldrige National Quality Award, 141–42
  overview, 131–32, 145
  pharmaceutical industry, 48–49
  process technicians and quality improvement, 144–45
  quality initiatives, 135–36
  quality movement and its pioneers, 132–35
    Dr. Walter Shewhart, 132–33
    Dr. W. Edwards Deming, 133–34
    F.W. Taylor, 132
    the Japanese influence, 135
    Joseph M. Juran, 134–35
    Philip B. Crosby, 135
  self-directed and self-managed teams, 141
  Six Sigma, 141

statistical process control and other analysis tools, 137–40
total quality management (TQM), 136
Quarry, 20, 31

R
Radiant section, 296, 298–99
Radiant tubes, 296, 298
Radiation, 158, 278
  definition, 149
Rankine, 156
Raw materials, 2, 6
RCRA (Resource Conservation and Recovery Act), 116
Reactants, 169, 174
Reaction furnaces, 212, 222
Reactive distillation, 318
Reactors, 220
  definition, 212
  fixed bed, 221
  fluidized bed, 222
  purpose, 220–22
  stirred tank, 221
  symbols, 224
  tubular, 222
Reboilers, 274, 280
Receiver, 338, 341
Reciprocating compressors, 238, 240
Reciprocating pumps, 227, 229
Recorders, 349, 355
Rectifying (fractionating) section, 314, 316
Refining
  definition, 20, 29
  oil
    conversion, 29
    distillation (fractionation), 29
    formulating and blending, 29
    treatment, 29
Reflux, 314, 316
Refractory lining, 296, 298
Refrigerant, 338, 341
Refrigeration systems, 340–41
Regulations affecting the process industries
  DOT CFR 49.173.1—Hazardous Materials—General Requirements for Shipments and Packaging, 116
  EPA 40 CFR Parts 239–282—Resource Conservation and Recovery Act (RCRA), 116
  EPA Clean Air and Clean Water Acts, 116–17
  Equal Employment Opportunity Act, 114
  OSHA 1910.1000 Air Contaminants, 116
  OSHA 1910.1200—Hazard Communication (HAZCOM), 115–16
  OSHA 1910.120-Hazardous Waste Operations and Emergency Response (HAZWOPER), 116
  OSHA 1910.132—Personal Protective Equipment (PPE), 115
  OSHA 1910.119—Process Safety Management (PSM), 114–15

Index

face protection, 123
eye protection, 123
examples, 122
ear protection, 123
body protection, 124
personal protective equipment (PPE), 121–24
overview, 108–9, 127–28
OSHA Voluntary Protection Program (VPP), 125
safety, health, and environmental, 120
moral and ethical, 120
legal, 120
economic, 121
non-compliance with regulations, consequences of
key terms, 107–8
ISO 14000 Standard, 125–26
ISO 9000 Standard, 125
information technology, 126
hazards found in the process industries, 109–10
fire triangle and fire tetrahedron, 118–20
fire classes, 119–20
engineering controls, 121
attitudes and behaviors that help prevent accidents, 117–18
administrative controls, 121
Safety, health, environment, and security, 11
Safety Data Sheet (SDS), 108, 115

S

Rotors, 259
Rotary pumps, 227–29
definition, 238
Rotary compressors, 239–40
Roots, Philander Higley, 243
Roots, Francis Marion, 243
Roots blower, 243
Roosevelt, Franklin Delano, 57
Rocks, 20, 30
Rock formations, 20, 30
Rock cycle, 20, 30
Roasting, 67, 69
Risers, 305, 308
Reverse osmosis, 67, 70
Respiratory protection, 123–24
Resource Conservation and Recovery Act (RCRA), 116, 328
Resistance (Ohms), 260
Resistance, 259–60
Remotes, 349, 355
Relief valve, 195, 204–5
Occupational Safety and Health Administration (OSHA), 16, 113
Nuclear Regulatory Commission (NRC), 16, 114
Environmental Protection Agency (EPA), 16, 112–13
Department of Transportation (DOT), 16, 113
Department of Homeland Security (DHS), 16
Alcohol, Tobacco, and Firearms (ATF), 16
Regulatory agencies and their responsibilities

foot protection, 124
hand protection, 124
head protection, 122–23
respiratory protection, 123–24
physical security and cybersecurity, 126–27
regulations affecting the process industries
DOT CFR 49.173.1—Hazardous Materials—General Requirements for Shipments and Packaging, 116
EPA 40 CFR Parts 239–282—Resource Conservation and Recovery Act (RCRA), 116
EPA Clean Air and Clean Water Acts, 116–17
Equal Employment Opportunity Act, 114
OSHA 1910.1000 Air Contaminants, 116
OSHA 1910.1200—Hazard Communication (HAZCOM), 115–16
OSHA 1910.120—Hazardous Waste Operations and Emergency Response (HAZWOPER), 116
OSHA 1910.132—Personal Protective Equipment (PPE), 115
OSHA 1910.119—Process Safety Management (PSM), 114–15
regulatory agencies and their responsibilities
Department of Transportation (DOT), 113
disasters prompted safety legislation and development of, 110–12
Environmental Protection Agency (EPA), 112–13
Nuclear Regulatory Commission (NRC), 114
Occupational Safety and Health Administration (OSHA), 113
role of process technicians, 117
Safety valve, 204, 205
definition, 195
Salem Witch Trials, 67
Salt water, 154
Scatter plots, 140
definition, 131
Schematics, 180, 190
Screw pump, 229
Security. See also Safety, health, environment, and security
cybersecurity, 107, 110, 126–27
hazards, 110
physical security, 108, 110, 126–27
Self-contained breathing apparatus (SCBA), 124, 330
Self-directed and self-managed teams, 141
Self-managed team, 92, 95
Semiconductors, 259, 260
Sensible heat, 149, 157
Sensors, 349, 355
Series circuits, 261
Service (raw) water, 324–26
Set point, 249, 255
Settleable solids, 76, 80
Shafts, 249, 253
Shell inlets, 274, 277
Shell outlets, 274, 277

Shells, 274, 277
Shewhart, Dr. Walter, 132–33
Shift work, 12–13
Shock bank, 299
   definition, 296
Shock tubes, 299
   definition, 296
Sieve tray, 317
Sight glass, tubular type, 352
Sillman, Benjamin Jr., 22
Simulations, 180, 191
Six Sigma, 131, 141
Skills, 8
Smith, Douglas, 94
Socket weld, 195, 198
Soda bottling process, 71
Soil pollution, 108, 110
Solar (CSP), 54, 61
Solar (PV), 54, 61
Solder, 195, 198
Solids, 150
   definition, 149
   dissolved, 76, 80
   settleable, 76, 80
   suspended, 76, 80
Soluble, 169, 175
Solutes, 169, 175
Solutions, 169, 175
Solvents, 169, 175
Sour water, 324, 325
Spacer rods, 274, 277
SPC (statistical process control), 137–38
   definition, 131
Specialty chemicals, 41
Specific gravity, 153–54
   definition, 149
Specific heat, 149, 157
Spherical tanks, 212, 215
Stacks, 296, 299
Stators, 259
Steam and condensate systems, 328–29, 332, 334
Steam drums, 305, 307
Steam generator. *See* Boilers
Steam turbines, 250
   components, 252–53
   definition, 249
   hazards, 254–55
   symbols, 255
Stereotyping, 92, 102
Stirred tank reactors, 212, 221
Storm sewer, 328
Stripping section, 314, 316
Suction, 238
Suction screens, 287, 290
Sump, 212, 216
Superheated steam, 305, 308
Superheaters, 305, 308

Surface mining, 32–33
Surging (compressors), 244
Suspended solids, 76, 80
Swan, Joseph, 56
Symbols, 181
   definition, 180
Synergy, 92, 95
Synthetic, definition of, 41
Synthetic materials, 42, 44

T

Taguchi, Dr. Genichi, 135
Tanks, 212, 213
   types, 214–15
Tape gauge, 352
Tasks, 92, 95
Taylor, F.W., 132
Team dynamics, 92, 95
Teams, 11–12, 92
   characteristics of high-performance teams
     composition, 94
     process, 95
     technique, 94–95
   comparison between work groups and team, 93
   definition, 93
   failure of, 97–98
   giving feedback, 101
   key terms, 91–92
   overview, 92, 103–4
   receiving feedback, 101
   resolving conflict
     consequences phase, 100
     contract phase, 100
     describe phase, 100
     express phase, 100
     specify phase, 100
   stages of team development
     stage 1: forming, 96, 97
     stage 3: norming, 96, 97
     stage 4: performing, 96, 97
     stage 2: storming, 96, 97
   synergy and team dynamics, 95
   team tasks *versus* process, 95–96
   types in the process industries, 93–94
   workforce diversity, 102–3
     respecting, 103
     terms associated with diversity, 102–3
   work groups, 92–93
Temperature, 149, 156–58
Temperature conversion formulas, 157
Temperature instrumentation, 351–52
Tesla, Nikola, 264
Thermal imaging camera, 351, 352
Threaded (screwed) pipe, 195, 197
Three Mile Island, 57
Throttle valves, 249, 253
Throttling, 195, 200

Index

T
Titanic, 298
Title blocks, 180, 182–83
Toasting, 67, 69
Toilet paper, invention of, 84
Total productive maintenance (TPM), 131, 143
Total quality management (TQM), 131, 136
Tower. *See* Column section
Towers cooling. *See* Cooling towers
Transducers, 349, 356
Transformers, 259, 263
Transmitters, 349, 355
Transportation, 27
 definition, 20
 oil and gas, 28–29
Tray downcomers, 314, 316
Trays, 314, 316–18
Tray-type distillation columns, 316–18
Treated water, 324, 325
Treatment
 definition, 20
 refining, 29
Trip valves, 249, 253
Tube bundles, 274, 277
Tube inlets, 274, 277
Tube outlets, 274, 277
Tube sheets, 274, 277
Tubular reactors, 222
 definition, 212
Turbidity, 76, 77
Turbines, 58
 definition, 54, 249
 key terms, 248–49
 monitoring and maintenance activities, 255, 255t
 operational hazards, 254–55, 254t–55t
 overview, 249, 256
 principles of operation, 253–54
 steam turbine components, 252–53
 steam turbine symbols, 255
 types, 249–52
 gas turbines, 250–51
 hydraulic turbines, 251
 steam turbines, 250
 wind turbines, 251–52
Turbulent flow, 279
 definition, 274

U
Underground mining, 33–34
Union of Japanese Scientists and Engineers, 134
Union *versus* non-union work environment, 13
Unit, 2, 6
U.S. Department of Agriculture (USDA), 72–73
Utility air, 324, 329
Utility flow diagrams (UFDs), 183, 186
 definition, 180

V
Vacuum pressure (psiv), 149, 160
Valve actuators, 205–6
Valve chatter, 203
Valve positioners, 350, 356
Valves, 195, 199–206. *See also* Piping and valves
Valve tray, 317
Vane separators, 212, 220
Van Helmont, Jan Baptista, 42
Van Syckel, Samuel, 23
Vapor pressure
 boiling point and, 160
 definition, 149
Vapor recovery systems, 218
 definition, 212
Variables, 131, 138
Velocity, 150, 155
Venturi tube, 353
Vessels, 212–13
 common components, 215–20
 definition, 212
 dikes, firewalls, and containment walls, 220
 key terms, 211–12
 monitoring and maintenance activities, 223–24, 223t
 operational hazards, 222–23, 223t
 overview, 212–13, 224
 purpose, 213–14
 reactors
 purpose and types, 220–22
 symbols, 224
 tank types, 214–15
 types, 213
Viscosity, 153
 definition, 150
Volta, Alessandro, 56
Voltmeter, 259–60
Volts, 54, 260
 definition, 259
 origin of name, 56
Voluntary Protection Program (VPP)
 definition, 108
 OSHA, 125
Volutes, 227, 230
Vortex, 212, 218
Vortex breaker, 212, 218

W
Wastewater, 324, 325, 328
Water. *See also* Water and wastewater treatment industry
 boiler feed, 326–27
 cooling water, 327
 fire water, 326
 introduction, 324–25
 operational hazards associated with process utility systems, 334

Water. (*Continued*)
　potable water, 326
　process water, 327–28
　service or raw water, 326
　treated water, 326
　utility in process industries, 324–25
　wastewater, 328
Water and wastewater treatment industry, 5
　environmental regulations and considerations, 81
　growth and development, 77–78
　history of wastewater treatment industry, 78
　industrial wastewater treatment process, 79–80
　key terms, 76
　municipal water treatment process, 78–79
　overview, 77, 81
　process technicians, 80–81
Water distribution header, 287, 290
Watermarks, 84
Water pollution, 108, 110
Water treatment plant, 79
Water tube boiler, 307, 310
Watt, James, 56

Watts, 54, 261
　definition, 259
　origin of name, 56
Weight, 150, 151
Weirs, 212, 219, 314, 316
Wilcox, Stephen, 308
Wind power, 54, 62
Wind turbines, 251–52
　definition, 249
Wiring diagrams, 180, 190
Workforce diversity, 102–3
　respecting, 103
　terms associated with diversity, 102–3
Work groups, 92
　definition, 92–93
　versus teams, 93
Working as team. *See* Teams
Workplace conditions and expectations, 10

Z
Zero defects, 131, 135